普通高等院校公共基础课程系列教材

离散数学简明教程

（双语版）

成科扬　肖　文　张建明　编著

清华大学出版社
北　京

内 容 简 介

本书是作者团队结合多年教学实践经验与科学研究成果，在力求通俗易懂、简明扼要的指导思想下编写而成的。本书共 11 章，内容包含数理逻辑、集合与关系、函数、代数结构、图和树等。本书体系严谨、文字精练、内容充实、例题丰富，配套丰富的教学资源，适合高校教学使用。除此之外，本书综合国内外离散数学的相关新资料，采用双语的形式，从而培养读者的外文科技文献阅读能力。

本书适合高等院校计算机及相关专业作为"离散数学"课程的教材使用，也可以作为对离散数学感兴趣的读者的入门参考用书。

图书在版编目（CIP）数据

离散数学简明教程：双语版：汉、英/成科扬，肖文，张建明编著. —北京：清华大学出版社，2023.8
普通高等院校公共基础课程系列教材
ISBN 978-7-302-64144-5

Ⅰ. ①离… Ⅱ. ①成… ②肖… ③张… Ⅲ. ①离散数学－高等学校－教材－汉、英 Ⅳ. ①O158

中国国家版本馆 CIP 数据核字(2023)第 131900 号

责任编辑：吴梦佳
封面设计：傅瑞学
责任校对：李　梅
责任印制：曹婉颖

出版发行：清华大学出版社
　　网　　址：https://www.tup.com.cn，https://www.wqxuetang.com
　　地　　址：北京清华大学学研大厦 A 座　　　　　　　邮　　编：100084
　　社　总　机：010-83470000　　　　　　　　　　　　邮　　购：010-62786544
　　投稿与读者服务：010-62776969，c-service@tup.tsinghua.edu.cn
　　质量反馈：010-62772015，zhiliang@tup.tsinghua.edu.cn
　　课件下载：https://www.tup.com.cn，010-83470410
印　装　者：三河市龙大印装有限公司
经　　销：全国新华书店
开　　本：185mm×260mm　　　　印　　张：22　　　　字　　数：557 千字
版　　次：2023 年 10 月第 1 版　　　　　　　　　　印　　次：2023 年 10 月第 1 次印刷
定　　价：69.00 元

产品编号：090509-01

前　言

离散数学是现代数学的重要分支,也是计算机科学与技术、网络工程、软件工程等专业的核心课程。它为计算机专业的数据结构、操作系统、编译原理、算法设计与分析、数字电路、密码学基础、人工智能等课程提供了必要的数学基础。

离散这个概念是和连续相对应的,高等数学中第一章是"函数与极限",这就是一个连续概念的体现。在现实中,我们发现连续的问题更常见,那么我们为什么要学习离散数学呢?因为计算机的运算是离散型(以二进制为基础)的,也就是说,计算机通常只能用离散的数据模仿连续的数据,而不能做到真正的连续,故而对计算机而言,离散就是它们理解数学的方式。简而言之,我们可以把离散数学理解为数学和计算机之间的桥梁。离散数学在计算机学科中有着许多的应用,比如,笛卡儿积在数据库中具有无可替代的作用,离散结构与算法思考反映出数据结构的结构知识,而近年大热的人工智能领域、以逻辑数学为推理基础的逻辑推理也是其重点研究的内容。离散数学在计算机学科中正发挥着越来越大的作用,以离散数学作为计算机学科研究的依据与方法,可以促进计算机学科逐渐趋于完善。因此,我们应该重视离散数学在计算机学科中的作用,学好离散数学对后续相关方面的研究有着十分重大的意义。

离散数学涉及内容较多,本书共分为数理逻辑、集合论、代数结构和图论4篇,共11章。第1章和第2章分别介绍命题逻辑和谓词逻辑的基本概念、推理理论及应用等;第3章和第4章主要介绍离散结构的集合表示,讨论了集合、关系和函数的各种运算、性质、表示方法及应用;第5章～第7章主要介绍代数结构,包括代数系统、群、格与布尔代数;8～11章主要介绍离散结构的图形表示,即图论,包括图、欧拉图、哈密顿图、平面图及树。本书结合大量图例、公式及例题,清晰地介绍并展示了离散数学中的概念与技术,行文流畅,通俗易懂,由浅入深,并在每章的最后配有相当数量的习题,帮助读者巩固所学知识。除此之外,为了与国际接轨,本书采用双语版本,方便读者对照阅读,从而培养读者的外语阅读与写作能力。

本书是作者团队结合多年教学实践与科学研究,在力求通俗易懂、简明扼要的指导思想下编写而成的。在编写过程中,我们着重突出以下四个特点。

(1) 易理解性。我们希望本书对读者来说是易读易懂的,所以注意突出重点,论证做到详细明了。每一个概念的抛出,都会给出一些相应的例题,方便读者理解学习。

(2) 独立且具灵活性。为了便于灵活使用,本书做了精心的设计,每章分成长度大致相等的若干节,每节又根据内容划分成若干小节以方便学习。同时各章节各有独立性,但尽可能保留各章节之间的联系,规范并统一了符号和术语。

(3) 实用性。在加强基本理论教学的同时,也注意了分析问题、解决问题的技能培养和训练。因此,本书在每章后面加入了大量不同类型的练习,不仅提供了足够多的简单练习用于巩固基础,还提供了大量中等难度的练习和许多具有挑战性的练习供读者灵活运用所学知识挑战自我。

(4) 国际化。计算机专业相较于其他专业对于国际化有着更高的要求,因为计算机相关的资料、论文及编程软件都是以英语为主的语言载体,所以专业化的英语术语表达对读者来说十分重要,这也是采用双语编写的初衷。

在编写过程中,我们参考了国内外多种版本的离散数学教材和相关的文献资料,从中吸取了许多优秀的思想与方法,在此向这些学者、专家致谢。另外,我们还要感谢万浩、司宇、蒋森林、梁赛、裴运申等在本书的编写、核对等方面作出的贡献,以及江苏大学研究生教材建设项目、来华留学全英文授课课程教材建设项目资助。

由于编者水平有限,书中不足之处在所难免,敬请读者批评、指正。

编　者

2023 年 3 月

Contents

Part Ⅰ Mathematical Logic

Part Ⅱ Set Theory

Part Ⅲ　The Algebraic Structure

Part Ⅳ Graph Theory

目　录

第1篇　数理逻辑

第2篇　集合论

第 3 篇 代 数 结 构

第 4 篇　图　　论

第1篇
Part I

数理逻辑
Mathematical Logic

Chapter 1   Propositional Logic

Propositional logic can also be called propositional calculus or statement calculus. It mainly studies the deducible relationship between the premise and the conclusion which is composed of the proposition as the basic unit. So what is a proposition? How to express a proposition? How to derive some conclusions from a set of premises? We will take you to discuss these core issues in detail in the following.

1.1   Propositions and Connectives

A proposition is expressed as a statement with a definite meaning. A proposition always has a value called truth value. There are only "True" and "False" values, which are called True and False, usually with 0 or F for "False" and 1 or T for "True".

Statement that represents a proposition must have two conditions: first, the statement is a declarative sentence; second, the statement has a uniquely determined true or false meaning. There are two steps to judge whether a given sentence is a proposition: first, whether it is a declarative sentence; second, whether it has a unique truth value.

There are two types of propositions: the first type is the proposition that cannot be decomposed into simpler statements, which is called atomic proposition; the second type of proposition consists of a combination of connectives, punctuation marks and atomic propositions.

All of these propositions should have certain truth values.

The atomic proposition, also known as a simple proposition, is an inference about a single property of an object, and it can also represent any true or false statement about a certain object or situation. Its semantic truth value is determined by objective facts.

Compound propositions, which can be decomposed into some simple propositions in grammatical structure, are formed by some simple propositions constructed by proposition connectors.

An example is given to illustrate the concept of the proposition.

(1) The sun rises in the east and sets in the west.

(2) The grass is green.

(3) Stand up!

(4) 71% of the earth's surface is water.

(5) Will it rain tomorrow?

(6) What a beautiful day!

(7) In the afternoon I go to the library, or to the laboratory.

(8) Both Zhang San and Li Si are postgraduates of Jiangsu University.

In the above examples, (1), (2), (4), (7), (8) are propositions. Among them, (1), (7), (8) are compound propositions. (3), (5), (6) are not propositions, because they aren't declarative sentences.

In mathematical logic, compound propositions are usually constituted by "connectives". Propositional connectives also known as propositional operators, have strict logical meanings to ensure the consistency between the semantics of the symbolic system and those of the original natural system. There are five logical connectives in common use: Negation Connective, Conjunction Connective, Disjunction Connective, Conditional Connective and Biconditional Connective.

1.1.1 Negation Connective

Definition 1.1.1 If P is a proposition, then the negation of P is a compound proposition, denoted as "$\neg P$" and pronounced as "not P" or "P's negation". If P is T, then $\neg P$ is F; If P is F, then the truth value of $\neg P$ is T.

The connective "$\neg$" can also be seen as a logical operator, which is a unary operation. The relationship between P and $\neg P$ is shown in Table 1.1.1, which is called the truth table of the negation connective "$\neg$".

Example 1.1.1 Negate the following propositions.

P: Zhang San is a college student.

$\neg P$: Zhang San is not a college student.

1.1.2 Conjunction Connective

Definition 1.1.2 If P and Q are both propositions, then the conjunction of P and Q is a compound proposition, denoted as $P \wedge Q$, pronounced as "P and Q" or "P conjunction Q". The value of $P \wedge Q$ is T if and only if both P and Q are T.

The connective "$\wedge$" can also be viewed as a binary logical operation. The truth value table of the connective "$\wedge$" is shown in Table 1.1.2.

Table 1.1.1

P	$\neg P$
F	T
T	F

Table 1.1.2

P	Q	$P \wedge Q$
F	F	F
F	T	F
T	F	F
T	T	T

Example 1.1.2 Suppose P: The COVID-19 epidemic was raging in 2020. Q: The COVID-19 broke out in many countries around the world. Then $P \wedge Q$: The COVID-19 epidemic was raging and broke out in many countries worldwide in 2020.

1.1.3 Disjunction Connective

Definition 1.1.3 Suppose both P and Q are the propositions, and the disjunction of P and Q is a compound proposition, as $P \lor Q$, pronounced "P or Q," or "P disjunction Q". If and only if P and Q are F, $P \lor Q$ is F.

Connective "$\lor$" can also be seen as a logic operation, and it is a binary logic operation. "$\lor$" similar to "or" in Chinese, but not the same. In Chinese, there are "inclusive or" and "exclusive or". The truth table of the conjunction "$\lor$" is shown in Table 1.1.3.

Example 1.1.3 Which of the following two propositions is inclusive or? Which is exclusive or?

(1) Watch the acrobatics on TV or in the theater. (exclusive or)

(2) The light bulb or switch is out of order. (inclusive or, "$\lor$" means inclusive or)

1.1.4 Conditional Connective

Definition 1.1.4 Suppose both P and Q are propositions, and the conditional proposition is a compound proposition, denoted as $P \rightarrow Q$, read "if P, then Q". If and only if P is T and Q is F, $P \rightarrow Q$ is F. P is the antecedent of the conditional proposition $P \rightarrow Q$, and Q is the decedent of the conditional proposition $P \rightarrow Q$.

The connective "$\rightarrow$" can also be viewed as a logical operation, which is a binary logical operation. The truth value table of the connective "$\rightarrow$" is shown in Table 1.1.4.

Table 1.1.3

P	Q	$P \lor Q$
F	F	F
F	T	T
T	F	T
T	T	T

Table 1.1.4

P	Q	$P \rightarrow Q$
F	F	T
F	T	T
T	F	F
T	T	T

Example 1.1.4 P: Xiao Yu studies hard.

Q: Xiao Yu does well in his studies.

$P \rightarrow Q$: If Xiao Yu studies hard, he will get good grades.

The connective "$\rightarrow$" and the Chinese "if…, then…" are similar, but different.

1.1.5 Biconditional Connective

Definition 1.1.5 Set both P and Q as propositions, the complex proposition $P \leftrightarrow Q$ can be called dual condition, which can be pronounced as "P dual condition Q" or "P can be if and only Q". It can be T if and only if the true value of P and Q are the same.

The connective "$\leftrightarrow$" can also be understood as a logical operation, which is binary logic.

The truth value table of the connective "↔" is shown in Table 1.1.5.

Table 1.1.5

P	Q	$P \leftrightarrow Q$	P	Q	$P \leftrightarrow Q$
F	F	T	T	F	F
F	T	F	T	T	T

As described above, a biconditional connective represents a sufficient and necessary relationship and can be determined only according to the definition of a connective, regardless of its antecedes.

Example   1.1.5   Suppose P: Li Si is a merit student.

Q: Li Si is excellent in moral, intellectual and physical qualities.

$P \leftrightarrow Q$: Li Si is a merit student if and only if he is excellent in moral, intellectual and physical qualities.

1.2   Propositional Formula and Translation

It has been mentioned before that propositions that do not contain other propositions as part of them are called atomic propositions. At the same time, atomic propositions cannot contain any connectives. Then a proposition containing at least one conjunction is called a compound proposition.

Let P and Q be any two propositions, then $\neg P$, $P \wedge Q$, $(P \wedge Q) \rightarrow (P \vee Q)$, $\neg P \rightarrow (Q \leftrightarrow \neg P)$ are compound propositions.

If P and Q are propositional variables, then all the above formulas are called propositional formulas, also called well-formed formulas. P and Q are called the components of a propositional formula.

Note: Propositional formulas contain propositional variables, so it is impossible to calculate their true values. Only when the propositional variables in a propositional formula are brought in with propositions that determine the truth values, a proposition can be obtained.

Definition 1.2.1   The symbol string formed according to the following rules is called the well-formed formulas of propositional calculus (wff), also known as propositional formula, or formula for short:

(1) A single propositional variable or constant is a well-formed formula.

(2) If A is a well-formed formula, then $\neg A$ is also a well-formed formula.

(3) If A and B are well-formed formulas, then $(A \wedge B)$, $(A \vee B)$, $(A \rightarrow B)$, $(A \leftrightarrow B)$ are well-formed formulas.

(4) If and only if (1), (2) and (3) are applied a finite number times, the resulting sign string is a well-formed formula.

The propositional formula is generally represented by the capital letters A, B, C, ...

The method of defining the formula is called inductive definition, which consists of three parts: basis, induction and boundary. Among them, (1) is called basis, (2) (3) is called induction, and(4) is called boundary.

For convenience, the involution formula is agreed as follows:

(1) To reduce the number of parentheses used, it is agreed that the outermost parentheses can be omitted.

(2) The order of priority of connectives is as follows: $\neg$, $\wedge$, $\vee$, $\rightarrow$, $\leftrightarrow$. According to this priority, if the brackets are removed and the operation order of the original formula can be omitted, then $P \wedge Q \rightarrow R$ is equivalent to$(P \wedge Q) \rightarrow R$, which is also a compound formula.

Example 1.2.1 Determine whether the following symbol string is a compound formula.

(1) $\neg (P \vee Q)$

(2) $(\neg P \rightarrow Q)$

(3) $(P \rightarrow (P \leftrightarrow Q))$

(4) $(P \leftrightarrow Q) \rightarrow (\vee Q)$

(5) $(P \wedge Q, (P \wedge Q) \wedge Q)$

Solution It can be seen from the above definitions that(1),(2),(3) are all well-formed formulas;(4) and(5) are not well-formed formulas. Because $\vee$ in(4) is binocular conjunction and a propositional argument is missing on the left; while there is a comma in(5), which does not meet the definition of the formula.

With the concept of well-formed formulas of connectives, well-formed formulas can be represented by compound formulas. This process is often called the symbolization of propositions. The whole process is equivalent to translating some sentences(propositions) in natural language into symbolic forms in mathematical logic. It can be carried out according to the following steps:

(1) Determines whether a given statement is a proposition.

(2) Find out the atomic propositions in the statement and determine the conjunctions corresponding to the conjunctions.

(3) In with or by grammar, the original proposition is expressed as a compound formula composed of the atomic proposition, connectives and parentheses.

Example 1.2.2 Symbolize the following proposition: The G124 train from Changsha to Zhenjiang leaves at 9:30 or 10:00 a.m.

Solution P: The G124 train from Changsha to Zhenjiang leaves at 9:30 a.m.

Q: The G124 train from Changsha to Zhenjiang leaves at 10:00 a.m.

In this example, the word "or" is exclusive or, which means is, the train can not leave at 9:30 and 10:00 a.m. at the same time, it can only be one of the two; while $\vee$ in mathematical logic is inclusive or, so we can't use $\vee$ directly. Then construct a truth table for analysis, as shown in Table 1.2.1.

It can be seen from the table the original proposition cannot, denoted by the five connectives separately, but can use the proposition and connectives, the propositional symbol can be turned into: $\neg(P \leftrightarrow Q)$.

Table　1.2.1

P	Q	original proposition	$P \leftrightarrow Q$	$\neg(P \leftrightarrow Q)$	P	Q	original proposition	$P \leftrightarrow Q$	$\neg(P \leftrightarrow Q)$
T	T	F	T	F	F	T	T	F	T
T	F	T	F	T	F	F	F	T	F

Example　1.2.3　Symbolizing the following proposition: He is not only gifted but also hardworking.

Solution　P: He is gifted.　　Q: He works hard.

According to "not only … but also …", we can know the result of this question is $P \wedge Q$.

Example　1.2.4　Symbolize the following proposition: He is gifted but not hard working.

Solution　The actual meaning of this sentence is that he is gifted and not hard working. So if the hypothesis is as follows:

P: He is gifted.　　$\neg Q$: He is not hard working.

The result of this question is $P \wedge \neg Q$.

Example　1.2.5　Symbolize the following proposition: unless you study hard, you will fail.

Solution　The actual meaning of this sentence is that if you don't work hard, you'll lose the course. So if the hypothesis is as follows:

P: You work hard.　　Q: you will fail.

The result of this question is $\neg P \rightarrow Q$.

Example　1.2.6　Symbolizing the following propositions: He won the gold or silver award in the International Collegiate Programming Contest.

Solution　Find out the atomic propositions and use propositional symbols to represent them:

A: He won the gold award in the International Collegiate Programming Contest.

B: He won the silver award in the International Collegiate Programming Contest.

So the result of this problem is $(A \wedge \neg B) \vee (\neg A \wedge B)$.

From the above examples, we can find that some connectives in natural language, such as "not only… but also…" "or" "although… but…" each have their specific meaning. Therefore, they need to be translated into appropriate logical connectives according to different situations. At the same time, to correctly express the relationship between propositions, sometimes we can list the "truth table" to further analyze each original proposition.

1.3 Truth Tables and Equivalent Formulas

From the above Example 1.2.2, when a proposition cannot be directly symbolized, the truth table is often used for auxiliary analysis, which will get the symbolic results faster. Therefore, we will further learn the related knowledge of the truth table.

Definition 1.3.1 Suppose P_1, P_2, ..., P_n are all propositional variables appearing in proposition formula A. Assigning a true value to P_1, P_2, ..., P_n is called an assignment or explanation of proposition formula A. If the assigned value makes the true value of A be T, then this assignment is called a real assignment of A. If the true value of A is set to F, then the assignment is called a false assignment of A.

Definition 1.3.2 In proposition formula A, a truth value of proposition formula A is determined by assigning a value to each value of proposition formula A. The truth value table of proposition formula A is formed by summarizing all truth values into a table.

Example 1.3.1 Construct the truth table of $P \lor \neg Q$.

Solution The results are shown in Table 1.3.1.

Example 1.3.2 Construct the truth table of $(P \land Q) \land \neg P$.

Solution The results are shown in Table 1.3.2.

Table 1.3.1

P	Q	$\neg Q$	$P \lor \neg Q$
T	T	F	T
T	F	T	T
F	T	F	F
F	F	T	T

Table 1.3.2

P	Q	$P \land Q$	$\neg P$	$(P \land Q) \land \neg P$
T	T	T	F	F
T	F	F	F	F
F	T	F	T	F
F	F	F	T	F

Example 1.3.3 Construct the truth table of $(P \lor Q) \lor (\neg P \land Q)$.

Solution The results are shown in Table 1.3.3.

Table 1.3.3

P	Q	$\neg P$	$P \lor Q$	$\neg P \land Q$	$(P \lor Q) \lor (\neg P \land Q)$
T	T	F	T	F	T
T	F	F	T	F	T
F	T	T	T	T	T
F	F	T	F	F	F

Example 1.3.4 Construct the truth table of $\neg (P \land Q) \leftrightarrow (\neg P \lor \neg Q)$.

Solution The results are shown in Table 1.3.4.

Table 1.3.4

P	Q	$P \wedge Q$	$\neg(P \wedge Q)$	$\neg P$	$\neg Q$	$\neg P \vee \neg Q$	$\neg(P \wedge Q) \leftrightarrow (\neg P \vee \neg Q)$
T	T	T	F	F	F	F	T
T	F	F	T	F	T	T	T
F	T	F	T	T	F	T	T
F	F	F	T	T	T	T	T

In the truth table, the number of truth values of the propositional formula depends on the number of components. For example, there are four possible assignments for a propositional formula composed of two propositional variables, while there are eight possible assignments for a propositional formula composed of three propositional variables. Therefore, in general, there are 2^n truth value cases for propositional formulas composed of n propositional variables.

It can be found from the truth table that the corresponding truth values of some propositional formulas are the same as those of other propositional formulas under all assignments of components, the corresponding truth values of $\neg(Q \rightarrow P)$ and $\neg P \wedge Q$ are the same, as shown in Table 1.3.5.

Table 1.3.5

P	Q	$\neg P$	$Q \rightarrow P$	$\neg(Q \rightarrow P)$	$\neg P \wedge Q$
T	T	F	T	F	F
T	F	F	T	F	F
F	T	T	F	T	T
F	F	T	T	F	F

Definition 1.3.3   Given two propositional formulas A and B, let P_1, P_2, P_3, ..., P_n be all atomic variables in A and B. If any group of truth values of P_1, P_2, P_3, ..., P_n are assigned, and the true values of A and B are the same, then A and B are said to be equivalent or logically equal. Recorded as $A \Leftrightarrow B$.

It can be proved that propositional formula equivalence has the following three properties.

(1) Reflexivity, for any proposition formula A, $A \Leftrightarrow A$.

(2) Symmetry, for any propositional formula A and B, if $A \Leftrightarrow B$, then $B \Leftrightarrow A$.

(3) Transitivity, for any propositional formula A, B and C, if $A \Leftrightarrow B$, $B \Leftrightarrow C$, then $A \Leftrightarrow C$.

Example 1.3.5   Prove $P \leftrightarrow Q \Leftrightarrow (P \rightarrow Q) \wedge (Q \rightarrow P)$.

Proof It can be seen from Table 1.3.6 that $P \leftrightarrow Q$ and $P \rightarrow Q \wedge (Q \rightarrow P)$ have the same truth value, so the proposition is proved.

Table 1.3.6

P	Q	$P \rightarrow Q$	$Q \rightarrow P$	$P \leftrightarrow Q$	$(P \rightarrow Q) \wedge (Q \rightarrow P)$
T	T	T	T	T	T
T	F	F	T	F	F
F	T	T	F	F	F
F	F	T	T	T	T

Example 1.3.6　Prove $P \lor (P \land Q) \Leftrightarrow P$, $P \land (P \lor Q) \Leftrightarrow P$.

Proof　The truth table is shown in Table 1.3.7, and the equivalent relationship between the two formulas can be clearly seen from the truth table.

Table 1.3.7

P	Q	$P \land Q$	$P \lor Q$	$P \lor (P \land Q)$	$P \land (P \lor Q)$
T	T	T	T	T	T
T	F	F	T	T	T
F	T	F	T	F	F
F	F	F	F	F	F

As shown in Table 1.3.8, common propositional equivalence theorems are listed.

Table 1.3.8

Number	The Law	Equivalence Relation
1	Involution law	$\neg \neg P \Leftrightarrow P$
2	Idempotent law	$P \lor P \Leftrightarrow P$, $P \land P \Leftrightarrow P$
3	Associative law	$(P \lor Q) \lor R \Leftrightarrow P \lor (Q \lor R)$, $(P \land Q) \land R \Leftrightarrow P \land (Q \land R)$
4	Exchange law	$P \lor Q \Leftrightarrow Q \lor P$, $P \land Q \Leftrightarrow Q \land P$
5	Distributive law	$P \lor (Q \land R) \Leftrightarrow (P \lor Q) \land (P \lor R)$, $P \land (Q \lor R) \Leftrightarrow (P \land Q) \lor (P \land R)$
6	Absorption law	$P \lor (P \land Q) \Leftrightarrow P$, $P \land (P \lor Q) \Leftrightarrow P$
7	De Morgan's law	$\neg (P \lor Q) \Leftrightarrow \neg P \land \neg Q$, $\neg (P \land Q) \Leftrightarrow \neg P \lor \neg Q$
8	Identity law	$P \lor F \Leftrightarrow P$, $P \land T \Leftrightarrow P$
9	Zero law	$P \lor T \Leftrightarrow T$, $P \land F \Leftrightarrow F$
10	Negation law	$P \lor \neg P \Leftrightarrow T$, $P \land \neg P \Leftrightarrow F$
11	Conditional equivalence	$A \to B \Leftrightarrow \neg A \lor B$
12	Two conditional equivalence	$A \leftrightarrow B \Leftrightarrow (A \to B) \land (B \to A)$
13	Hypothetical translocation	$A \to B \Leftrightarrow \neg B \to \neg A$
14	Double conditional negation equivalence	$A \leftrightarrow B \Leftrightarrow \neg A \leftrightarrow \neg B$
15	Domestication of conjunctions	$\neg (A \leftrightarrow B) \Leftrightarrow A \leftrightarrow \neg B$

Example 1.3.7　Prove De Morgan's law $\neg (P \lor Q) \Leftrightarrow \neg P \land \neg Q$ by the truth table.

Proof　Table 1.3.9 is the truth table of $\neg (P \lor Q)$ and $\neg P \land \neg Q$. From the table, we can find that De Morgan's law is correct.

Table 1.3.9

P	Q	$\neg P$	$\neg Q$	$P \lor Q$	$\neg (P \lor Q)$	$\neg P \land \neg Q$
T	T	F	F	T	F	F
T	F	F	T	T	F	F
F	T	T	F	T	F	F
F	F	T	T	F	T	T

In a propositional formula, if a part of a proposition is replaced by a formula, a new formula will be produced. For example, if $(P \to Q)$ is replaced in $P \to (P \lor (P \to Q))$ with $(P \land Q)$, then $P \to (P \lor (P \land Q))$ is different from the original formula. To ensure that the substituted formula is equivalent to the original formula, it is necessary to make some provisions for the replacement.

Definition 1.3.4　If Y is a part of well-formed formulas A, and Y itself is a well-formed formula, then Y is called a subformula of formula A. For example, if $A \Leftrightarrow Q \rightarrow (P \vee (P \wedge Q))$, $Y \Leftrightarrow P \wedge Q$, then Y is a subformula of A.

Theorem 1.3.1　Let Y be a subformula of A. If $Y \Leftrightarrow X$, replacing Y in A with X, the obtained formula B is equivalent to formula A, that is $A \Leftrightarrow B$.

Proof　Because under any assignment of the corresponding argument, the true values of Y and X are the same, after replacing Y with X, the true values of formula B and formula A must be the same under the corresponding assignment of truth values, and then $A \Leftrightarrow B$.

The permutation satisfying the conditions of Theorem 1.3.1 is called an equivalent permutation.

Example 1.3.8　Prove $P \leftrightarrow Q \Leftrightarrow (P \wedge Q) \vee (\neg P \wedge \neg Q)$ with equivalent permutation and truth table.

Proof

$$P \leftrightarrow Q \Leftrightarrow (P \rightarrow Q) \wedge (Q \rightarrow P)$$
$$\Leftrightarrow (\neg P \vee Q) \wedge (\neg Q \vee P)$$
$$\Leftrightarrow (\neg P \wedge \neg Q) \vee (\neg P \wedge P) \vee (Q \wedge \neg Q) \vee (Q \wedge P)$$
$$\Leftrightarrow (\neg P \wedge \neg Q) \vee F \vee (Q \wedge P)$$
$$\Leftrightarrow (\neg P \wedge \neg Q) \vee (Q \wedge P)$$
$$\Leftrightarrow (P \wedge Q) \vee (\neg P \wedge \neg Q)$$
$$P \leftrightarrow Q \Leftrightarrow (P \wedge Q) \vee (\neg P \wedge \neg Q)$$

The equivalence is verified by the truth table, as shown in Table 1.3.10.

Table　1.3.10

P	Q	$\neg P$	$\neg Q$	$\neg P \wedge \neg Q$	$P \wedge Q$	$P \leftrightarrow Q$	$(P \wedge Q) \vee (\neg P \wedge \neg Q)$
T	T	F	F	F	T	T	T
T	F	F	T	F	F	F	F
F	T	T	F	F	F	F	F
F	F	T	T	T	F	T	T

Example 1.3.9　Prove $(P \wedge Q) \vee (P \wedge \neg Q) \Leftrightarrow P$ with equivalent permutations.

Proof

$$(P \wedge Q) \vee (P \wedge \neg Q) \Leftrightarrow P \wedge (Q \vee \neg Q)$$
$$\Leftrightarrow P \wedge T$$
$$\Leftrightarrow P$$

Example 1.3.10　Prove $Q \vee \neg ((P \rightarrow Q) \wedge P) \Leftrightarrow T$ with equivalent permutations.

Proof

$$Q \vee \neg ((P \rightarrow Q) \wedge P) \Leftrightarrow Q \vee \neg ((\neg P \vee Q) \wedge P)$$

$$\Leftrightarrow Q \vee \neg((\neg P \wedge P) \vee (Q \wedge P))$$

$$\Leftrightarrow Q \vee \neg(Q \wedge P)$$

$$\Leftrightarrow Q \vee (\neg Q \vee \neg P)$$

$$\Leftrightarrow (Q \vee \neg Q) \vee \neg P$$

$$\Leftrightarrow T \vee \neg P$$

$$\Leftrightarrow T$$

Example 1.3.11 Prove $((P \vee Q) \wedge \neg(\neg P \wedge (\neg Q \vee \neg R))) \vee (\neg P \wedge \neg Q) \vee (\neg P \wedge \neg R) \Leftrightarrow T$ with equivalent permutations.

Proof

The left side of the original formula $\Leftrightarrow ((P \vee Q) \wedge \neg(\neg P \wedge \neg(Q \wedge R))) \vee \neg(P \vee Q) \vee \neg(P \vee R)$

$$\Leftrightarrow ((P \vee Q) \wedge (P \vee (Q \wedge R))) \vee \neg(P \vee Q) \vee \neg(P \vee R)$$

$$\Leftrightarrow ((P \vee Q) \wedge ((P \vee Q) \wedge (P \vee R))) \vee \neg((P \vee Q) \wedge (P \vee R))$$

$$\Leftrightarrow ((P \vee Q) \wedge (P \vee R)) \vee \neg((P \vee Q) \wedge (P \vee R))$$

$$\Leftrightarrow T$$

1.4 Tautology and Implication

According to the truth table and proposition in Section 1.3, some propositions have true or false values no matter how they are assigned. The formula can be divided according to the value of the formula.

Definition 1.4.1 Let A be any propositional formula.

(1) If the value of A is true under any assignments, then A is called a tautology.

(2) If the value of A is true under any assignments, then A is called contradictory or absurdity.

(3) If A is not contradictory, then A is called satisfiable paradox.

According to the above definition, we can summarize the following properties.

(1) There is at least one real assignment in satisfiability.

(2) Tautology must be satisfiable, but satisfiable formula is not necessarily tautology.

Theorem 1.4.1 Any combination or disjunction of two tautologies is still a tautology.

Inference Any combination or disjunction of two paradoxes is still a paradox.

Theorem 1.4.2 A tautology is replaced by the same formula for every occurrence of the same component, and the result is still tautology.

Inference A paradox is replaced by the same formula in every place of the same component, and the result is still contradictory.

In the truth table, if the last column is all T, the formula is a tautology; if the last column is all F, the formula is contradictory; if at least one of the last columns is T, the formula is satisfiable.

Example 1.4.1 Determine the type of the propositional formula.

(1) $P \wedge \neg P$, (2) $P \vee \neg P$, (3) $\neg P \vee Q$.

Solution Table 1.4.1 gives the truth table of the three formulas.

Table 1.4.1

P	Q	$P \wedge \neg P$	$P \vee \neg P$	$\neg P \vee Q$
F	F	F	T	T
F	T	F	T	T
T	F	F	T	F
T	T	F	T	T

From Table 1.4.1, the values of (1) are all F, which is contradictory; the values of (2) are all T, which is tautologies; at least one of the values of (3) is T, the equation can be satisfied.

From the above discussion, it can be seen that a truth table can not only give the true value and false value of the formula, but also judge the type of formula.

Definition 1.4.2 Let A and B be propositional formulas. If and only if $A \rightarrow B$ is a tautology, $A \rightarrow B$ is called the implication of $A \Rightarrow B$.

$P \rightarrow Q$ can also be understood as that Q is a necessary condition for P.

Theorem 1.4.3 Implication is an important tool of logical reasoning. Here are some important implications, where A, B, C, D are arbitrary propositional formulas.

(1) Law of additional: $A \Rightarrow A \vee B$

(2) Law of reduction: $A \wedge B \Rightarrow A$

(3) Hypothetical reasoning: $A \wedge (A \rightarrow B) \Rightarrow B$

(4) modus tollendo ponens: $\neg B \wedge (A \rightarrow B) \Rightarrow \neg A$

(5) Disjunctive syllogism: $\neg B \wedge (A \vee B) \Rightarrow A$

(6) Hypothetical syllogism: $(A \rightarrow B) \wedge (B \rightarrow C) \Rightarrow (A \rightarrow C)$

(7) Equivalent syllogism: $(A \leftrightarrow B) \wedge (B \leftrightarrow C) \Rightarrow (A \leftrightarrow C)$

(8) Structural dilemma:

$$(A \vee C) \wedge (A \rightarrow B) \wedge (C \rightarrow D) \Rightarrow B \vee D$$

$$(A \vee \neg A) \wedge (A \rightarrow B) \wedge (\neg A \rightarrow B) \Rightarrow B$$

(9) Destructive dilemma:

$$(\neg B \vee \neg D) \wedge (A \rightarrow B) \wedge (C \rightarrow D) \Rightarrow (\neg A \vee \neg C)$$

Theorem 1.4.4 Let A and B be any two propositional formulas, The necessary and sufficient condition of $A \Leftrightarrow B$ is $A \Rightarrow B$ and $B \Rightarrow A$.

Theorem 1.4.5 Let A, B and C be the combined formula.

(1) $A \Rightarrow A$ (Implication is reflexive).

(2) If $A \Rightarrow B$ and A is a tautology, then B must be a tautology.

(3) If $A \Rightarrow B$ and $B \Rightarrow C$, then $A \Rightarrow C$ (Implication is transitive).

(4) If $A \Rightarrow B$ and $A \Rightarrow C$, then $A \Rightarrow B \wedge C$.

(5) If $A \Rightarrow B$ and $C \Rightarrow B$, then $A \vee C \Rightarrow B$.

(6) If $A \Rightarrow B$, C is an arbitrary formula, then $A \wedge C \Rightarrow B \wedge C$.

1.5 Duality and Normal Form

From Section 1.4, we can see that except for involution law, propositional laws appear in pairs, but the difference is that $\wedge$ and $\vee$ are interchangeable. We call such a formula dual law.

Definition 1.5.1   In the propositional formula A which contains only the connectives $\neg$, $\vee$, $\wedge$, the conjunctions $\vee$, $\wedge$, F and T are replaced by $\wedge$, $\vee$, T, F respectively. The formula obtained is called the dual formula of formula A, which is denoted as A^*.

Let A^* be the duality of A, if we change $\vee$, $\wedge$, F, T in A^* into $\wedge$, $\vee$, T, F respectively, A will be obtained. That is, A is the dual of A^*, $(A^*)^* \Leftrightarrow A$. Therefore, A^* and A are dual to each other.

Example 1.5.1   Write the duality of the following expression.

(1) $(A \vee B) \wedge C$

(2) $(A \wedge B) \vee T$

(3) $\neg B \wedge (A \vee B) \wedge (A \vee C)$

Solution   The corresponding dual formula of the above expression.

(1) $(A \wedge B) \vee C$

(2) $(A \vee B) \wedge F$

(3) $\neg B \vee (A \wedge B) \vee (A \wedge C)$

Theorem 1.5.1   Let A^* be the duality of A, p_1, p_2, ..., p_n is the atomic argument in A and A^*, then

$$\neg A(p_1, p_2, ..., p_n) \Leftrightarrow A^*(\neg p_1, \neg p_2, ..., \neg p_n)$$

$$A(\neg p_1, \neg p_2, ..., \neg p_n) \Leftrightarrow \neg A^*(p_1, p_2, ..., p_n)$$

Theorem 1.5.2   Let p_1, $p_2 ..., p_n$, be the atomic argument in A and B, if $A \Leftrightarrow B$ then $A^* \Leftrightarrow B^*$.

Definition 1.5.2   A disjunctive form consisting of some propositional arguments or their negations is called basic sum, also called simple disjunctive. A single argument or its negation is a fundamental sum.

For example, $\neg p \vee q$, $p \vee \neg q$, $p \vee q$, $\neg q$, $\neg p$, q are all fundamental sums.

Definition 1.5.3   A disjunctive form consisting of some propositional arguments or their negations is called a basic product, also called a simple conjunction. A single argument or its negation is a fundamental product.

For example, $\neg p \wedge q$, $p \wedge \neg q$, $p \wedge q$, $\neg p$, $\neg q$, p are all fundamental products.

Note that an argument is both a simple disjunctive and simple conjunction.

Definition 1.5.4 The formula composed of the conjunction of the basic sum is called conjunctive normal form. It is agreed that a single basic sum is a conjunctive normal form.

Definition 1.5.5 The formula composed of disjunctive fundamental product is called disjunctive normal form. A single elementary product is a disjunctive normal form.

Disjunctive and conjunctive paradigms are called paradigms. Disjunctive and conjunctive paradigms have the following properties.

(1) A disjunctive normal form is a contradiction if and only if every simple conjunction of it is a contradiction.

(2) A conjunctive paradigm is a tautology if and only if every simple disjunctive is a tautology.

For any propositional formula, the steps to find its conjunctive normal form and disjunctive normal form are as follows.

(1) Eliminate conjunctions "→" and "↔" in the formula which are transformed into $\land$, $\lor$ and $\neg$, using implication equivalence and equivalent equivalence.

Example 1.5.2 $A \rightarrow B \Leftrightarrow \neg A \lor B$

$A \leftrightarrow B \Leftrightarrow (A \rightarrow B) \land (B \rightarrow A)$

(2) The negative conjunction "$\neg$" is eliminated by using the double negative law or moved before the argument of each proposition by using the de Morgan law.

Example 1.5.3 $\neg \neg A \Leftrightarrow A$

$\neg (A \lor B) \Leftrightarrow \neg A \land \neg B$

$\neg (A \land B) \Leftrightarrow \neg A \lor \neg B$

(3) The formula is reduced to conjunctive normal form and disjunctive normal form by using distributive law and combining law.

Example 1.5.4 $A \land (B \lor C) \Leftrightarrow (A \land B) \lor (A \land C)$

$A \lor (B \land C) \Leftrightarrow (A \lor B) \land (A \lor C)$

Example 1.5.5 Finding the conjunctive and disjunctive paradigms of $(p \lor q) \leftrightarrow p$.

Solution (1) Find the conjunctive normal form

$(p \lor q) \leftrightarrow p \Leftrightarrow ((p \lor q) \rightarrow p) \land (p \rightarrow (p \lor q))$ (Eliminate↔)

$\Leftrightarrow (\neg (p \land q) \lor p) \land (\neg p \lor (p \lor q))$

$\Leftrightarrow ((\neg p \land \neg q) \lor p) \land (\neg p \lor (p \lor q))$ ($\neg$ Introversion)

$\Leftrightarrow (\neg p \lor p) \land (\neg q \lor p) \land (\neg p \lor p \lor q)$ (Distributive law, Conjunctive paradigm)

$\Leftrightarrow T \land (\neg q \lor p) \land (T \lor q)$ (The law of excluded middle)

$\Leftrightarrow T \land (\neg q \lor p) \land T$ (Zero law, Conjunctive paradigm)

$\Leftrightarrow (\neg q \lor p)$ (Identity, Conjunctive paradigm)

(2) Find disjunctive normal form.

$(p \lor q) \leftrightarrow p \Leftrightarrow ((p \lor q) \land p) \lor (\neg (p \lor q) \land \neg p)$ (Eliminate↔)

$\Leftrightarrow ((p \lor q) \land p) \lor ((\neg p \land \neg q) \land \neg p)$ ($\neg$ Introversion)

$\Leftrightarrow p \lor (\neg p \land \neg q \land \neg p)$ (Law of absorption, Disjunctive normal form)

$\Leftrightarrow p \lor (\neg p \land \neg p \land \neg q)$ (Exchange law)

$\Leftrightarrow p \lor (\neg p \land \neg q)$ (Idempotent law, Disjunctive normal form)

From this example, we can see that the disjunctive normal form of the propositional formula is not unique.

Definition 1.5.6   In the basic product, each variable and its negation do not exist at the same time, but one of them must appear and only appear once. Such a basic product is called Boolean conjunction, also called minor term or minimal term. The minimal term of p, q is $p \wedge q, p \wedge \neg q, \neg p \wedge q, \neg p \wedge \neg q$.

Generally, n propositional variables have 2^n minima.

Table 1.5.1 is the truth table of the minima of two variables p and q. The minima have the following properties.

(1) There is only one real assignment for each minimal term, and the real assignment of each minimal term is different from each other. There is a one-to-one correspondence between the minimal term and its real assignment. We can encode the minimal term with the real assignment, and use the code as the subscript of m to represent the minimal item, which is called the name of the minimal item. The minimal term, real assignment and names of the two propositional arguments are shown in Table 1.5.2. The minimal term, real assignment and name of three propositional arguments are shown in Table 1.5.3. From Table 1.5.2 and Table 1.5.3, it can be seen that the corresponding relationship between the minimal term and its real assignment is: the argument corresponds to T, and the negative of the argument corresponds to F.

Table   1.5.1

p	q	$p \wedge q$	$p \wedge \neg q$	$\neg p \wedge q$	$\neg p \wedge \neg q$
F	F	F	F	F	T
F	T	F	F	T	F
T	F	F	T	F	F
T	T	T	F	F	F

Table   1.5.2

Minterm	Real assignment	Name	Minterm	Real assignment	Name
$\neg p \wedge \neg q$	FF	m_0	$p \wedge \neg q$	TF	m_2
$\neg p \wedge q$	FT	m_1	$p \wedge q$	TT	m_3

Table   1.5.3

Minterm	Real assignment	Name	Minterm	Real assignment	Name
$\neg p \wedge \neg q \wedge \neg r$	FFF	m_0	$p \wedge \neg q \wedge \neg r$	TFF	m_4
$\neg p \wedge \neg q \wedge r$	FFT	m_1	$p \wedge \neg q \wedge r$	TFT	m_5
$\neg p \wedge q \wedge \neg r$	FTF	m_2	$p \wedge q \wedge \neg r$	TTF	m_6
$\neg p \wedge q \wedge r$	FTT	m_3	$p \wedge q \wedge r$	TTT	m_7

(2) The conjunction of any two different minimal terms is an eternal false form. For example:

$$m_{001} \wedge m_{100} \Leftrightarrow (\neg p \wedge \neg q \wedge r) \wedge (p \wedge \neg q \wedge \neg r)$$
$$\Leftrightarrow \neg p \wedge \neg q \wedge r \wedge p \wedge \neg q \wedge \neg r \Leftrightarrow F$$

(3) The disjunctive form of all minimal terms is the eternal truth form. Recorded as:

$$\sum_{i=0}^{2^n-1} m_i \Leftrightarrow m_0 \lor m_1 \lor \ldots \lor m_{2^n-1} \Leftrightarrow T$$

Definition 1.5.7　For a given propositional formula, if there is an equivalent formula that only consists of disjunction of minimal terms, the formula is called the main disjunctive normal form of the original formula.

Any propositional formula has its equivalent principal disjunctive normal form. The principal disjunctive normal form of a propositional formula can be obtained by the following two methods.

(1) The steps for finding the principal disjunctive normal form with an equivalent algorithm are as follows.

① Reduction to disjunctive paradigm.

② Remove all permanent and false fundamental products from disjunctive normal forms.

③ In the basic product, the repeated conjunctions are merged with the same variables.

④ In the basic product, add a proposition variable that does not appear, that is, add $\land (p \lor \lnot p)$, expand it with the distributive law, and finally merge the same minimal term.

Example　1.5.6　Find the principal disjunctive normal form of $(p \land q) \lor (\lnot p \land r) \lor (q \land r)$ by equivalent algorithm.

Solution　$(p \land q) \lor (\lnot p \land r) \lor (q \land r) \Leftrightarrow (p \land q \land (r \lor \lnot r)) \lor (\lnot p \land r \land (q \lor \lnot q))$
$\lor (q \land r \land (p \lor \lnot p))$
$\Leftrightarrow (p \land q \land r) \lor (p \land q \land \lnot r) \lor (\lnot p \land q \land r) \lor$
$(\lnot p \land \lnot q \land r) \lor (p \land q \land r) \lor (\lnot p \land q \land r)$
$\Leftrightarrow (p \land q \land r) \lor (p \land q \land \lnot r) \lor (\lnot p \land q \land r) \lor$
$(\lnot p \land \lnot q \land r)$
$\Leftrightarrow m_{111} \lor m_{110} \lor m_{011} \lor m_{001}$
$\Leftrightarrow m_7 \lor m_6 \lor m_3 \lor m_1 \Leftrightarrow \Sigma 1,3,6,7$

(2) Truth table method: using a truth table to find principal disjunctive normal form.

① Construct a truth table of propositional formula.

② Find the minimal term corresponding to the real assignment of the formula.

③ The disjunction of these minimal terms is the main disjunctive normal form of this formula.

Example　1.5.7　The principal disjunctive normal form of $(p \to q) \to r$ is obtained by the truth table method.

Solution　Table 1.5.4 is the truth table of $(p \to q) \to r$.

Table　1.5.4

p	q	r	$p \to q$	$(p \to q) \to r$	p	q	r	$p \to q$	$(p \to q) \to r$
0	0	0	1	0	1	0	0	0	1
0	0	1	1	1	1	0	1	0	1
0	1	0	1	0	1	1	0	1	0
0	1	1	1	1	1	1	1	1	1

According to Table 1.5.4, the minimum term corresponding to the real value of the formula is

$\neg p \wedge \neg q \wedge r$ (Real assignment is 001)

$\neg p \wedge q \wedge r$ (Real assignment is 011)

$p \wedge \neg q \wedge \neg r$ (Real assignment is 100)

$p \wedge \neg q \wedge r$ (Real assignment is 101)

$p \wedge q \wedge r$ (Real assignment is 111)

The main disjunctive normal form with the real value of $(p \rightarrow q) \rightarrow r$ is as follows.

$(p \wedge q \wedge r) \vee (p \wedge \neg q \wedge r) \vee (p \wedge \neg q \wedge \neg r) \vee (\neg p \wedge q \wedge r) \vee (\neg p \wedge \neg q \wedge r)$

$\Leftrightarrow m_{111} \vee m_{101} \vee m_{100} \vee m_{011} \vee m_{001} \Leftrightarrow m_7 \vee m_5 \vee m_4 \vee m_3 \vee m_1 \Leftrightarrow \Sigma 1,3,4,5,7$

Therefore, the principal disjunctive normal form does not contain any minimal term. The principal disjunctive normal form of the contradiction is marked as 0. Since tautology has no false assignment, the main disjunctive normal form contains 2^n (n is the number of propositional arguments in the formula). As for satisfiability, the number of minimal terms in its principal disjunctive normal form must be less than or equal to 2^n.

Definition 1.5.8   In the basic sum, each argument and its negation do not exist at the same time, but one of them must appear and only appear once. Such a basic sum is called Boolean disjunction, also known as a major term or maximal term.

The maximum term of two variables p and q is:

$$p \vee q \qquad p \vee \neg q \qquad \neg p \vee q \qquad \neg p \vee \neg q$$

In general, n variables have 2^n maximal terms.

The maximal term has the following three properties.

(1) There is only one false assignment for each maximal term, and the false assignment varies with the maximum term. The maximum term and its false assignment constitute a one-to-one correspondence. Therefore, the maximum term can be encoded by false assignment, and the maximum term is represented by the code as the subscript of M, which is called the name of the maximum term.

For example, the maximum term $\neg p \vee \neg q$ of the two arguments p, q has a false value of 11, which is expressed as M_{11}. 11 is understood as a binary number, and its decimal system is expressed as 3, so M_{11} is expressed as M_3.

See Table 1.5.5 for the maximum term, false assignment and name of two propositional arguments, and Table 1.5.6 for the maximum term, false assignment and name of the three propositional variables.

Table   1.5.5

Maximal term	False assignment	Name
$p \vee q$	00	M_0
$p \vee \neg q$	01	M_1
$\neg p \vee q$	10	M_2
$\neg p \vee \neg q$	11	M_3

Table   1.5.6

Maximal term	False assignment	Name	Maximal term	False assignment	Name
$p \vee q \vee r$	000	M_0	$\neg p \vee q \vee r$	100	M_4
$p \vee q \vee \neg r$	001	M_1	$\neg p \vee q \vee \neg r$	101	M_5
$p \vee \neg q \vee r$	010	M_2	$\neg p \vee \neg q \vee r$	110	M_6
$p \vee \neg q \vee \neg r$	011	M_3	$\neg p \vee \neg q \vee \neg r$	111	M_7

It can be found from Table 1.5.5 and Table 1.5.6 that the corresponding relationship between the maximum term and the false assignment is: the argument corresponds to 0, and the negative of the argument corresponds to 1.

(2) The disjunctive form of any two different maximal terms is an eternal truth form.

(3) The conjunctive form of all the maximal items is eternal false. As follows:

$$\prod_{i=0}^{2^n-1} M_i \Leftrightarrow M_0 \wedge M_1 \wedge \ldots \wedge M_{2^n-1} \Leftrightarrow 0$$

Definition 1.5.9   For a given propositional formula, if there is an equivalent formula that only consists of the conjunction of the maximal term, the equivalent formula is called the main conjunctive normal form of the original formula.

Any propositional formula has its equivalent principal conjunctive normal form. The principal conjunctive normal form can also be obtained by the following two methods.

(1) Equivalent algorithm: that is, it is derived from the basic equivalent formula. The calculation steps are as follows.

① It is reduced to the conjunctive paradigm.

② Remove all the real basic sums.

③ Merges recurring disjunctions and the same arguments in the base.

④ Add propositional arguments that do not appear in the basic sum. That is to add $\vee (p \wedge \neg p)$, then expand the formula by applying the distributive law, and finally, merge the same maximal terms.

Example   1.5.8   The principal conjunctive normal form of $(p \rightarrow q) \rightarrow r$ is obtained by equivalent calculus.

$$(p \rightarrow q) \rightarrow r \Leftrightarrow \neg(\neg p \vee q) \vee r \Leftrightarrow (p \wedge \neg q) \vee r \Leftrightarrow (p \vee r) \wedge (\neg q \vee r)$$
$$\Leftrightarrow (p \vee r \vee (q \wedge \neg q)) \wedge (\neg q \vee r \vee (p \wedge \neg p))$$
$$\Leftrightarrow (p \vee r \vee q) \wedge (p \vee r \vee \neg q) \wedge (p \vee \neg q \vee r) \wedge (\neg p \vee \neg q \vee r)$$
$$\Leftrightarrow (p \vee q \vee r) \wedge (p \vee \neg q \vee r) \wedge (\neg p \vee \neg q \vee r)$$
$$\Leftrightarrow M_{000} \wedge M_{010} \wedge M_{110} \Leftrightarrow M_0 \wedge M_2 \wedge M_6 \Leftrightarrow \prod_{0,2,6}$$

(2) Truth table method: use the truth table to find the principal conjunction normal form. The calculation steps are as follows.

① Construct truth table of propositional formula.

② Find the maximum term corresponding to the false assignment of the formula.

③ The disjunction of these maxima is the main conjunctive normal form of this formula.

Example **1.5.9** The principal conjunctive normal form of $(p \to q) \to r$ is obtained by the truth table method. The truth table of $(p \to q) \to r$ is Table 1.5.4. The large term corresponding to the false assignment of the formula is as follows.

$p \lor q \lor r$ (False assignment is 000)

$p \lor \neg q \lor r$ (False assignment is 010)

$\neg p \lor \neg q \lor r$ (False assignment is 110)

The principal conjunctive paradigm is:

$$(p \lor q \lor r) \land (p \lor \neg q \lor r) \land (\neg p \lor \neg q \lor r)$$
$$\Leftrightarrow M_{000} \land M_{010} \land M_{110} \Leftrightarrow M_0 \land M_2 \land M_6 \Leftrightarrow \prod_{0,2,6}$$

Therefore, the principal conjunctive normal form of the contradiction contains 2^n maxima (n is the number of propositional arguments in the formula). However, tautology has no false assignment, so the principal conjunctive normal form does not contain any maximum term. Mark the main conjunctive paradigm of tautology as 1. As for satisfiability, the number of maximal terms in their principal conjunctive normal form must be less than 2^n.

1.6 The Reasoning Theory of Propositional Calculus

Logic is the study of the forms and laws of thought. One of its main tasks is to provide a set of rules for reasoning. The process of deducing a conclusion from a set of premises and using the rules of inference provided is often called deductive or formal proof.

In general, empirically, if the premises are true, the conclusions derived from the rules of reasoning provided should also be true. In mathematics, these premises are often called axioms, and the resulting conclusion is called a theorem. This process of reasoning is called the process of proof.

In mathematical logic, attention is focused on the study of inference rules. First, some reasoning rules are given. Starting from some premises and based on the reasoning rules provided, a conclusion is derived. Such a conclusion is called a valid conclusion. In determining the validity of an argument, we do not care about the actual truth value of the premises. In other words, in mathematical logic, we study the validity of deduction, not the correctness. A valid conclusion is true only if the premises are all true, and since the premises are not always true, its valid conclusion is not always true. So when the premises are false, the premises contain the conclusion whether or not the conclusion is true. This is different from the reasoning usually applied in practice.

1.6.1 Effective Conclusion

Definition 1.6.1 Let A and B be two propositional formulae, if and only if $A \to B$ is an eternally true formula, namely $A \Leftrightarrow B$, then B is A valid conclusion of A, or B can be deduced logically from A.

This definition can be extended to n premises.

Suppose $H_1, H_2, \ldots, H_n, B$ is a propositional formula, if and only if $H_1 \wedge H_2 \wedge \ldots \wedge H_n \Rightarrow B$ says that B is a valid conclusion of the premise set $\{H_1, H_2, \ldots, H_n\}$, or that $H_1, H_2, \ldots, H_n$ logically leads to B.

1.6.2　A Method for Proving a Valid Conclusion

1. Truth Table Method

`Example` `1.6.1`　An error in a statistical form is due either to unreliable material or to faulty calculations. The error of this statistical table is not due to the unreliable materials, so the error of this statistical table is due to a calculation error.

`Solution`　Suppose P: the error in a statistical table is due to unreliable materials; Q: An error in a statistical table is due to an error in calculation.

The original proposition can be symbolized as:

$$(P \vee Q) \wedge \neg P \Rightarrow Q$$

List the truth tables as shown in Table 1.6.1.

Table　1.6.1

P	Q	$P \vee Q$	$\neg P$	$(P \vee Q) \wedge \neg P$
T	T	T	F	F
T	F	T	F	F
F	T	T	T	T
F	F	F	T	F

As shown in Table 1.6.1, the ante-post $(P \vee Q) \wedge \neg P$ of the third exercise is T, the ante-post Q is also T. So Q is a valid conclusion for the premise $(P \vee Q) \wedge \neg P$.

`Example` `1.6.2`　If Mr. Zhang comes, the question can be answered, and if Mr. Li comes, the question can be answered too. In a word, here comes Mr. Zhang or Mr. Li, the question can be answered.

`Solution`　P: Here comes Mr. Zhang, Q: Here comes Mr. Li, R: The question can be answered. Then the original proposition is expressed as:

$$(P \to R) \wedge (Q \to R) \wedge (P \vee Q) \Rightarrow R$$

List the truth tables as shown in Table 1.6.2.

Table　1.6.2

P	Q	R	$P \to R$	$Q \to R$	$P \vee Q$	P	Q	R	$P \to R$	$Q \to R$	$P \vee Q$
T	T	T	T	T	T	F	T	T	T	T	T
T	T	F	F	F	T	F	T	F	T	F	T
T	F	T	T	T	T	F	F	T	T	T	F
T	F	F	F	T	T	F	F	F	T	T	F

As you can see from the table, there are 3 situations which $P \rightarrow R$, $Q \rightarrow R$, $P \lor Q$ truth values are all T, and the truth values of R arc all T. So

$$(P \lor Q) \land (Q \rightarrow R) \land (P \lor Q) \Rightarrow R$$

Example 1.6.3 Today is Sunday or it is sunny; It's Sunday today. So, it's sunny today.

Solution Suppose P: Today is Sunday, Q: It's sunny today. The original proposition can then be symbolized as:

$$(P \lor Q) \land P \Rightarrow Q$$

The truth table is listed as shown in Table 1.6.3.

Table 1.6.3

P	Q	$P \lor Q$	$(P \lor Q) \land Q$	Q
T	T	T	T	T
T	F	T	T	F
F	T	T	F	T
F	F	F	F	F

As can be seen from truth Table 1.6.3, the truth value of the premise in the second row is T and the truth value of the corresponding conclusion is F. Therefore, this conclusion is not a valid conclusion of the premise.

2. Direct Proofs

Direct proof is a group of premises, using some recognized rules of reasoning, according to the known equivalent or implication formula, deducing effective conclusions.

Rule P: premise can be introduced and used at any time in the derivation process. Rule T: in derivation, if one or more formulas and tautology contain formula S, formula S can be introduced into derivation.

The commonly used implication and equivalence expressions are listed in Table 1.6.4 and Table 1.6.5.

Table 1.6.4

Name	Implication expression	Name	Implication expression
I_1	$P \land Q \Rightarrow P$	I_4	$Q \Rightarrow P \lor Q$
I_2	$P \land Q \Rightarrow Q$	I_5	$\neg P \Rightarrow P \rightarrow Q$
I_3	$P \Rightarrow P \lor Q$		

Table 1.6.5

Name	Equivalence expression	Name	Equivalence expression
E_1	$\neg \neg P \Leftrightarrow P$	E_8	$\neg (P \land Q) \Leftrightarrow \neg P \lor \neg Q$
E_2	$P \land Q \Leftrightarrow Q \land P$	E_9	$\neg (P \lor Q) \Leftrightarrow \neg P \land \neg Q$
E_3	$P \lor Q \Leftrightarrow Q \lor P$	E_{10}	$P \lor P \Leftrightarrow P$
E_4	$(P \land Q) \land R \Leftrightarrow P \land (Q \land R)$	E_{11}	$P \land P \Leftrightarrow P$
E_5	$(P \lor Q) \lor R \Leftrightarrow P \lor (Q \lor R)$	E_{12}	$R \lor (P \land \neg P) \Leftrightarrow R$
E_6	$P \land (Q \lor R) \Leftrightarrow (P \land Q) \lor (P \land R)$	E_{13}	$R \land (P \lor \neg P) \Leftrightarrow R$
E_7	$P \lor (Q \land R) \Leftrightarrow (P \lor Q) \land (P \lor R)$	E_{14}	$R \lor (P \lor \neg P) \Leftrightarrow T$

续表

Name	Equivalence expression	Name	Equivalence expression
E_{15}	$R \wedge (P \wedge \neg P) \Leftrightarrow F$	E_{19}	$P \rightarrow (Q \rightarrow R) \Leftrightarrow (P \wedge Q) \rightarrow R$
E_{16}	$P \rightarrow Q \Leftrightarrow \neg P \vee Q$	E_{20}	$P \rightleftarrows Q \Leftrightarrow (P \rightarrow Q) \wedge (Q \rightarrow P)$
E_{17}	$\neg(P \rightarrow Q) \Leftrightarrow P \wedge \neg Q$	E_{21}	$P \rightleftarrows Q \Leftrightarrow (P \wedge Q) \vee (\neg P \wedge \neg Q)$
E_{18}	$P \rightarrow Q \Leftrightarrow \neg Q \rightarrow \neg P$	E_{22}	$\neg(P \rightleftarrows Q) \Leftrightarrow P \rightleftarrows \neg Q$

Example 1.6.4 Prove:

$$(P \vee Q) \wedge (P \rightarrow R) \wedge (Q \rightarrow S) \Rightarrow S \vee R$$

Proof 1
(1) $P \vee Q$ P
(2) $\neg P \rightarrow Q$ T(1)E
(3) $Q \rightarrow S$ P
(4) $\neg P \rightarrow S$ T(2),(3)I
(5) $\neg S \rightarrow P$ T(4)E
(6) $P \rightarrow R$ P
(7) $\neg S \rightarrow R$ T(5),(6)I
(8) $S \vee R$ P

Proof 2
(1) $P \rightarrow R$ P
(2) $P \vee Q \rightarrow R \vee Q$ T(1)I
(3) $Q \rightarrow S$ P
(4) $Q \vee R \rightarrow S \vee R$ T(3)I
(5) $P \vee Q \rightarrow S \vee R$ T(2),(4)I
(6) $P \vee Q$ P
(7) $S \vee R$ T(5),(6)I

Example 1.6.5 Prove:

$$(W \vee R) \rightarrow V, \ V \rightarrow C \vee S, \ S \rightarrow U, \ \neg C \wedge \neg U \Rightarrow \neg W$$

Proof
(1) $\neg C \wedge \neg U$ P
(2) $\neg U$ T(1)I
(3) $S \rightarrow U$ P
(4) $\neg S$ T(2),(3)I
(5) $\neg C$ T(1)I
(6) $\neg C \wedge \neg S$ T(4),(5)I
(7) $\neg(C \vee S)$ T(6)E
(8) $(W \vee R) \rightarrow V$ P
(9) $V \rightarrow (C \vee S)$ P
(10) $(W \vee R) \rightarrow (C \vee S)$ T(8),(9)I
(11) $\neg(W \vee R)$ T(7),(10)I
(12) $\neg W \wedge \neg R$ T(11)E
(13) $\neg W$ T(12)I

3. Indirect Demonstration

Definition 1.6.2 If the propositional argument in formula H_1, H_2, ..., H_m is P_1, P_2, ..., P_n, and some truth assignments of P_1, P_2, ..., P_n can make the truth value of $H_1 \wedge H_2 \wedge ... \wedge H_m$ T, then formula H_1, H_2, ..., H_m is said to be compatible. Formula H_1, H_2, ..., H_m is said to be incompatible if the truth assignment of H_1, H_2, ..., H_m for each set of P_1, P_2, ..., P_n causes the truth value of H_1, H_2, ..., H_m to be F.

Incompatible concepts can now be applied to the proof of propositional formulas.

A group has H_1, H_2, ..., H_m premise, I want to derive conclusion C, which is the proof of $H_1 \wedge H_2 \wedge ... \wedge H_m \Rightarrow C$, named $S \Rightarrow C$, which $\neg C \rightarrow \neg S$ is true, or $C \vee \neg S$ is true, so $\neg C \wedge S$ is false. So if we want to prove $H_1 \wedge H_2 \wedge ... \wedge H_m \Rightarrow C$, we only need to prove $H_1 \wedge H_2 \wedge ... \wedge H_m$ and $\neg C$ are incompatible.

Example 1.6.6 Prove: $A \rightarrow B$, $\neg(B \vee C)$ can logically derive $\neg A$.

Proof
(1) $A \rightarrow B$ P
(2) A P
(3) $\neg(B \vee C)$ P
(4) $\neg B \wedge \neg C$ T(3)E
(5) B T(1), (2)I
(6) $\neg B$ T(4)I
(7) $B \wedge \neg B$ T(5), (6)I

Example 1.6.7 Prove:
$$(P \vee Q) \wedge (P \rightarrow R) \wedge (Q \rightarrow S) \Rightarrow S \vee R$$

Proof
(1) $\neg(S \vee R)$ P(Additional premise)
(2) $\neg S \wedge \neg R$ T(1)E
(3) $P \vee Q$ P
(4) $\neg P \rightarrow Q$ T(3)E
(5) $Q \rightarrow S$ P
(6) $\neg P \rightarrow S$ T(4), (5)I
(7) $\neg S \rightarrow P$ T(6)E
(8) $(\neg S \wedge \neg R) \rightarrow (P \wedge \neg R)$ T(7)I
(9) $P \wedge \neg R$ T(2), (8)I
(10) $P \rightarrow R$ P
(11) $\neg P \vee R$ T(10)E
(12) $\neg(P \wedge \neg R)$ T(11)E
(13) $(P \wedge \neg R) \wedge \neg(P \wedge \neg R)$ T(9), (12)I

1.7 Application of Propositional Logic

Logic has many important applications in mathematics, computer science and many other subjects. Propositional logic and its rules can be used to design computer circuits, construct computer programs, verify the correctness of programs, construct expert systems, and analyze and solve many familiar puzzles.

In digital circuits, the logic gate symbols are as follows.

AND gate:

OR gate:

NOT gate:

NOR gate:

AND OR NOT gate:

XOR gate:

XNOR gate:

Application in inference questions: for example, M company to improve the staff's professional workability, therefore, decided to send a few excellent staff to study abroad, A, B, C, D and E are selected from many applicants company several more qualified personnel, but in order not to affect the normal work need, their five not all go, so this time send must meet the following conditions:

① If A is sent abroad to study, B will also be sent abroad to study;

② Some of D and E will be sent to study abroad;

③ Only one of the B and C students was sent to study abroad;

④ Both C and D were sent, or neither;

⑤ If E is sent out to study, then A and B will also be sent out.

To arrange the work reasonably, how should the company choose from these five people and who should study abroad according to these conditions?

Solution 1 General inference method.

Analysis: This question is to deduce how many people should be sent out, who should be sent out, and certainly someone was sent out, but the number and character are not fixed, then we must assume one by one, to see what kind of assumption is reasonable.

(1) Suppose that if A must be sent, then according to condition ①, we can infer that B must also be sent. According to condition ③, we can infer that C will not go, and from condition ④, we can infer that C and D will not go. Then according to condition②, we can infer that e will go. After the above inference, we can come to a selection scheme: if A, B and E are sent to study, C and D will not be sent.

(2) Suppose that if B must be sent out, and according to condition 3, C will not be sent out, then condition ④ can infer that C and D will not be sent out, and condition ② can infer that e will be sent out, then condition ⑤can infer that a and B will also be sent out. After the above inference, we can also get a selection scheme: A, B and e will be sent out to study, C and D will not be sent.

(3) Suppose that if C must be dispatched and B is not dispatched according to condition ③, then from condition ④ it can be inferred that D must also be dispatched, and from condition ② it can be inferred that E may or may not be dispatched. Suppose that if E is dispatched, then from condition ⑤ it can be inferred that a and B must also be dispatched, which is in contradiction with the previous reasoning, so E must not be dispatched. Therefore, from the above reasoning, we can get a selection scheme: C and D are dispatched, and A, B and E are not dispatched.

(4) Suppose that if D must be dispatched, then it can be inferred from condition ④ that C must be dispatched, from condition ③ that B must not be dispatched, and from conditions ① and ⑤ that A and E must not be dispatched, so a dispatch scheme can be deduced from the above inference: C and D are dispatched, and a, B and E are not dispatched.

(5) Suppose that if E must be sent out, then it can be inferred from condition ⑤ that A and B must also be sent out to learn, and from ③ ④ that C and D must not be sent out, so from the above inference, we can get a scheme: A, B and E are sent to study, while C and D are not sent.

From the above analysis, we can conclude that the company has only two dispatch schemes: one is that A, B and E are sent to study, while C and D are not sent; The other is that C and D are dispatched, while A, B and E are not.

Solution 2 Use the propositional logic inference method.

Analysis: If we want to infer who will go and what kind of sending scheme they have, we only need to symbolize the necessary proposition, and then write the proposition formula F_1, F_2, F_3, F_4 and F_5 according to the given conditions. Then we only need to transform $T \Leftrightarrow F_1 \wedge F_2 \wedge F_3 \wedge F_4 \wedge F_5$ into the disjunctive normal form, then the problem can be solved.

To symbolize the proposition, make

p: A must be sent to study;

q: B must be sent to study;

r: C must be sent to study;

s: D must have been sent to study;

t: E must be sent to study.

So the conditions can be symbolized:

① $F_1 \Leftrightarrow p \rightarrow q$;

② $F_2 \Leftrightarrow s \vee t$;

③ $F_3 \Leftrightarrow (q \wedge \neg r) \vee (\neg q \wedge r)$;

④ $F_4 \Leftrightarrow (r \wedge s) \vee (\neg r \wedge \neg s)$;

⑤ $F_5 \Leftrightarrow t \rightarrow (p \wedge q)$

Since the conditions are known, their conjunction must be true. Set:

$$T = F_1 \wedge F_2 \wedge F_3 \wedge F_4 \wedge F_5$$
$$= (p \rightarrow q) \wedge (s \vee t) \wedge ((q \wedge \neg r) \vee (\neg q \wedge r)) \wedge ((r \wedge s) \vee (\neg r \wedge \neg s)) \wedge (t \rightarrow (p \wedge q))$$

To get the dispatched scheme, only the disjunctive formula of $T = F_1 \wedge F_2 \wedge F_3 \wedge F_4 \wedge F_5$ is required. Then the main steps to obtain the disjunctive formula are as follows:

$$T \Leftrightarrow (\neg p \vee q) \wedge (s \vee t) \wedge ((q \wedge \neg r) \vee (\neg q \wedge r)) \wedge ((r \wedge s) \vee (\neg r \wedge \neg s)) \wedge (\neg t \vee (p \wedge q))$$
$$\Leftrightarrow ((\neg p \wedge s) \vee (\neg p \wedge t) \vee (q \wedge s) \vee (q \wedge t)) \wedge ((q \wedge \neg r \wedge \neg s) \vee (\neg q \wedge r \wedge s) \wedge$$
$$(\neg t \vee (p \wedge q))$$
$$\Leftrightarrow (\neg p \wedge \neg q \wedge r \wedge s) \vee (\neg p \wedge q \wedge \neg r \wedge \neg s \wedge t) \vee (q \wedge \neg r \wedge \neg s \wedge t) \wedge (\neg t \vee (p \wedge q))$$
$$\Leftrightarrow (\neg p \wedge \neg q \wedge r \wedge s \wedge \neg t) \vee (p \wedge q \wedge \neg r \wedge \neg s \wedge t)$$

From the above inference, it can be easily concluded that there are only two schemes: one is that A, B and E are sent out to study, while C and D are not sent out; The other is that C and D are sent out, but A, B and E are not.

Propositional deductive reasoning is used in reasoning problems, and most of the reasoning rules are based on reasoning theorems.

Exercises

1. Indicate which of the following statements are propositions and which are not. If they are propositions, indicate their truth values.

(1) Discrete mathematics is a required course in computer science and technology.

(2) Is the computer available?

(3) I'm going to the cinema tomorrow.

(4) No spitting!

(5) There is no maximum prime number.

(6) If I master English and French, it will be much easier to learn other European languages.

(7) $11 + 6 < 10$.

(8) $y = 4$.

(9) We must study hard.

2. Let P stands for the proposition "Snow under the sky", Q stands for "I'm going to town", R stands for the proposition "I have time".

Write the following propositions in symbolic form.

(1) If it doesn't snow and I have time, then I'll go to town.

(2) I will go to town only when I have time.

(3) It doesn't snow.

(4) If it snows, then I won't go to town.

3. Symbolize the following propositions.

(1) Wang Qiang is in good health and gets good grades.

(2) Xiao Li listened to music while reading a book.

(3) The weather is good or hot.

(4) If a and b are even, then a plus b is even.

(5) Quadrilateral $ABCD$ is a parallelogram, if and only if its opposite side is parallel.

4. Determine which of the following is the formula and which is not.

(1) $(Q \rightarrow R \wedge S)$

(2) $(P \rightleftarrows (R \rightarrow S))$

(3) $((\neg P \rightarrow Q) \rightarrow (Q \rightarrow P))$

(4) $(RS \rightarrow T)$

(5) $((P \rightarrow (Q \rightarrow R)) \rightarrow ((P \rightarrow Q) \rightarrow (P \rightarrow R)))$

5. Try to explain that the following formula is a fit formula according to the definition of the fit formula.

(1) $(A \rightarrow (A \vee B))$

(2) $((\neg A \wedge B) \wedge A)$

(3) $((\neg A \rightarrow B) \rightarrow (B \rightarrow A))$

(4) $((A \rightarrow B) \vee (B \rightarrow A))$

6. Which of the following equations are obtained by substitution?

(1) $(P \rightarrow (Q \rightarrow P))$

(2) $((((P \rightarrow Q) \wedge (R \rightarrow S)) \wedge (P \vee R)) \rightarrow (Q \vee S))$

(3) $(Q \rightarrow ((P \rightarrow P) \rightarrow Q))$

(4) $(P \rightarrow ((P \rightarrow (Q \rightarrow P)) \rightarrow P))$

(5) $(((R \rightarrow S) \wedge (Q \rightarrow P) \wedge (R \vee Q)) \rightarrow (S \vee P))$

7. Try to express atomic propositions as P, Q, R, etc. Then translate the following sentences with symbols.

(1) Either you haven't written to me, or it's lost in the route.

(2) If Zhang San and Li Si don't go, he will.

(3) We can't row and run.

(4) If you come, whether he sings or not depends on whether you accompany him.

8. Find the truth table of the following complex propositions.

(1) $P \rightarrow (Q \vee R)$

(2) $(P \wedge R) \vee (P \rightarrow Q)$

(3) $(P \vee Q) \rightleftarrows (Q \vee P)$

(4) $(P \vee \neg Q) \wedge R$

9. Try to find the truth table of the following propositions and explain the results.

(1) $(P \rightarrow Q) \wedge (Q \rightarrow P)$

(2) $(P \wedge Q) \rightarrow P$

(3) $Q \rightarrow (P \vee Q)$

(4) $(P \rightarrow Q) \rightleftarrows (\neg P \vee Q)$

(5) $(\neg P \vee Q) \wedge (\neg (P \wedge \neg Q))$

10. Prove the following equation.

(1) $A \rightarrow (B \rightarrow A) \Leftrightarrow \neg A \rightarrow (A \rightarrow \neg B)$

(2) $\neg (A \rightleftarrows B) \Leftrightarrow (A \vee B) \wedge \neg (A \wedge B)$

(3) $\neg (A \rightarrow B) \Leftrightarrow A \wedge \neg B$

(4) $\neg (A \rightleftarrows B) \Leftrightarrow (A \wedge \neg B) \vee (\neg A \wedge B)$

(5) $A \rightarrow (B \vee C) \Leftrightarrow (A \wedge \neg B) \rightarrow C$

(6) $(A \rightarrow D) \wedge (B \rightarrow D) \Leftrightarrow (A \vee B) \rightarrow D$

11. Prove that the followings are tautology.

(1) $(P \wedge (P \rightarrow Q)) \rightarrow Q$

(2) $\neg P \rightarrow (P \rightarrow Q)$

(3) $((P \rightarrow Q) \wedge (Q \rightarrow R)) \rightarrow (P \rightarrow R)$

(4) $(a \wedge b) \vee (b \wedge c) \vee (c \wedge a) \rightleftarrows (a \vee b) \wedge (b \vee c) \wedge (c \vee a)$

12. Try to prove the following implication without constructing a truth table.

(1) $(P \rightarrow Q) \Rightarrow P \rightarrow (P \wedge Q)$

(2) $(P \rightarrow Q) \rightarrow Q \Rightarrow P \vee Q$

(3) $(Q \rightarrow (P \wedge \neg P)) \rightarrow (R \rightarrow (R \rightarrow (P \wedge \neg P))) \Rightarrow R \rightarrow Q$

13. Find the disjunctive and conjunctive normal form of the formula $P \wedge (P \rightarrow Q)$.

14. Convert the following into disjunctive paradigms.

(1) $(\neg P \wedge Q) \rightarrow R$

(2) $P \rightarrow ((Q \wedge R) \rightarrow S)$

(3) $\neg (P \vee \neg Q) \wedge (S \rightarrow T)$

(4) $(P \rightarrow Q) \rightarrow R$

(5) $\neg (P \wedge Q) \wedge (P \vee Q)$

15. Convert the following into conjunctive paradigms.

(1) $P \vee (\neg P \wedge Q \wedge R)$

(2) $\neg (P \rightarrow Q) \vee (P \vee Q)$

(3) $\neg (P \rightarrow Q)$

(4) $(P \rightarrow Q) \rightarrow R$

(5) $(\neg P \wedge Q) \vee (P \wedge \neg Q)$

16. Use rules of inference to prove the following.

(1) $\neg (P \wedge \neg Q), \neg Q \vee R, \neg R \Rightarrow \neg P$

(2) $J \rightarrow (M \vee N), (H \vee G) \rightarrow J, H \vee G \Rightarrow M \vee N$

(3) $B \wedge C$, $(B \rightleftarrows C) \rightarrow (H \vee G) \Rightarrow G \vee H$

(4) $P \rightarrow Q$, $(\neg Q \vee R) \wedge \neg R$, $\neg(\neg P \wedge S) \Rightarrow \neg S$

17. Use only rules P and T to deduce the following formula.

(1) $\neg A \vee B$, $C \rightarrow \neg B \Rightarrow A \rightarrow \neg C$

(2) $A \vee B \rightarrow C \wedge D$, $D \vee E \rightarrow F \Rightarrow A \rightarrow F$

(3) $A \rightarrow (B \rightarrow C)$, $(C \wedge D) \rightarrow E$,

　　$\neg F \rightarrow (D \wedge \neg E) \Rightarrow A \rightarrow (B \rightarrow F)$

(4) $A \rightarrow (B \wedge C)$, $\neg B \vee D$,

　　$(E \rightarrow \neg F) \rightarrow \neg D$, $B \rightarrow (A \wedge \neg E) \Rightarrow B \rightarrow E$

18. Please use the CP rule to prove(1), (2), (3) and(4) of exercise 17.

第1章 命 题 逻 辑

命题逻辑又可以称为命题演算,或语句演算。它主要研究以命题为基本单位构成的前提和结论之间的可推导关系。那么究竟什么是命题?如何表示一个命题?如何由一组前提推导一些结论?下面带大家一起详细地讨论这些核心问题。

1.1 命题与联结词

命题是具有确定真假意义的陈述句。一个命题总是具有一个"值",称为真值。真值只有"真"和"假"两种,记作 True(真)和 False(假),通常使用 0 或 F 表示"假",用 1 或 T 表示"真"。

命题必须具备两个条件:其一,语句是陈述句;其二,语句有唯一确定的真假意义。判断给定句子是否为命题,应该分两步:首先判定它是否为陈述句;其次判断它是否有唯一的真值。

命题可分为以下两种类型:第一种类型是不能分解为更简单的陈述语句的命题,称作原子命题;第二种类型是由联结词、标点符号和原子命题复合构成的命题,称作复合命题。

所有这些命题,都应具有确定的真值。

原子命题又称为简单命题,是对某个事物的某个性质进行推断,也可以表示任何有关某个事物或情况的真假陈述。它的语义真值由客观事实决定。

复合命题是从语法结构上可分解成若干简单命题的命题,由若干简单命题通过命题联结词组合而成。

下面给出实例,说明命题的概念。

(1) 太阳东升西落。

(2) 草是绿的。

(3) 全体起立!

(4) 地球表面 71% 是水。

(5) 明天是否下雨呢?

(6) 今天天气真好呀!

(7) 下午我去图书馆或者去实验室。

(8) 张三和李四都是江苏大学的研究生。

在上面的这些例子中,(1)、(2)、(4)、(7)、(8)是命题,其中(1)、(7)、(8)是复合命题。(3)、(5)、(6)都不是命题,因为它们都不是陈述句。

在数理逻辑中,通常通过"联结词"构成复合命题。命题联结词也称为命题运算符,具有严格的逻辑含义,以求保证符号系统的语义与其原有自然系统语义的一致性。常用的逻辑联结词有五种:否定联结词、合取联结词、析取联结词、条件联结词和双条件联结词。

1.1.1 否定联结词

定义 1.1.1 ▶ 设 P 为命题,则 P 的否定命题是一个复合命题,记作 $\neg P$,读作"非 P"或"P 的否定"。若 P 为 T,则 $\neg P$ 为 F;若 P 为 F,则 $\neg P$ 为 T。

联结词"$\neg$"也可以看作逻辑运算,它是一元运算。P 和 $\neg P$ 的关系如表 1.1.1 所示,表 1.1.1 叫作否定联结词"$\neg$"的真值表。

例 1.1.1 否定下列命题。

P:张三是一名大学生。

$\neg P$:张三不是一名大学生。

1.1.2 合取联结词

定义 1.1.2 ▶ 设 P 和 Q 均为命题,则 P 和 Q 的合取是一个复合命题,记作 $P \wedge Q$,读作"P 与 Q"或"P 合取 Q"。当且仅当 P 和 Q 均为 T 时,$P \wedge Q$ 才为 T。

联结词"$\wedge$"也可以看成逻辑运算,它是二元逻辑运算。联结词"$\wedge$"的真值表如表 1.1.2 所示。

表 1.1.1

P	$\neg P$
F	T
T	F

表 1.1.2

P	Q	$P \wedge Q$
F	F	F
F	T	F
T	F	F
T	T	T

例 1.1.2 设 P:2020 年新冠肺炎来势汹汹,Q:全球很多国家都暴发了疫情,则 $P \wedge Q$:2020 年新冠肺炎来势汹汹并且全球很多国家都暴发了疫情。

1.1.3 析取联结词

定义 1.1.3 ▶ 设 P 和 Q 均为命题,P 和 Q 的析取是一个复合命题,记作 $P \vee Q$,读作"P 或 Q"或者"P 析取 Q"。当且仅当 P 和 Q 均为 F 时,$P \vee Q$ 才为 F。

联结词"$\vee$"也可以看成逻辑运算,它是二元逻辑运算。"$\vee$"与汉语中的"或"相似,但又不相同。汉语中的"或"有可兼或与不可兼或(排斥或)的区别。联结词"$\vee$"的真值表如表 1.1.3 所示。

例 1.1.3 下列两个命题中的"或",哪个是可兼或?哪个是不可兼或?

(1) 在电视上看这场杂技或在剧场里看这场杂技。(不可兼或)

(2) 灯泡有故障或开关有故障。(可兼或)

1.1.4 条件联结词

定义 1.1.4 ▶ 设 P 和 Q 均为命题,其条件命题是一个复合命题,记作 $P \rightarrow Q$,读作"如果 P,那么 Q"或"若 P,则 Q"。当且仅当 P 为 T,Q 为 F 时,$P \rightarrow Q$ 才为 F。P 称为条件命题 $P \rightarrow Q$ 的前件,Q 称为条件命题 $P \rightarrow Q$ 的后件。

联结词"→"也可以看成逻辑运算,它是二元逻辑运算。联结词"→"的真值表如表 1.1.4 所示。

表　1.1.3		
P	Q	$P \lor Q$
F	F	F
F	T	T
T	F	T
T	T	T

表　1.1.4		
P	Q	$P \to Q$
F	F	T
F	T	T
T	F	F
T	T	T

例 1.1.4　　P:小宇努力学习。

Q:小宇学习成绩优秀。

$P \to Q$:如果小宇努力学习,那么他的学习成绩就优秀。

联结词"→"与汉语中的"如果……,那么……"或"若……,则……"相似,但又不完全相同。

1.1.5　双条件联结词

定义 1.1.5 ▶ 设 P 和 Q 均为命题,其复合命题 $P \leftrightarrow Q$ 称为双条件命题,读作"P 双条件 Q"或"P 当且仅当 Q"。当且仅当 P 和 Q 的真值相同时,$P \leftrightarrow Q$ 为 T。

联结词"↔"也可以理解成逻辑运算,它是二元逻辑运算。联结词"↔"的真值表如表 1.1.5 所示。

表 1.1.5

P	Q	$P \leftrightarrow Q$	P	Q	$P \leftrightarrow Q$
F	F	T	T	F	F
F	T	F	T	T	T

与前面所述相同,双条件联结词表示的是一种充分必要关系,只根据联结词的定义来确定其真值。

例 1.1.5　设 P:李四是三好学生。

Q:李四在德、智、体方面全都是优秀的。

$P \leftrightarrow Q$:李四是三好学生当且仅当他在德、智、体方面全都是优秀的。

1.2　命题公式与翻译

此前已经提到过,不包含其他命题作为其组成部分的命题,即在结构上不能再分解为更简单的陈述语句的命题叫作原子命题。原子命题不能包含任何联结词。至少包含一个联结词的命题叫作复合命题。

设 P 和 Q 是任意两个命题,则 $\neg P$,$P \land Q$,$(P \land Q) \to (P \lor Q)$,$\neg P \to (Q \leftrightarrow \neg P)$ 等都是复合命题。

若 P 和 Q 是命题变元,则上述各式均称作命题公式,也叫合式公式。P 和 Q 称作命题公式的分量。

需注意:命题公式中包含命题变元,因此无法计算其真值,仅当一个命题公式中的命题变元用确定真值的命题代入时,才得到一个命题。

定义 1.2.1 ▶ 按下列规则构成的符号串称为命题演算的合式公式(wff),也称为命题公式,简称公式。

(1) 单个命题变元或者命题常量是合式公式。

(2) 如果 A 是合式公式,那么 $\neg A$ 也是合式公式。

(3) 如果 A 和 B 是合式公式,那么 $(A \wedge B),(A \vee B),(A \rightarrow B),(A \leftrightarrow B)$ 都是合式公式。

(4) 当且仅当能够有限次地应用(1)、(2)、(3)所得到的包含命题变元、联结词和括号的符号串是合式公式。

命题公式一般用大写的英文字母 $A,B,C,\cdots$ 表示。

这个合式公式定义的方法称为归纳定义,它包含三个部分:基础、归纳和界限。其中(1)称为基础,(2)和(3)称为归纳,(4)称为界限。

为方便起见,对合式公式的约定如下。

(1) 为减少圆括号的使用数量,约定最外层的圆括号可以省略。

(2) 规定联结词的优先级由高到低依次为 $\neg$、$\wedge$、$\vee$、$\rightarrow$、$\leftrightarrow$。按此优先级,如果夫掉括号不改变原公式运算顺序,也可以省略这些括号,即 $P \wedge Q \rightarrow R$ 等同于 $(P \wedge Q) \rightarrow R$,这也是合式公式。

例 1.2.1 判断下列符号串是否是合式公式。

(1) $\neg(P \vee Q)$

(2) $(\neg P \rightarrow Q)$

(3) $(P \rightarrow (P \leftrightarrow Q))$

(4) $(P \leftrightarrow Q) \rightarrow (\vee Q)$

(5) $(P \wedge Q,(P \wedge Q) \wedge Q)$

解 由以上定义可知,(1)、(2)、(3)是合式公式;(4)、(5)不是合式公式。因为(4)中 $\vee$ 为双目联结词,所以其左侧缺少一个命题变元;而(5)中多出一个逗号,不满足合式公式的定义。

有了命题合式公式的概念,就可以用合式公式表示复合命题,常将这个过程称为命题的符号化。整个过程等同于把自然语言中的某些语句(命题)翻译成为数理逻辑中的符号形式。其可按以下步骤进行。

(1) 确定给定的语句是否为命题。

(2) 找出语句中的原子命题,同时确定语句中连词对应的联结词。

(3) 用正确的语法将原命题表示成由原子命题、联结词和圆括号组成的合式公式。

例 1.2.2 将下列命题符号化:长沙到镇江的 G124 次列车是上午九点半或十点开。

解 P:长沙到镇江的 G124 次列车是上午九点半开。

Q:长沙到镇江的 G124 次列车是上午十点开。

在本例中,汉语的"或"是不可兼或,也就是说列车不可能同时在上午九点半和十点开,只可能是其二者之一,而数理逻辑中的析取是可兼或,所以不能简单地将 P 和 Q 析取。应构造真值表进行分析,如表 1.2.1 所示。

从表中可以看出原命题不能用前述五个联结词单独表示出来,但是将命题和联结词组合就可以把原命题符号化为 $\neg(P \leftrightarrow Q)$。

表 1.2.1

P	Q	原命题	$P\leftrightarrow Q$	$\neg(P\leftrightarrow Q)$	P	Q	原命题	$P\leftrightarrow Q$	$\neg(P\leftrightarrow Q)$
T	T	F	T	F	F	T	T	F	T
T	F	T	F	T	F	F	F	T	F

例 1.2.3 将下列命题符号化:他不仅天赋高还很努力。

解 P:他天赋高。Q:他很努力。

在自然语言中这个"不仅……还……"显然与数理逻辑中的合取所表示的"且"一致,所以本题可以符号化为 $P\wedge Q$。

例 1.2.4 将下列命题符号化:他虽然天赋高但是不努力。

解 这里"虽然……但是……"这个词组不能直接用前述联结词表示,需稍加转换,其实际的意义是:他天赋高且不努力。所以,设

P:他天赋高,Q:他不努力。

那么本例题可以符号化为 $P\wedge\neg Q$。

例 1.2.5 将下列命题符号化:除非你用功,否则你将挂科。

解 这个命题的实际意义是:如果你不用功,那么你将挂科。所以可以设

P:你不用功,Q:你将挂科。

那么本例题可以符号化为 $\neg P\rightarrow Q$。

例 1.2.6 将下列命题符号化:他在国际大学生程序设计竞赛上获得了金奖或银奖。

解 找出各个原子命题,并用命题符号表示。

A:他在国际大学生程序设计竞赛上获得了金奖。

B:他在国际大学生程序设计竞赛上获得了银奖。

那么本例题可符号化为$(A\wedge\neg B)\vee(\neg A\wedge B)$。

从上面的例子可以看出,自然语言中的一些联结词,如"不仅……还……""或""虽然……但是……"等各有其具体的含义,因此需要根据不同的情况将其翻译成适当的逻辑联结词。同时为便于正确地表达命题间的相互关系,有时也可以用列出"真值表"的方法进一步分析各个原命题。

1.3 真值表与等价公式

通过上面的例题 1.2.2 可知,当无法直接对一个命题进行符号化时往往可以借助真值表进行辅助分析,从而更快地得到符号化的结果。下面我们将进一步学习真值表的相关知识。

定义 1.3.1 ▶ 设 $P_1,P_2,\cdots,P_n$ 是出现在命题公式 A 中的全部命题变元,给 $P_1,P_2,\cdots,P_n$ 各指定一个真值,称为对命题公式 A 的赋值或者解释。若指定的赋值使 A 的真值为 T,则称这个赋值为 A 的成真赋值;若指定的赋值使 A 的真值为 F,则称这个赋值为 A 的成假赋值。

定义 1.3.2 ▶ 在命题公式 A 中,对命题公式 A 的每一个赋值,就确定了 A 的一个真值,把所有的真值汇成表,就是命题公式 A 的真值表。

例 1.3.1 构造 $P\vee\neg Q$ 的真值表。

解 结果如表 1.3.1 所示。

例 1.3.2 构造 $(P \wedge Q) \wedge \neg P$ 的真值表。

解 结果如表 1.3.2 所示。

表 1.3.1

P	Q	$\neg Q$	$P \vee \neg Q$
T	T	F	T
T	F	T	T
F	T	F	F
F	F	T	T

表 1.3.2

P	Q	$P \wedge Q$	$\neg P$	$(P \wedge Q) \wedge \neg P$
T	T	T	F	F
T	F	F	F	F
F	T	F	T	F
F	F	F	T	F

例 1.3.3 构造 $(P \vee Q) \vee (\neg P \wedge Q)$ 的真值表。

解 结果如表 1.3.3 所示。

表 1.3.3

P	Q	$\neg P$	$P \vee Q$	$\neg P \wedge Q$	$(P \vee Q) \vee (\neg P \wedge Q)$
T	T	F	T	F	T
T	F	F	T	F	T
F	T	T	T	T	T
F	F	T	F	F	F

例 1.3.4 构造 $\neg(P \wedge Q) \leftrightarrow (\neg P \vee \neg Q)$ 的真值表。

解 结果如表 1.3.4 所示。

表 1.3.4

P	Q	$P \wedge Q$	$\neg(P \wedge Q)$	$\neg P$	$\neg Q$	$\neg P \vee \neg Q$	$\neg(P \wedge Q) \leftrightarrow (\neg P \vee \neg Q)$
T	T	T	F	F	F	F	T
T	F	F	T	F	T	T	T
F	T	F	T	T	F	T	T
F	F	F	T	T	T	T	T

在真值表中,命题公式真值的取值数目取决于分量的个数。例如,由两个命题变元组成的命题公式共有 4 种可能的赋值;由 3 个命题变元组成的命题公式共有 8 种可能的赋值。所以,n 个命题变元组成的命题公式共有 2^n 种真值情况。

从真值表中可以看出,有些命题公式在分量的所有指派情况下,其对应的真值与另一命题公式完全相同,如 $\neg(Q \rightarrow P)$ 与 $\neg P \wedge Q$ 的对应真值相同,如表 1.3.5 所示。

表 1.3.5

P	Q	$\neg P$	$Q \rightarrow P$	$\neg(Q \rightarrow P)$	$\neg P \wedge Q$
T	T	F	T	F	F
T	F	F	T	F	F
F	T	T	F	T	T
F	F	T	T	F	F

定义 1.3.3 ▶ 给定两个命题公式 A 和 B,设 $P_1, P_2, P_3, \cdots, P_n$ 为所有出现在 A 和 B 中的原子变元,若给 $P_1, P_2, P_3, \cdots, P_n$ 任一组真值指派,A 和 B 的真值都相同,则称 A 和 B 是

等价的或者二者逻辑相等,记作 $A \Leftrightarrow B$。

可以证明,命题公式等价有下面三条性质。

(1) 自反性,即对任意命题公式 A 都有 $A \Leftrightarrow A$。

(2) 对称性,即对任意命题公式 A 和 B,若 $A \Leftrightarrow B$,则 $B \Leftrightarrow A$。

(3) 传递性,即对任意命题公式 A,B 和 C,若 $A \Leftrightarrow B$,$B \Leftrightarrow C$,则 $A \Leftrightarrow C$。

例 1.3.5 证明 $P \leftrightarrow Q \Leftrightarrow (P \rightarrow Q) \wedge (Q \rightarrow P)$。

证明 列出真值表,如表 1.3.6 所示。

由表 1.3.6 可知,$P \leftrightarrow Q$ 与 $(P \rightarrow Q) \wedge (Q \rightarrow P)$ 的真值相同,所以命题得证。

表 1.3.6

P	Q	$P \rightarrow Q$	$Q \rightarrow P$	$P \leftrightarrow Q$	$(P \rightarrow Q) \wedge (Q \rightarrow P)$
T	T	T	T	T	T
T	F	F	T	F	F
F	T	T	F	F	F
F	F	T	T	T	T

例 1.3.6 证明 $P \vee (P \wedge Q) \Leftrightarrow P$,$P \wedge (P \vee Q) \Leftrightarrow P$。

证明 其真值表如表 1.3.7 所示。从真值表中可以清晰地看出二者之间的等价关系。

表 1.3.7

P	Q	$P \wedge Q$	$P \vee Q$	$P \vee (P \wedge Q)$	$P \wedge (P \vee Q)$
T	T	T	T	T	T
T	F	F	T	T	T
F	T	F	T	F	F
F	F	F	F	F	F

如表 1.3.8 所示,常用的命题等价定理都可以用真值表给予验证。

表 1.3.8

序号	定 律	等 价 关 系
1	对合律	$\neg \neg P \Leftrightarrow P$
2	幂等律	$P \vee P \Leftrightarrow P$,$P \wedge P \Leftrightarrow P$
3	结合律	$(P \vee Q) \vee R \Leftrightarrow P \vee (Q \vee R)$,$(P \wedge Q) \wedge R \Leftrightarrow P \wedge (Q \wedge R)$
4	交换律	$P \vee Q \Leftrightarrow Q \vee P$,$P \wedge Q \Leftrightarrow Q \wedge P$
5	分配律	$P \vee (Q \wedge R) \Leftrightarrow (P \vee Q) \wedge (P \vee R)$,$P \wedge (Q \vee R) \Leftrightarrow (P \wedge Q) \vee (P \wedge R)$
6	吸收律	$P \vee (P \wedge Q) \Leftrightarrow P$,$P \wedge (P \vee Q) \Leftrightarrow P$
7	德·摩根定律	$\neg (P \vee Q) \Leftrightarrow \neg P \wedge \neg Q$,$\neg (P \wedge Q) \Leftrightarrow \neg P \vee \neg Q$
8	同一律	$P \vee F \Leftrightarrow P$,$P \wedge T \Leftrightarrow P$
9	零律	$P \vee T \Leftrightarrow T$,$P \wedge F \Leftrightarrow F$
10	否定律	$P \vee \neg P \Leftrightarrow T$,$P \wedge \neg P \Leftrightarrow F$
11	条件等价式	$A \rightarrow B \Leftrightarrow \neg A \vee B$
12	双条件等价式	$A \leftrightarrow B \Leftrightarrow (A \rightarrow B) \wedge (B \rightarrow A)$
13	假言易位式	$A \rightarrow B \Leftrightarrow \neg B \rightarrow \neg A$
14	双条件否定等价式	$A \leftrightarrow B \Leftrightarrow \neg A \leftrightarrow \neg B$
15	联结词归化	$\neg (A \leftrightarrow B) \Leftrightarrow A \leftrightarrow \neg B$

例 1.3.7 用真值表证明德·摩根定律$\neg(P \vee Q) \Leftrightarrow \neg P \wedge \neg Q$。

证明 表 1.3.9 是$\neg(P \vee Q)$和$\neg P \wedge \neg Q$的真值表,从表中可以看出德·摩根定律是正确的。

表 1.3.9

P	Q	$\neg P$	$\neg Q$	$P \vee Q$	$\neg(P \vee Q)$	$\neg P \wedge \neg Q$
T	T	F	F	T	F	F
T	F	F	T	T	F	F
F	T	T	F	T	F	F
F	F	T	T	F	T	T

在一个命题公式中,如果用公式置换命题的某个部分,一般会产生某种新的公式。例如,$P \rightarrow (P \vee (P \rightarrow Q))$中以$(P \wedge Q)$取代$(P \rightarrow Q)$,则$P \rightarrow (P \vee (P \wedge Q))$就与原式不同。为保证取代后的公式与原始公式是等价的,需对置换做出一些规定。

定义 1.3.4 如果Y是合式公式A中的一部分,并且Y本身也是一个合式公式,则称Y为公式A的子公式。例如,$A \Leftrightarrow Q \rightarrow (P \vee (P \wedge Q))$,$Y \Leftrightarrow P \wedge Q$,则$Y$是$A$的子公式。

定理 1.3.1 设Y是合式公式A的子公式,若$Y \Leftrightarrow X$,将A中的Y用X置换,所得到的公式B与公式A是等价的,即$A \Leftrightarrow B$。

证明 因为在相应变元的任一种指派下,Y与X的真值相同,故用X取代Y之后,公式B与公式A在相应的真值指派情况下,其真值必相同,故$A \Leftrightarrow B$。

满足定理 1.3.1 条件的置换称为等价置换(等价代换)。

例 1.3.8 用等价置换及真值表证明$P \leftrightarrow Q \Leftrightarrow (P \wedge Q) \vee (\neg P \wedge \neg Q)$。

证明
$$
\begin{aligned}
P \leftrightarrow Q &\Leftrightarrow (P \rightarrow Q) \wedge (Q \rightarrow P) \\
&\Leftrightarrow (\neg P \vee Q) \wedge (\neg Q \vee P) \\
&\Leftrightarrow (\neg P \wedge \neg Q) \vee (\neg P \wedge P) \vee (Q \wedge \neg Q) \vee (Q \wedge P) \\
&\Leftrightarrow (\neg P \wedge \neg Q) \vee F \vee (Q \wedge P) \\
&\Leftrightarrow (\neg P \wedge \neg Q) \vee (Q \wedge P) \\
&\Leftrightarrow (P \wedge Q) \vee (\neg P \wedge \neg Q) \\
P \leftrightarrow Q &\Leftrightarrow (P \wedge Q) \vee (\neg P \wedge \neg Q)
\end{aligned}
$$

也可用真值表验证,如表 1.3.10 所示。

表 1.3.10

P	Q	$\neg P$	$\neg Q$	$\neg P \wedge \neg Q$	$P \wedge Q$	$P \leftrightarrow Q$	$(P \wedge Q) \vee (\neg P \wedge \neg Q)$
T	T	F	F	F	T	T	T
T	F	F	T	F	F	F	F
F	T	T	F	F	F	F	F
F	F	T	T	T	F	T	T

例 1.3.9 用等价置换证明$(P \wedge Q) \vee (P \wedge \neg Q) \Leftrightarrow P$。

证明
$$
\begin{aligned}
(P \wedge Q) \vee (P \wedge \neg Q) &\Leftrightarrow P \wedge (Q \vee \neg Q) \\
&\Leftrightarrow P \wedge T \\
&\Leftrightarrow P
\end{aligned}
$$

例 1.3.10 用等价置换证明 $Q \vee \neg((P \rightarrow Q) \wedge P) \Leftrightarrow T$。

证明 $Q \vee \neg((P \rightarrow Q) \wedge P) \Leftrightarrow Q \vee \neg((\neg P \vee Q) \wedge P)$
$$\Leftrightarrow Q \vee \neg((\neg P \wedge P) \vee (Q \wedge P))$$
$$\Leftrightarrow Q \vee \neg(Q \wedge P)$$
$$\Leftrightarrow Q \vee (\neg Q \vee \neg P)$$
$$\Leftrightarrow (Q \vee \neg Q) \vee \neg P$$
$$\Leftrightarrow T \vee \neg P$$
$$\Leftrightarrow T$$

例 1.3.11 用等价置换证明 $((P \vee Q) \wedge \neg(\neg P \wedge (\neg Q \vee \neg R))) \vee (\neg P \wedge \neg Q) \vee (\neg P \wedge \neg R) \Leftrightarrow T$。

证明 原式左边 $\Leftrightarrow ((P \vee Q) \wedge \neg(\neg P \wedge \neg(Q \wedge R))) \vee \neg(P \vee Q) \vee \neg(P \vee R)$
$$\Leftrightarrow ((P \vee Q) \wedge (P \vee (Q \wedge R))) \vee \neg(P \vee Q) \vee \neg(P \vee R)$$
$$\Leftrightarrow ((P \vee Q) \wedge ((P \vee Q) \wedge (P \vee R))) \vee \neg((P \vee Q) \wedge (P \vee R))$$
$$\Leftrightarrow ((P \vee Q) \wedge (P \vee R)) \vee \neg((P \vee Q) \wedge (P \vee R))$$
$$\Leftrightarrow T$$

1.4 重言式与蕴含式

根据 1.3 节真值表和命题的等价公式推证可以发现,有些命题无论如何赋值,其取值均为真或假。根据公式在各种赋值下的取值情况,可以将公式进行分类。

定义 1.4.1 ▶ 设 A 为任一命题公式。

(1) 若 A 在任意赋值下的取值均为真,则称 A 为重言式或永真式。

(2) 若 A 在任意赋值下的取值均为假,则称 A 为矛盾式或永假式。

(3) 若 A 不是矛盾式,则称 A 为可满足式。

根据以上定义,我们可以总结出以下性质。

(1) 可满足式至少存在一个成真赋值。

(2) 重言式一定是可满足式,但可满足式不一定是重言式。

定理 1.4.1 ▶ 任何两个重言式的合取或析取仍然是一个重言式。

推论 任何两个矛盾式的合取或析取仍然是一个矛盾式。

定理 1.4.2 ▶ 一个重言式,对同一分量出现的每一处都用同一合式公式置换,其结果仍是重言式。

推论 一个矛盾式,对同一分量出现的每一处都用同一合式公式置换,其结果仍是矛盾式。

在真值表中,若最后一列全为 T,则公式为重言式;若最后一列全为 F,则公式为矛盾式;若最后一列至少有一个为 T,则公式为可满足式。

例 1.4.1 判断下列命题公式的类型。

(1) $P \wedge \neg P$;(2) $P \vee \neg P$;(3) $\neg P \vee Q$。

解 表 1.4.1 给出了三个公式的真值表。

表 1.4.1

P	Q	$P \land \neg P$	$P \lor \neg P$	$\neg P \lor Q$
F	F	F	T	T
F	T	F	T	T
T	F	F	T	F
T	T	F	T	T

(1)的取值均为 F,为矛盾式;(2)的取值均为 T,为重言式;(3)的取值中至少一个为 T,为可满足式。

从以上讨论可以看出,真值表不仅可以给出公式的成真赋值和成假赋值,同时还可以判断公式的类型。

定义 1.4.2 ▶ 设 A 和 B 是命题公式,当且仅当 $A \to B$ 为重言式时,$A \to B$ 称为 A 与 B 的蕴含式,记作 $A \Rightarrow B$。

$P \to Q$ 也可以理解为 Q 是 P 的必要条件。

定理 1.4.3 ▶ 蕴含式是逻辑推理的重要工具。下面是一些重要的蕴含式,其中 A,B,C,D 是任意的命题公式。

(1) 附加律:$A \Rightarrow A \lor B$

(2) 化简律:$A \land B \Rightarrow A$

(3) 假言推理:$A \land (A \to B) \Rightarrow B$

(4) 拒取式:$\neg B \land (A \to B) \Rightarrow \neg A$

(5) 析取三段论:$\neg B \land (A \lor B) \Rightarrow A$

(6) 假言三段论:$(A \to B) \land (B \to C) \Rightarrow (A \to C)$

(7) 等价三段论:$(A \leftrightarrow B) \land (B \leftrightarrow C) \Rightarrow (A \leftrightarrow C)$

(8) 构造性二难:$(A \lor C) \land (A \to B) \land (C \to D) \Rightarrow B \lor D$
$$(A \lor \neg A) \land (A \to B) \land (\neg A \to B) \Rightarrow B$$

(9) 破坏性二难:$(\neg B \lor \neg D) \land (A \to B) \land (C \to D) \Rightarrow (\neg A \lor \neg C)$

定理 1.4.4 ▶ 设 A、B 为任意两个命题公式,$A \Leftrightarrow B$ 的充分必要条件是 $A \Rightarrow B$ 且 $B \Rightarrow A$。

定理 1.4.5 ▶ 设 A、B、C 为合式公式。

(1) $A \Rightarrow A$(即蕴含是自反的)。

(2) 若 $A \Rightarrow B$ 且 A 为重言式,则 B 必为重言式。

(3) 若 $A \Rightarrow B$ 且 $B \Rightarrow C$,则 $A \Rightarrow C$(即蕴含是传递的)。

(4) 若 $A \Rightarrow B$ 且 $A \Rightarrow C$,则 $A \Rightarrow B \land C$。

(5) 若 $A \Rightarrow B$ 且 $C \Rightarrow B$,则 $A \lor C \Rightarrow B$。

(6) 若 $A \Rightarrow B$,C 是任意公式,则 $A \land C \Rightarrow B \land C$。

1.5 对偶与范式

从 1.4 节可看到命题定律除对合律外都是成对出现的,其不同的只是 $\land$ 和 $\lor$ 互换,我们称其具有对偶规律。

定义 1.5.1 ▶ 在仅含联结词 $\neg$,$\land$,$\lor$ 的命题公式 A 中,将联结词 $\lor$,$\land$,F,T 分别换成

∧,∨,T,F 所得的公式称为公式 A 的对偶式,记为 A^*。

设 A^* 是 A 的对偶式,将 A^* 中的 ∨,∧,F,T 分别换成 ∧,∨,T,F,就会得到 A。即 A 是 A^* 的对偶式,$(A^*)^* \Leftrightarrow A$。所以说,$A^*$ 和 A 互为对偶式。

例 1.5.1 写出下列表达式的对偶式。

(1) $(A \lor B) \land C$

(2) $(A \land B) \lor T$

(3) $\lnot B \land (A \lor B) \land (A \lor C)$

解 上述表达式对应的对偶式如下。

(1) $(A \land B) \lor C$

(2) $(A \lor B) \land F$

(3) $\lnot B \lor (A \land B) \lor (A \land C)$

定理 1.5.1 ▶ 设 A^* 是 A 的对偶式,$p_1, p_2, \cdots, p_n$ 是出现在 A 和 A^* 中的原子变元,则
$$\lnot A(p_1, p_2, \cdots, p_n) \Leftrightarrow A^*(\lnot p_1, \lnot p_2, \cdots, \lnot p_n)$$
$$A(\lnot p_1, \lnot p_2, \cdots, \lnot p_n) \Leftrightarrow \lnot A^*(p_1, p_2, \cdots, p_n)$$

定理 1.5.2 ▶ (对偶原理)设 $p_1, p_2, \cdots, p_n$ 是出现在公式 A 和 B 中的所有原子变元,如果 $A \Leftrightarrow B$,则 $A^* \Leftrightarrow B^*$。

定义 1.5.2 ▶ 由一些命题变元或其否定构成的析取式称为基本和,也叫简单析取式。约定单个变元或其否定都是基本和。

例如,$\lnot p \lor q$、$p \lor \lnot q$、$p \lor q$、$\lnot q$、$\lnot p$、q 都是基本和。

定义 1.5.3 ▶ 由一些命题变元或其否定构成的合取式称为基本积,也叫简单合取式。约定单个变元或其否定都是基本积。

例如,$\lnot p \land q$、$p \land \lnot q$、$p \land q$、$\lnot p$、$\lnot q$、p 都是基本积。

注意,单个变元既是简单析取式也是简单合取式。

定义 1.5.4 ▶ 由基本和的合取构成的公式叫作合取范式。约定单个基本和都是合取范式。

定义 1.5.5 ▶ 由基本积的析取构成的公式叫作析取范式。约定单个基本积都是析取范式。

析取范式与合取范式统称为范式。析取范式和合取范式具有以下性质。

(1) 一个析取范式是矛盾式,当且仅当它的每个简单合取式都是矛盾式。

(2) 一个合取范式是重言式,当且仅当它的每个简单析取式都是重言式。

任何一个命题公式,求其合取范式和析取范式的步骤如下。

(1) 消去公式中的联结词"→"和"↔",利用蕴含等值式和等价等值式将其转换成 ∧、∨、¬。

例 1.5.2 $A \to B \Leftrightarrow \lnot A \lor B$
$$A \leftrightarrow B \Leftrightarrow (A \to B) \land (B \to A)$$

(2) 利用双重否定律消去否定联结词"¬",或利用德·摩根定律将否定联结词"¬"移到各命题变元前(¬内移)。

例 1.5.3 $\lnot \lnot A \Leftrightarrow A$
$$\lnot (A \lor B) \Leftrightarrow \lnot A \land \lnot B$$

$$\neg(A \land B) \Leftrightarrow \neg A \lor \neg B$$

（3）利用分配律、结合律将公式归约为合取范式和析取范式。

例 1.5.4　$A \land (B \lor C) \Leftrightarrow (A \land B) \lor (A \land C)$

$A \lor (B \land C) \Leftrightarrow (A \lor B) \land (A \lor C)$

例 1.5.5　求命题公式$(p \lor q) \leftrightarrow p$的合取范式和析取范式。

解　（1）求合取范式。

$(p \lor q) \leftrightarrow p \Leftrightarrow ((p \lor q) \to p) \land (p \to (p \lor q))$　（消去$\leftrightarrow$）

$\Leftrightarrow (\neg(p \land q) \lor p) \land (\neg p \lor (p \lor q))$

$\Leftrightarrow ((\neg p \land \neg q) \lor p) \land (\neg p \lor (p \lor q))$　（$\neg$内移）

$\Leftrightarrow (\neg p \lor p) \land (\neg q \lor p) \land (\neg p \lor p \lor q)$　（分配律，合取范式）

$\Leftrightarrow T \land (\neg q \lor p) \land (T \lor q)$　（排中律）

$\Leftrightarrow T \land (\neg q \lor p) \land T$　（零律，合取范式）

$\Leftrightarrow (\neg q \lor p)$　（同一律，合取范式）

从上面这个例子可以看出，公示的合区范式并不唯一。

（2）求析取范式。

$(p \lor q) \leftrightarrow p \Leftrightarrow ((p \lor q) \land p) \lor (\neg(p \lor q) \land \neg p)$　（消去$\leftrightarrow$）

$\Leftrightarrow ((p \lor q) \land p) \lor ((\neg p \land \neg q) \land \neg p)$　（$\neg$内移）

$\Leftrightarrow p \lor (\neg p \land \neg q \land \neg p)$　（吸收律，析取范式）

$\Leftrightarrow p \lor (\neg p \land \neg p \land \neg q)$　（交换律）

$\Leftrightarrow p \lor (\neg p \land \neg q)$　（幂等律，析取范式）

由此例可以看出，命题公式的析取范式也不唯一。

定义 1.5.6　在基本积中，每个变元及其否定不同时存在，但两者之一必须出现且仅出现一次，这样的基本积叫作布尔合取，也叫小项或极小项。p, q的极小项为$p \land q$，$p \land \neg q$，$\neg p \land q$，$\neg p \land \neg q$。

一般地，n个命题变元共有2^n个极小项。

表1.5.1是两个变元p和q的极小项的真值表。极小项的性质如下。

（1）每个极小项只有一个成真赋值，且各极小项的成真赋值互不相同。极小项和它的成真赋值一一对应。可用成真赋值为极小项进行编码，并将编码作为名称m的下标来表示该极小项，叫作该极小项的名称。两个命题变元的极小项、成真赋值和名称如表1.5.2所示。三个命题变元的极小项、成真赋值和名称如表1.5.3所示。从表1.5.2和表1.5.3中可以看出，极小项与其成真赋值的对应关系为变元对应T，而变元的否定对应F。

表 1.5.1

p	q	$p \land q$	$p \land \neg q$	$\neg p \land q$	$\neg p \land \neg q$
F	F	F	F	F	T
F	T	F	F	T	F
T	F	F	T	F	F
T	T	T	F	F	F

表　1.5.2

极小项	成真赋值	名称	极小项	成真赋值	名称
$\neg p \wedge \neg q$	FF	m_0	$p \wedge \neg q$	TF	m_2
$\neg p \wedge q$	FT	m_1	$p \wedge q$	TT	m_3

表　1.5.3

极小项	成真赋值	名称	极小项	成真赋值	名称
$\neg p \wedge \neg q \wedge \neg r$	FFF	m_0	$p \wedge \neg q \wedge \neg r$	TFF	m_4
$\neg p \wedge \neg q \wedge r$	FFT	m_1	$p \wedge \neg q \wedge r$	TFT	m_5
$\neg p \wedge q \wedge \neg r$	FTF	m_2	$p \wedge q \wedge \neg r$	TTF	m_6
$\neg p \wedge q \wedge r$	FTT	m_3	$p \wedge q \wedge r$	TTT	m_7

（2）任意两个不同的极小项的合取式为永假式。例如：

$$m_{001} \wedge m_{100} \Leftrightarrow (\neg p \wedge \neg q \wedge r) \wedge (p \wedge \neg q \wedge \neg r)$$
$$\Leftrightarrow \neg p \wedge \neg q \wedge r \wedge p \wedge \neg q \wedge \neg r \Leftrightarrow F$$

（3）全体极小项的析取式为永真式，记为

$$\sum_{i=0}^{2^n-1} m_i \Leftrightarrow m_0 \vee m_1 \vee \cdots \vee m_{2^n-1} \Leftrightarrow T$$

定义 1.5.7 ▶ 对给定的命题公式，如果有一个它的等价公式仅由极小项的析取组成，称该公式为原公式的主析取范式。

任何命题公式都存在与之等价的主析取范式。一个命题公式的主析取范式可以用以下两种方法求得。

（1）用等价演算法求主析取范式。其步骤如下。

① 化归为析取范式。

② 除去析取范式中所有永假的基本积。

③ 在基本积中合并重复出现的合取项和相同变元。

④ 在基本积中补入没有出现的命题变元，即添加 $\wedge (p \vee \neg p)$，再用分配律展开，最后合并相同的极小项。

例 1.5.6　用等价演算法求 $(p \wedge q) \vee (\neg p \wedge r) \vee (q \wedge r)$ 的主析取范式。

解　$(p \wedge q) \vee (\neg p \wedge r) \vee (q \wedge r) \Leftrightarrow (p \wedge q \wedge (r \vee \neg r)) \vee (\neg p \wedge r \wedge (q \vee \neg q)) \vee (q \wedge r \wedge (p \vee \neg p))$

$\Leftrightarrow (p \wedge q \wedge r) \vee (p \wedge q \wedge \neg r) \vee (\neg p \wedge q \wedge r) \vee (\neg p \wedge \neg q \wedge r) \vee (p \wedge q \wedge r) \vee (\neg p \wedge q \wedge r)$

$\Leftrightarrow (p \wedge q \wedge r) \vee (p \wedge q \wedge \neg r) \vee (\neg p \wedge q \wedge r) \vee (\neg p \wedge \neg q \wedge r)$

$\Leftrightarrow m_{111} \vee m_{110} \vee m_{011} \vee m_{001}$

$\Leftrightarrow m_7 \vee m_6 \vee m_3 \vee m_1 \Leftrightarrow \Sigma 1,3,6,7$

（2）真值表法，即用真值表求主析取范式。

① 构造命题公式的真值表。

② 找出公式的成真赋值对应的极小项。

③ 这些极小项的析取就是此公式的主析取范式。

例 1.5.7 用真值表法,求 $(p \to q) \to r$ 的主析取范式。

解 $(p \to q) \to r$ 的真值表如表 1.5.4 所示。

表 1.5.4

p	q	r	$p \to q$	$(p \to q) \to r$	p	q	r	$p \to q$	$(p \to q) \to r$
0	0	0	1	0	1	0	0	0	1
0	0	1	1	1	1	0	1	0	1
0	1	0	1	0	1	1	0	1	0
0	1	1	1	1	1	1	1	1	1

根据表 1.5.4,公式的成真赋值对应的极小项为

$\neg p \wedge \neg q \wedge r$ (成真赋值为 001)

$\neg p \wedge q \wedge r$ (成真赋值为 011)

$p \wedge \neg q \wedge \neg r$ (成真赋值为 100)

$p \wedge \neg q \wedge r$ (成真赋值为 101)

$p \wedge q \wedge r$ (成真赋值为 111)

$(p \to q) \to r$ 的主析取范式为

$$(p \wedge q \wedge r) \vee (p \wedge \neg q \wedge r) \vee (p \wedge \neg q \wedge \neg r) \vee (\neg p \wedge q \wedge r) \vee (\neg p \wedge \neg q \wedge r)$$

$\Leftrightarrow m_{111} \vee m_{101} \vee m_{100} \vee m_{011} \vee m_{001} \Leftrightarrow m_7 \vee m_5 \vee m_4 \vee m_3 \vee m_1 \Leftrightarrow \sum 1,3,4,5,7$

矛盾式无成真赋值,因此主析取范式不含任何极小项,将矛盾式的主析取范式记为 0。而重言式无成假赋值,因此主析取范式含 2^n (n 为公式中命题变元的个数)个极小项。至于可满足式,它的主析取范式中极小项的个数一定小于或等于 2^n。

定义 1.5.8 ▶ 在基本和中,每个变元及其否定不同时存在,但两者之一必须出现且仅出现一次,这样的基本和叫作布尔析取,也叫大项或极大项。

两个变元 p,q 构成的极大项为

$$p \vee q \qquad p \vee \neg q \qquad \neg p \vee q \qquad \neg p \vee \neg q$$

和极小项相同的是,一般地,n 个变元共有 2^n 个极大项。

极大项有以下三个性质。

(1) 每个极大项只有一个成假赋值,极大项不同,成假赋值也不同。极大项和它的成假赋值一一对应,故可用成假赋值为该极大项进行编码,并将编码作为名称 M 的下标来表示该极大项,叫作极大项的名称。

例如,两个变元 p,q 的极大项 $\neg p \vee \neg q$ 的成假赋值是 11,表示为 M_{11}。把 11 理解为二进制数,用 10 进制表示为 3,所以 M_{11} 又可表示为 M_3。

两个命题变元的极大项的成假赋值及名称如表 1.5.5 所示,三个命题变元的极大项的成假赋值及名称如表 1.5.6 所示。

表 1.5.5

极大项	成假赋值	名称
$p \vee q$	00	M_0
$p \vee \neg q$	01	M_1
$\neg p \vee q$	10	M_2
$\neg p \vee \neg q$	11	M_3

表　1.5.6

极大项	成假赋值	名称	极大项	成假赋值	名称
$p \vee q \vee r$	000	M_0	$\neg p \vee q \vee r$	100	M_4
$p \vee q \vee \neg r$	001	M_1	$\neg p \vee q \vee \neg r$	101	M_5
$p \vee \neg q \vee r$	010	M_2	$\neg p \vee \neg q \vee r$	110	M_6
$p \vee \neg q \vee \neg r$	011	M_3	$\neg p \vee \neg q \vee \neg r$	111	M_7

从表 1.5.5 和表 1.5.6 中可以看出,极大项与成假赋值的对应关系:变元对应 0,变元的否定对应 1。

(2) 任意两个不同的极大项的析取式为永真式。

(3) 全体极大项的合取式为永假式,记作

$$\prod_{i=0}^{2^n-1} M_i \Leftrightarrow M_0 \wedge M_1 \wedge \cdots \wedge M_{2^n-1} \Leftrightarrow 0$$

定义 1.5.9 ▶对给定的命题公式,如果有一个它的等价公式仅由极大项的合取组成,则该等价式称为原公式的主合取范式。

任何命题公式都存在与之等价的主合取范式。主合取范式也可以由以下两种方法求得。

(1) 等价演算法:用基本等价公式推出。其演算步骤如下。

① 化归为合取范式。

② 除去所有永真的基本和。

③ 在基本和中合并重复出现的析取项和相同的变元。

④ 在基本和中补充没有出现的命题变元,即增加 $\vee(p \wedge \neg p)$,然后应用分配律展开公式,最后合并相同的极大项。

例 1.5.8　用等价演算法求 $(p \to q) \to r$ 的主合取范式。

$(p \to q) \to r \Leftrightarrow \neg(\neg p \vee q) \vee r \Leftrightarrow (p \wedge \neg q) \vee r \Leftrightarrow (p \vee r) \wedge (\neg q \vee r)$

$\qquad \Leftrightarrow (p \vee r \vee (q \wedge \neg q)) \wedge (\neg q \vee r \vee (p \wedge \neg p))$

$\qquad \Leftrightarrow (p \vee r \vee q) \wedge (p \vee r \vee \neg q) \wedge (p \vee \neg q \vee r) \wedge (\neg p \vee \neg q \vee r)$

$\qquad \Leftrightarrow (p \vee q \vee r) \wedge (p \vee \neg q \vee r) \wedge (\neg p \vee \neg q \vee r)$

$\qquad \Leftrightarrow M_{000} \wedge M_{010} \wedge M_{110} \Leftrightarrow M_0 \wedge M_2 \wedge M_6 \Leftrightarrow \prod_{0,2,6}$

(2) 真值表法:用真值表求主合取范式。其演算步骤如下。

① 构造命题公式的真值表。

② 找出公式的成假赋值对应的极大项。

③ 这些极大项的析取就是此公式的主合取范式。

例 1.5.9　用真值表法求 $(p \to q) \to r$ 的主合取范式。$(p \to q) \to r$ 的真值表是表 1.5.4。

公式的成假赋值对应的极大项为

$\qquad p \vee q \vee r$　　　（成假赋值为 000）

$\qquad p \vee \neg q \vee r$　　　（成假赋值为 010）

$\qquad \neg p \vee \neg q \vee r$　　　（成假赋值为 110）

主合取范式为

$$(p \vee q \vee r) \wedge (p \vee \neg q \vee r) \wedge (\neg p \vee \neg q \vee r)$$

$$\Leftrightarrow M_{000} \wedge M_{010} \wedge M_{110} \Leftrightarrow M_0 \wedge M_2 \wedge M_6 \Leftrightarrow \prod_{0,2,6}$$

矛盾式无成真赋值,因此矛盾式的主合取范式含 2^n(n 为公式中命题变元的个数)个极大项。而重言式无成假赋值,因此其主合取范式不含任何极大项。将重言式的主合取范式记为 1。可满足式的主合取范式中极大项的个数一定小于 2^n。

1.6 命题演算的推理理论

逻辑学是研究思维形式及其规律的科学,其主要任务之一是提供一套关于推理的规则。给出一些前提,利用所提供的推理规则,推导出一个结论,这个过程通常称为演绎或形式证明。

一般来说,根据经验,如果前提是真的,利用所提供的推理规则推导出的结论也应该是真的。在数学中,这些前提通常称为公理,推导出的结论称为定理。这样的推理过程称为证明过程。

在数理逻辑中,注意力集中在推理规则的研究。先给出一些推理规则,从一些前提出发,根据所提供的推理规则推导出结论,这种推导出来的结论称为有效结论,这种论证过程称为有效论证。在确定论证的有效性时,并不关心前提的实际真值。也就是说,在数理逻辑中,研究的是演绎的有效性,而不是通常所说的正确性。只有在假定的前提都为真命题时,由此推导出的有效结论才为真命题。由于通常作为前提的命题并非全是永真式,所以它的有效结论并不一定都是真命题。因此当前提为假时,无论结论是否为真,前提都是蕴含结论的。这一点与通常实际中应用的推理是不同的。

1.6.1 有效结论

定义 1.6.1 ▶ 设 A 和 B 是两个命题公式,当且仅当 $A \rightarrow B$ 是永真式,即 $A \Rightarrow B$,才说 B 是 A 的有效结论,或 B 可由 A 逻辑地推出。

这个定义可以推广到有 n 个前提的情况。

设 $H_1, H_2, \cdots, H_n, B$ 是命题公式,当且仅当 $H_1 \wedge H_2 \wedge \cdots \wedge H_n \Rightarrow B$,才说 B 是前提集合 $\{H_1, H_2, \cdots, H_n\}$ 的有效结论,或由 $H_1, H_2, \cdots, H_n$ 可逻辑地推出 B。

1.6.2 证明有效结论的方法

1. 真值表法

例 1.6.1 统计表格的错误或者是由于材料不可靠,或者是由于计算有错误。这份统计表格的错误不是由于材料不可靠,所以这份统计表格的错误是由于计算有错误。

解 设 P:统计表格的错误是由于材料不可靠;Q:统计表格的错误是由于计算有错误。则原命题可符号化为

$$(P \vee Q) \wedge \neg P \Rightarrow Q$$

列出真值表如表 1.6.1 所示。

表 1.6.1

P	Q	$P \vee Q$	$\neg P$	$(P \vee Q) \wedge \neg P$
T	T	T	F	F
T	F	T	F	F
F	T	T	T	T
F	F	F	T	F

由表 1.6.1 可知,只有第三行使前件$(P \lor Q) \land \neg P$ 为 T,此时后件 Q 也为 T,所以 Q 是前提$(P \lor Q) \land \neg P$ 的有效结论。

例 1.6.2　如果张老师来了,这个问题就可以得到解答。如果李老师来了,这个问题也可以得到解答。总之,无论来的是张老师还是李老师,这个问题都可以得到解答。

解　P:张老师来了,Q:李老师来了,R:这个问题可以解答。则原命题可以表示为

$$(P \to R) \land (Q \to R) \land (P \lor Q) \Rightarrow R$$

列出真值表如表 1.6.2 所示。

表　1.6.2

P	Q	R	$P \to R$	$Q \to R$	$P \lor Q$	P	Q	R	$P \to R$	$Q \to R$	$P \lor Q$
T	T	T	T	T	T	F	T	T	T	T	T
T	T	F	F	F	T	F	T	F	T	F	T
T	F	T	T	T	T	F	F	T	T	T	F
T	F	F	F	T	T	F	F	F	T	T	F

从表中可以看出,$P \to R$,$Q \to R$,$P \lor Q$ 的真值都为 T 的情况有三种,其中 R 的真值均为T。故

$$(P \lor Q) \land (Q \to R) \land (P \lor Q) \Rightarrow R$$

例 1.6.3　今天是星期天或者今天是晴天;今天是星期天。所以,今天是晴天。

解　设 P:今天是星期天,Q:今天是晴天。于是原命题可符号化为

$$(P \lor Q) \land P \Rightarrow Q$$

列出真值表如表 1.6.3 所示。

表　1.6.3

P	Q	$P \lor Q$	$(P \lor Q) \land Q$	Q
T	T	T	T	T
T	F	T	T	F
F	T	T	F	T
F	F	F	F	F

从真值表 1.6.3 中可以看出,第二行前提的真值为 T 而相应结论的真值为 F,所以,此结论不是前提的有效结论。

2. 直接证法

直接证法就是由一组前提,利用一些公认的推理规则,根据已知的等价或蕴含公式,推演得到有效的结论。

P 规则:前提在推导过程中的任何时候都可以引入使用。T 规则:在推导中,如果有一个或多个公式、重言蕴含公式 S,则公式 S 可以引入推导中。

现将常用的蕴含式和等价式列入表 1.6.4 和表 1.6.5 中。

表 1.6.4

名称	蕴含式	名称	蕴含式
I_1	$P \wedge Q \Rightarrow P$	I_4	$Q \Rightarrow P \vee Q$
I_2	$P \wedge Q \Rightarrow Q$	I_5	$\neg P \Rightarrow P \rightarrow Q$
I_3	$P \Rightarrow P \vee Q$		

表 1.6.5

名称	等 价 式	名称	等 价 式
E_1	$\neg \neg P \Leftrightarrow P$	E_{12}	$R \vee (P \wedge \neg P) \Leftrightarrow R$
E_2	$P \wedge Q \Leftrightarrow Q \wedge P$	E_{13}	$R \wedge (P \vee \neg P) \Leftrightarrow R$
E_3	$P \vee Q \Leftrightarrow Q \vee P$	E_{14}	$R \vee (P \vee \neg P) \Leftrightarrow T$
E_4	$(P \wedge Q) \wedge R \Leftrightarrow P \wedge (Q \wedge R)$	E_{15}	$R \wedge (P \wedge \neg P) \Leftrightarrow F$
E_5	$(P \vee Q) \vee R \Leftrightarrow P \vee (Q \vee R)$	E_{16}	$P \rightarrow Q \Leftrightarrow \neg P \vee Q$
E_6	$P \wedge (Q \vee R) \Leftrightarrow (P \wedge Q) \vee (P \wedge R)$	E_{17}	$\neg (P \rightarrow Q) \Leftrightarrow P \wedge \neg Q$
E_7	$P \vee (Q \wedge R) \Leftrightarrow (P \vee Q) \wedge (P \vee R)$	E_{18}	$P \rightarrow Q \Leftrightarrow \neg Q \rightarrow \neg P$
E_8	$\neg (P \wedge Q) \Leftrightarrow \neg P \vee \neg Q$	E_{19}	$P \rightarrow (Q \rightarrow R) \Leftrightarrow (P \wedge Q) \rightarrow R$
E_9	$\neg (P \vee Q) \Leftrightarrow \neg P \wedge \neg Q$	E_{20}	$P \rightleftarrows Q \Leftrightarrow (P \rightarrow Q) \wedge (Q \rightarrow P)$
E_{10}	$P \vee P \Leftrightarrow P$	E_{21}	$P \rightleftarrows Q \Leftrightarrow (P \wedge Q) \vee (\neg P \wedge \neg Q)$
E_{11}	$P \wedge P \Leftrightarrow P$	E_{22}	$\neg (P \rightleftarrows Q) \Leftrightarrow P \rightleftarrows \neg Q$

例 1.6.4 证明：

$$(P \vee Q) \wedge (P \rightarrow R) \wedge (Q \rightarrow S) \Rightarrow S \vee R$$

证明 1

(1) $P \vee Q$ P

(2) $\neg P \rightarrow Q$ T(1)E

(3) $Q \rightarrow S$ P

(4) $\neg P \rightarrow S$ T(2),(3)I

(5) $\neg S \rightarrow P$ T(4)E

(6) $P \rightarrow R$ P

(7) $\neg S \rightarrow R$ T(5),(6)I

(8) $S \vee R$ P

证明 2

(1) $P \rightarrow R$ P

(2) $P \vee Q \rightarrow R \vee Q$ T(1)I

(3) $Q \rightarrow S$ P

(4) $Q \vee R \rightarrow S \vee R$ T(3)I

(5) $P \vee Q \rightarrow S \vee R$ T(2),(4)I

(6) $P \vee Q$ P

(7) $S \vee R$ T(5),(6)I

例 1.6.5 证明：

$$(W \vee R) \rightarrow V, V \rightarrow C \vee S, S \rightarrow U, \neg C \wedge \neg U \Rightarrow \neg W$$

证明

(1) $\neg C \wedge \neg U$	P
(2) $\neg U$	T(1)I
(3) $S \rightarrow U$	P
(4) $\neg S$	T(2),(3)I
(5) $\neg C$	T(1)I
(6) $\neg C \wedge \neg S$	T(4),(5)I
(7) $\neg (C \vee S)$	T(6)E
(8) $(W \vee R) \rightarrow V$	P
(9) $V \rightarrow (C \vee S)$	P
(10) $(W \vee R) \rightarrow (C \vee S)$	T(8),(9)I
(11) $\neg (W \vee R)$	T(7),(10)I
(12) $\neg W \wedge \neg R$	T(11)E
(13) $\neg W$	T(12)I

3. 间接证法

定义 1.6.2 ▶ 假设公式 $H_1, H_2, \cdots, H_m$ 中的命题变元为 $P_1, P_2, \cdots, P_n$,对 $P_1, P_2, \cdots,$ P_n 的一些真值指派,如果能使 $H_1 \wedge H_2 \wedge \cdots \wedge H_m$ 的真值为 T,则称公式 $H_1, H_2, \cdots, H_m$ 是相容的。如果对 $P_1, P_2, \cdots, P_n$ 的每一组真值指派使 $H_1 \wedge H_2 \wedge \cdots \wedge H_m$ 的真值均为 F,则称公式 $H_1, H_2, \cdots, H_m$ 是不相容的。

下面把不相容概念应用于命题公式的证明中。

设有一组前提 $H_1, H_2, \cdots, H_m$,要推出结论 C,即证 $H_1 \wedge H_2 \wedge, \cdots, \wedge H_m \Rightarrow C$,记作 $S \Rightarrow$ C,即 $\neg C \vee \neg S$ 为永真,或 $C \vee \neg S$ 为永真,故 $\neg C \wedge S$ 为永假。因此要证明 $H_1 \wedge H_2 \wedge \cdots \wedge$ $H_m \Rightarrow C$,只需证明 $H_1 \wedge H_2 \wedge \cdots \wedge H_m$ 与 $\neg C$ 是不相容的即可。

例 1.6.6　证明:$A \rightarrow B, \neg (B \vee C)$ 可逻辑推出 $\neg A$。

证明

(1) $A \rightarrow B$	P
(2) A	P(附加前提)
(3) $\neg (B \vee C)$	P
(4) $\neg B \wedge \neg C$	T(3)E
(5) B	T(1),(2)I
(6) $\neg B$	T(4)I
(7) $B \wedge \neg B$	T(5),(6)I

例 1.6.7　证明:

$$(P \vee Q) \wedge (P \rightarrow R) \wedge (Q \rightarrow S) \Rightarrow S \vee R$$

证明

(1) $\neg (S \vee R)$	P(附加前提)
(2) $\neg S \wedge \neg R$	T(1)E
(3) $P \vee Q$	P

(4) $\lnot P \to Q$	T(3)E
(5) $Q \to S$	P
(6) $\lnot P \to S$	T(4),(5)I
(7) $\lnot S \to P$	T(6)E
(8) $(\lnot S \land \lnot R) \to (P \land \lnot R)$	T(7)I
(9) $P \land \lnot R$	T(2),(8)I
(10) $P \to R$	P
(11) $\lnot P \lor R$	T(10)E
(12) $\lnot(P \land \lnot R)$	T(11)E
(13) $(P \land \lnot R) \land \lnot(P \land \lnot R)$	T(9),(12)I

1.7 命题逻辑的应用

逻辑在数学、计算机科学和其他许多学科中有许多重要的应用。命题逻辑及其规则可用于设计计算机电路、构造计算机程序、验证程序的正确性及构造专家系统,还可以用于分析和求解许多熟悉的谜题。

数字电路中,逻辑门符号如下。

与门:

或门:

非门:

或非门:

与或非门:

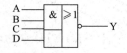

异或门:

同或门：

下面通过一个例子具体介绍命题逻辑在推断题中的应用。

M 公司为提高员工的专业能力，决定派出几名优秀的员工出国考察学习，甲、乙、丙、丁、戊是公司从众多申请者中选出的几名比较优秀的人员，但为不影响正常工作需要，他们五个不能都去，所以这次选派必须满足以下条件：

① 如果甲被派出国学习，则乙也要被派出国学习；

② 丁、戊两人中必有人被选派出国学习；

③ 乙、丙两人中有且仅有一人被选派出国学习；

④ 丙、丁两人要么都派去，要么都不去；

⑤ 若戊被派出国学习，那么甲、乙也将被派出国学习。

为合理安排工作，根据以上条件，公司应该怎样从这五人中挑选谁出国学习呢？

解法 1　一般的推断方法。

分析：此题是要推断出派出几个人，应该派出谁，而且肯定有人被派出，但派出人数和人物不固定，那么我们就必须一个一个假设，看怎样的假设才是合理的。

（1）假设甲一定被派出，那么根据条件①可以推断出乙一定也被派出，根据条件③推出丙不被派出，又由④可以推断出丙、丁都不被派出，那么根据条件②又可推断出戊一定去，经过以上推断可得出一种选派方案：甲、乙、戊被派出学习，丙、丁不会被派出学习。

（2）假设乙一定被派出，根据条件③推出丙不被派出，那么由条件④可以推断出丙、丁一定不被派出，而又由②可以推断出戊一定被派出，那么由条件⑤又可以推断出甲、乙也将会派出，经过以上推断我们也可得到一种选派方案：甲、乙、戊被派出学习，丙、丁不会被派出学习。

（3）假设丙一定被派出，根据条件③推出乙不被派出，那么由条件④可以推断出丁一定也被派出，由条件可以推出乙一定不被派出，而又由条件②可以推出戊可能被派出，也可能不被派出。如果戊被派出，由条件⑤可以推出甲、乙一定也被派出，这与前面的推理相矛盾，所以戊一定不被派出，因此由以上推断可以得到一种选派方案：丙、丁被派出学习，甲、乙、戊不被派出学习。

（4）假设丁一定被派出，那么由条件④可以推断出丙一定被派出，根据条件③推出乙一定不被派出，又由条件①⑤可以推出甲、戊一定也不被派出，所以由以上推断可得出一种选派方案：丙、丁被派出学习，甲、乙、戊不被派出学习。

（5）假设戊一定被派出，那么由条件⑤可以推断出甲、乙一定也被派出，又由条件③④可以推断出丙、丁一定不被派出，所以由以上的推断可以得出一种选派方案：甲、乙、戊被派出学习，丙、丁不会被派出学习。

由以上分析可以得出该公司只有两种选派方案：一种是甲、乙、戊被派出学习，丙、丁不被派出学习；另一种是丙、丁被派出学习，甲、乙、戊不被派出学习。

解法 2　用命题逻辑推断的方法。

分析：如果要推断他们五人谁去，又有怎样的派出方案，只需将必要的命题符号化，再根据所给条件写出各个命题公式 F_1、F_2、F_3、F_4、F_5，然后只需要将 $T \Leftrightarrow F_1 \wedge F_2 \wedge F_3 \wedge F_4 \wedge F_5$ 变换成析取范式的形式，就可以解决这个问题。

将命题符号化,令

p:甲被派出学习;

q:乙被派出学习;

r:丙被派出学习;

s:丁被派出学习;

t:戊被派出学习。

那么各条件可符号化为

① $F_1 \Leftrightarrow p \rightarrow q$;

② $F_2 \Leftrightarrow s \lor t$;

③ $F_3 \Leftrightarrow (q \land \neg r) \lor (\neg q \land r)$;

④ $F_4 \Leftrightarrow (r \land s) \lor (\neg r \land \neg s)$;

⑤ $F_5 \Leftrightarrow t \rightarrow (p \land q)$

由于各条件都是已知的,所以它们的合取式一定为真。设

$T \Leftrightarrow F_1 \land F_2 \land F_3 \land F_4 \land F_5$

$\Leftrightarrow (p \rightarrow q) \land (s \lor t) \land ((q \land \neg r) \lor (\neg q \land r)) \land ((r \land s) \lor (\neg r \land \neg s)) \land (t \rightarrow (p \land q))$

要想得到派出方案,只需求出 $T = F_1 \land F_2 \land F_3 \land F_4 \land F_5$ 的析取式。求其析取式的步骤为

$T \Leftrightarrow (\neg p \lor q) \land (s \lor t) \land ((q \land \neg r) \lor (\neg q \land r)) \land ((r \land s) \lor (\neg r \land \neg s))$
$\quad \land (\neg t \lor (p \land q))$

$\Leftrightarrow ((\neg p \land s) \lor (\neg p \land t) \lor (q \land s) \lor (q \land t)) \land ((q \land \neg r \land \neg s) \lor (\neg q \land r \land s) \land$
$\quad (\neg t \lor (p \land q))$

$\Leftrightarrow (\neg p \land \neg q \land r \land s) \lor (\neg p \land q \land \neg r \land \neg s \land t) \lor (q \land \neg r \land \neg s \land t) \land (\neg t \lor (p \land q))$

$\Leftrightarrow (\neg p \land \neg q \land r \land s \land \neg t) \lor (p \land q \land \neg r \land \neg s \land t)$

由上面的推断可以很容易得出只有两种方案:一种是甲、乙、戊被派出学习,而丙、丁不被派出学习;另一种是丙、丁被派出学习,而甲、乙、戊不被派出学习。

在推理题中用命题演绎推理,而其中应用的各种推理规则大多是依靠推理定理完成的。

习　题

1. 指出下列语句哪些是命题,哪些不是命题。如果是命题,指出它的真值。

(1) 离散数学是计算机科学与技术专业的一门必修课。

(2) 计算机有空吗?

(3) 明天我去看电影。

(4) 请勿随地吐痰!

(5) 不存在最大质数。

(6) 如果我掌握了英语和法语,那么学习其他欧洲语言就容易得多。

(7) $11 + 6 < 10$。

(8) $y = 4$。

(9) 我们要努力学习。

2. 设 P 表示命题"天下雪"，Q 表示命题"我将去镇上"，R 表示命题"我有时间"。以符号形式写出下列命题。

(1) 如果天不下雪且我有时间，那么我将去镇上。

(2) 我将去镇上，仅当我有时间时。

(3) 天不下雪。

(4) 天下雪，那么我不去镇上。

3. 将下列命题符号化。

(1) 王强身体很好，成绩也很好。

(2) 小李一边看书，一边听音乐。

(3) 气候很好或者很热。

(4) 如果 a 和 b 都是偶数，则 $a+b$ 是偶数。

(5) 四边形 $ABCD$ 是平行四边形，当且仅当它的对边平行时。

4. 判断下列哪些是合式公式，哪些不是合式公式。

(1) $(Q \rightarrow R \wedge S)$

(2) $(P \rightleftarrows (R \rightarrow S))$

(3) $((\neg P \rightarrow Q) \rightarrow (Q \rightarrow P))$

(4) $(RS \rightarrow T)$

(5) $((P \rightarrow (Q \rightarrow R)) \rightarrow ((P \rightarrow Q) \rightarrow (P \rightarrow R)))$

5. 根据合式公式的定义，说明下列公式是合式公式。

(1) $(A \rightarrow (A \vee B))$

(2) $((\neg A \wedge B) \wedge A)$

(3) $((\neg A \rightarrow B) \rightarrow (B \rightarrow A))$

(4) $((A \rightarrow B) \vee (B \rightarrow A))$

6. 下列的式子中有哪些是由其他式子经过代换得到的。

(1) $(P \rightarrow (Q \rightarrow P))$

(2) $((((P \rightarrow Q) \wedge (R \rightarrow S)) \wedge (P \vee R)) \rightarrow (Q \vee S))$

(3) $(Q \rightarrow ((P \rightarrow P) \rightarrow Q))$

(4) $(P \rightarrow ((P \rightarrow (Q \rightarrow P)) \rightarrow P))$

(5) $(((R \rightarrow S) \wedge (Q \rightarrow P) \wedge (R \vee Q)) \rightarrow (S \vee P))$

7. 试把原子命题表示为 P, Q, R 等，然后用符号译出下列各句子。

(1) 或者你没有给我写信，或者它在途中丢失了。

(2) 如果张三和李四不去，他就去。

(3) 我们不能既划船又跑步。

(4) 如果你来了，那么他唱不唱歌将由你是否伴奏而定。

8. 求下列各复合命题的真值表。

(1) $P \rightarrow (Q \vee R)$

(2) $(P \wedge R) \vee (P \rightarrow Q)$

(3) $(P \vee Q) \rightleftarrows (Q \vee P)$

(4) $(P \vee \neg Q) \wedge R$

9. 试求下列各命题的真值表并解释其结果。

(1) $(P \rightarrow Q) \wedge (Q \rightarrow P)$

(2) $(P \wedge Q) \rightarrow P$

(3) $Q \rightarrow (P \vee Q)$

(4) $(P \rightarrow Q) \rightleftarrows (\neg P \vee Q)$

(5) $(\neg P \vee Q) \wedge (\neg (P \wedge \neg Q))$

10. 证明下列等式。

(1) $A \rightarrow (B \rightarrow A) \Leftrightarrow \neg A \rightarrow (A \rightarrow \neg B)$

(2) $\neg (A \rightleftarrows B) \Leftrightarrow (A \vee B) \wedge \neg (A \wedge B)$

(3) $\neg (A \rightarrow B) \Leftrightarrow A \wedge \neg B$

(4) $\neg (A \rightleftarrows B) \Leftrightarrow (A \wedge \neg B) \vee (\neg A \wedge B)$

(5) $A \rightarrow (B \vee C) \Leftrightarrow (A \wedge \neg B) \rightarrow C$

(6) $(A \rightarrow D) \wedge (B \rightarrow D) \Leftrightarrow (A \vee B) \rightarrow D$

11. 试证明下列各式为重言式。

(1) $(P \wedge (P \rightarrow Q)) \rightarrow Q$

(2) $\neg P \rightarrow (P \rightarrow Q)$

(3) $((P \rightarrow Q) \wedge (Q \rightarrow R)) \rightarrow (P \rightarrow R)$

(4) $(a \wedge b) \vee (b \wedge c) \vee (c \wedge a) \rightleftarrows (a \vee b) \wedge (b \vee c) \wedge (c \vee a)$

12. 不构造真值表证明下列蕴含式。

(1) $(P \rightarrow Q) \Rightarrow P \rightarrow (P \wedge Q)$

(2) $(P \rightarrow Q) \rightarrow Q \Rightarrow P \vee Q$

(3) $(Q \rightarrow (P \wedge \neg P)) \rightarrow (R \rightarrow (R \rightarrow (P \wedge \neg P))) \Rightarrow R \rightarrow Q$

13. 求公式 $P \wedge (P \rightarrow Q)$ 的析取范式和合取范式。

14. 把下列各式转化为析取范式。

(1) $(\neg P \wedge Q) \rightarrow R$

(2) $P \rightarrow ((Q \wedge R) \rightarrow S)$

(3) $\neg (P \vee \neg Q) \wedge (S \rightarrow T)$

(4) $(P \rightarrow Q) \rightarrow R$

(5) $\neg (P \wedge Q) \wedge (P \vee Q)$

15. 把下列各式转化为合取范式。

(1) $P \vee (\neg P \wedge Q \wedge R)$

(2) $\neg (P \rightarrow Q) \vee (P \vee Q)$

(3) $\neg (P \rightarrow Q)$

(4) $(P \rightarrow Q) \rightarrow R$

(5) $(\neg P \wedge Q) \vee (P \wedge \neg Q)$

16. 用推理规则证明下列各式。

(1) $\neg (P \wedge \neg Q), \neg Q \vee R, \neg R \Rightarrow \neg P$

(2) $J \rightarrow (M \vee N), (H \wedge G) \rightarrow J, H \wedge G \Rightarrow M \vee N$

(3) $B \wedge C, (B \rightleftarrows C) \rightarrow (H \vee G) \Rightarrow G \vee H$

(4) $P \rightarrow Q, (\neg Q \vee R) \wedge \neg R, \neg (\neg P \wedge S) \Rightarrow \neg S$

17. 仅用规则 P 和 T，推证以下公式。

(1) $\neg A \vee B, C \rightarrow \neg B \Rightarrow A \rightarrow \neg C$

(2) $A \vee B \rightarrow C \wedge D, D \vee E \rightarrow F \Rightarrow A \rightarrow F$

(3) $A \rightarrow (B \rightarrow C), (C \wedge D) \rightarrow E, \neg F \rightarrow (D \wedge \neg E) \Rightarrow A \rightarrow (B \rightarrow F)$

(4) $A \rightarrow (B \wedge C), \neg B \vee D, (E \rightarrow \neg F) \rightarrow \neg D, B \rightarrow (A \wedge \neg E) \Rightarrow B \rightarrow E$

18. 用 CP 规则推证上题。

Chapter 2　Predicate Logic

In the propositional logic of Chapter 1, we mainly study propositions and propositional calculus, and atomic propositions, as their indecomposable constituent units, play a very important role in studying the relationship between propositions.

However, we can make further analyses on atomic propositions. Especially, there are some commonalities between different atomic propositions. To describe the logical structure of propositions, we should start to study predicate logic.

Besides, propositional logic can not solve all inferences, such as Socrates' syllogism. All of these make us have to do in-depth research on the internal relations of propositions.

2.1　Predicate and Quantifier

A proposition is a sentence that reflects a judgment, and a sentence that does not reflect judgment is not a proposition. Generally speaking, the sentence reflecting judgment is composed of subject and predicate. For example, mobile phones are communication tools. Among them, "mobile phone" is the subject and "is the communication tool" is the predicate. The subject is generally an individual, and the individual refers to the concrete or abstract object that can exist independently in the research object. It can be an independent person or thing, or an abstract concept, such as "Marxism Leninism" "Capitalism" "Xiao Wang" "Teacher", etc. Objects are usually represented by lower case letters or lower case English letters with subscripts, which are called individual identifiers.

The identifier that represents a specific or specific individual is called an individual constant, which is usually expressed in lowercase English letters $a, b, c, \ldots$ or these English letters with subscripts. For example,

a: Li Lin　　b: Xiao Gao　　c: Zhang Yun　　d: Xiao Hua

a, b, c, d are individual constants.

The identifier representing any individual or a class of individuals is called an individual variable, which is often expressed as x, y, z or these English letters have subscripts.

The predicate is used to describe the nature or relationship of an individual. For example, Zhang San is a worker and Li Ming is a worker. These two propositions can be represented by different symbols P and Q, but the predicates of P and Q have the same attribute "a worker". Therefore, a symbol is introduced to indicate "is a worker", and then a method is introduced to represent the name of an individual. In this way, the essential attribute of the object "×× is a worker" is described. For example:

(a) He is a student.　(b) 11 is a prime number.　(c) Xiao Zhang is taller than Xiao Liu.

In the above sentence, "is a student", "is a prime number", and "... is taller than ..." are all predicates. The first two are individual predicates, and the last one is a predicate indicating the relationship between two individuals.

We usually use capital letters to represent predicates, and lowercase letters or lowercase letters with subscripts to represent individuals. For example, A stands for "is a teacher", c is for Xiao Zhang, and d is for Xiao Bai. Then $A(c)$ and $A(d)$ respectively indicate "Xiao Zhang is a teacher" and "Xiao Bai is a teacher".

When a proposition is expressed by a predicate, it must include two parts: the individual and the predicate. For example, the following three propositional predicates indicate.

$F(a)$: Zhang San is an excellent member of the Communist Youth League.

$G(b, c)$: Xiao Pei is taller than Xiao Li.

$H(d, e, f)$: Xiao Shan sits between Xiao Ning and Xiao Ming.

We call $F(a)$ a unary predicate, $G(b, c)$ a binary predicate, $H(d, e, f)$ a ternary predicate, and so on.

The formula obtained by filling the predicate with the associated individual constant is called predicate filling. $F(a)$, $G(b, c)$, $H(d, e, f)$ are predicate fillers. The predicate filler represents a proposition.

Generally speaking, n-ary predicates require n individual names to be inserted into fixed positions. If A is an n-ary predicate, a_1, a_2, ..., a_n is the name of an individual, then $A(a_1, a_2, ..., a_n)$ becomes a proposition.

In general, unitary predicates express the properties of individuals, while multivariate predicates express the relationships between individuals.

In order to explain the concept of propositional function, the relation between proposition and predicate will be explained first.

Let T be the predicate "be able to complete the task", a denotes the individual name Zhang San, b represents the individual name Li Si, and c represents the car, then $T(a)$, $T(b)$, $T(c)$ respectively represent different propositions, but they all have a common form, namely $T(x)$. When x takes a, b and c respectively, it means "Zhang San can complete the task", "Li Si can complete the task", "The car can complete the task".

Similarly, if $H(x, y)$ means "x is greater than y", then $H(3, 2)$ represents a true proposition "3 is greater than 2". $H(4, 5)$ is a false proposition: "4 is greater than 5".

It can be seen from the above that $T(x)$, $H(x, y)$ are not propositions in themselves, because their true values cannot be determined. Only when the variables x, y, take specific individuals, can a proposition be determined.

Definition 2.1.1　An expression composed of a predicate and some individual variables is called a simple propositional function.

According to this definition, n-ary predicate is a propositional function $P(x_1, x_2, ...,$

x_n) with n individual variables is called n-ary propositional function ($n \geqslant 1$). For example, $T(x)$, $H(x, y)$, $P(x_1, x_2 x_n)$ is called one variable proposition function, two variable proposition function and n variable proposition function ($n \geqslant 1$). In particular, when $n = 0$, it is called a 0-ary propositional function, it is a proposition itself, so the proposition is also the case when $n = 0$ in the n-ary predicate. $F(a)$, $G(b, c)$, $H(d, e, f)$ are all 0-ary propositional function. Therefore, propositions in propositional logic can be expressed as 0-ary propositional functions(predicate filling)in predicate logic, and propositions become special cases of propositional functions.

Definition 2.1.2 The expression composed of one or n simple propositional functions and connectives($\neg$、$\wedge$、$\vee$、$\rightarrow$、$\leftrightarrow$) is called compound a propositional function.

Example 2.1.1 Let $A(x)$ denote "x is serious in class", and $B(x)$ is "x is good at learning". Then $\neg A(x)$ means "x is not serious in class". $A(x) \wedge B(x)$ means "x is serious in class, and his grades are good." . $A(x) \rightarrow B(x)$ means "if x is serious in class, then x will get good grades".

Example 2.1.2 Let $H(x, y)$ denote that "x is lighter than y". When x and y refer to people or things, it is a proposition, but when x and y refer to real numbers, $H(x, y)$ is not a proposition.

A propositional function is not a proposition, only when the individual variable takes a specific name can it become a proposition. But the range in which the individual variable takes a specific value has a great influence on whether it becomes a proposition and the truth value of a proposition.

Example 2.1.3 $Q(x)$ means "x is a middle school teacher". If the scope of discussion of x is the teacher in a middle school, then $Q(x)$ is a tautology.

If the discussion scope of x is a teacher in a university, then $Q(x)$ is an absurdity.

If the discussion scope of x is the audience in a cinema, there may be middle school teachers in the audience, or there may not be, then $Q(x)$ is true for some audiences, and false for others.

Example 2.1.4 $(R(x, y) \wedge R(y, z)) \rightarrow R(x, z)$

If $R(x, y)$ is interpreted as "x is greater than y", when x, y, z are all values in the range of real numbers, then this formula is expressed as "if x is greater than y and y is greater than z, then x is greater than z". This is a tautology.

If $R(x, y)$ is interpreted as "x is the father of y", when the values of x, y and z are all human beings, then this formula is interpreted as "if x is the father of y and y is the father of z, then x is the father of z". Obviously, this is an absurdity.

If $R(x, y)$ is interpreted as "x is 10 meters away from y", If x, y, z are housed on the ground, then "x is 10 meters away from y and y is 10 meters away from z, then x is 10 meters away from z". The truth of this proposition will depend on the position of x, y, z, which may be T or F.

From the above examples, it can be seen that the determination of propositional functions as propositions is actually related to the scope of discussion of individual variables. This scope is called individual domain or universe. The individual domain can be a finite set or an infinite set. The individual domain containing any individual domain is called the total individual domain. It is a set composed of all objects in the universe. In this book, unless otherwise specified, the total individual domain is used.

Using some of the concepts mentioned above can not fully use symbols to express various propositions in daily life. For example, $R(x)$ means "x is married", when the value range of the individual field is a middle school teacher. Then $R(x)$ can be understood as that all the teachers in this middle school are married, or some teachers in this middle school are married. In order to avoid this kind of confusion, the concept of the quantifier is introduced to describe the different concepts of "all" and "some".

Then according to "all" and "there are some", we can know that quantifiers can be divided into two types, namely, the universal quantifier and the existential quantifier.

The commonly used words in daily life and mathematics, such as "all", "each" and "arbitrary", are collectively known as a universal quantifier, and they are symbolized as "$\forall$". We use $(\forall x)$, $(\forall y)$ and so on to represent all individuals in the individual domain, while $(\forall x)F(x)$ and $(\forall y)G(y)$ are used to indicate that all individuals in the domain have property F and property G respectively.

In daily life and mathematics, the words "being", "there is one", "some", "at least one" are called existential quantifiers, and they are symbolized as "$\exists$". We use $(\exists x)$, $(\exists y)$, etc to denote some individuals in the individual domain, while $(\exists x)F(x)$ and $(\exists y)G(y)$ are used to indicate that there are individuals with property F and individuals with property G in the domain.

Example 2.1.5　The following propositions are symbolized, and the truth values of propositions are examined in the three individual domains: ①, ②, and ③.

Proposition: (1) All numbers are greater than 2.

(2) At least one number is greater than 2.

Individual domain: ① $\{2, 1, 3, -2, 4\}$. ② $\{20, 30, 3\}$. ③ $\{2, -3, 1\}$.

Solution　Let $R(x)$: x is greater than 5.

(1) "All numbers are greater than 2." It is symbolized as $(\forall x)R(x)$.

In the individual domain, the true values of ①, ② and ③ are false, true and false.

(2) "At least one number is greater than two." It is symbolized as $(\exists x)R(x)$.

In the individual domain, the truth values of ①, ② and ③ are true, true and false.

After individual variables are quantified, propositional functions become propositions, and their true value is related to the selection of individual fields, which brings certain difficulties to the research of propositional functions. At the same time, when discussing the propositional function with a quantifier, we must determine its individual field, so to unify, we will use the total individual field in the future. After the total individual field is used, the variation range of each individual variable is limited by the characteristic predicate. The method of adding attribute predicates is as follows.

(1) For universal quantifiers, characteristic predicates are added as antecedents of conditional propositions.

(2) For existential quantifiers, characteristic predicates are added as conjunctions.

Example 2.1.6 Symbolize the following propositions in two individual domains: ① and ②.

Proposition: (1) All the students have to take an exam.

(2) There is a student to test.

Individual domain: ① A collection of all students. ② Total individual domain.

Solution Let $R(x)$: x takes an exam.

When the individual domain is ①, it can be symbolized as follows.

(1) Symbolized as $(\forall x)R(x)$.

(2) Symbolized as $(\exists x)R(x)$.

When the individual field is ②, let the characteristic predicate $A(x)$: x is the student.

(1) Symbolized as $(\forall x)(A(x) \to R(x))$.

(2) Symbolized as $(\exists x)(A(x) \wedge R(x))$.

2.2 Predicate Formula and Translation

Through the previous study, we can understand that simple propositional functions can form some predicate expressions by combining them with propositional connectives. With the concept of quantifier and predicate, predicate expression can easily describe the proposition of daily life. However, what kind of predicate formula can become a predicate formula and carry out predicate calculus? The following is a direct introduction to the concept of the predicate formula.

We call $P(x_1, x_2, \ldots, x_n)$ the atomic formula of predicate calculus, where $x_1, x_2, \ldots, x_n$ is called an individual variable, so the atomic formula includes various special examples in the following forms. Such as $A(x)$, $A(x, y)$, $A(f(x), y)$, $A(x, y, z)$, $A(a, y)$, etc. In general, we call propositions, propositional variables, predicate fillers and propositional functions all atomic formulas of predicate calculus.

Definition 2.2.1 The expression formed according to the following rules is called the well-formed formula of predicate calculus.

(1) The atomic formula of predicate calculus is a well-formed formula.

(2) If A is a well-formed formula, then $\neg A$ is a well-formed formula too.

(3) If A and B are well-formed formulas, then $(A \wedge B)$, $(A \vee B)$, $(A \to B)$ and $(A \leftrightarrow B)$ are well-formed formulas.

(4) If A is a well-formed formula and x is any individual variable in A, then $(\forall x)A$, $(\exists x)A$ are well-formed formulas.

(5) Only the formula obtained by applying (1), (2), (3), (4) finitely is a well-formed formula.

The following examples illustrate how to use predicate formulas to express some propositions in natural language.

Example 2.2.1 Not every real number is a rational number.

Solution Let $R(x)$: x be a real number. $Q(x)$: x is a rational number.

The proposition is symbolized as follows: $\neg(\forall x)(R(x)\rightarrow Q(x))$.

Example 2.2.2 No one does not make mistakes.

Solution Let $M(x)$: x is human. $F(x)$: x makes mistakes.

This proposition can be understood as: everyone is to make mistakes. In this case, the symbol is: $(\forall x)(M(x)\rightarrow F(x))$.

This proposition can also be understood as: It's not that some people don't make mistakes. In this case, it is symbolized as: $\neg(\exists x)(M(x)\wedge\neg F(x))$.

Example 2.2.3 Not all rabbits run faster than all turtles.

Solution Let $F(x)$: x is a rabbit. $G(x)$: x is the tortoise. $H(x,y)$: x runs faster than y.

The proposition is symbolized as: $\neg(\forall x)(\forall y)(F(x)\wedge G(y)\rightarrow H(x,y))$.

2.3 Constraints on Variables

All individual variables in a predicate formula are quantified and become propositions. To analyze a predicate formula and see whether it becomes a proposition or not, we must see how it is quantified.

Definition 2.3.1 If A is a part of predicate formula B and is a predicate formula, A is said to be a subformula of B.

Definition 2.3.2 The smallest formula following the quantifier is called the scope or scope of the quantifier.

Definition 2.3.3 The x in the quantifier $(\forall x)$ and $(\exists x)$ is called the quantified variable or the action variable of the quantifier.

Definition 2.3.4 All occurrences of x in the scope of quantifiers $(\forall x)$ and $(\exists x)$ are called bound appearance, and x is called bound variable; the occurrence of other variables other than the bound variable is called free occurrence, and the variable that appears freely is called free variable.

Example 2.3.1 Explain the scope of the following formulas and the constraints of the variables.

(1) $(\forall x)P(x)\rightarrow Q(y)$

(2) $(\forall x)(P(x)\vee(\exists y)Q(x,y))$

(3) $(\forall x)(\forall y)(P(x,y)\vee Q(y,z))\wedge(\exists x)P(x,y)$

(4) $(\forall x)(P(x)\vee(\exists x)Q(x,z)\rightarrow(\exists y)R(x,y))\vee Q(x,y)$

Solution (1) The scope of $(\forall x)$ is $P(x)$, x is a bound variable, and y is a free variable.

(2) The scope of $(\forall x)$ is $(P(x) \vee (\exists y)Q(x, y))$, and the scope of $(\exists y)$ is $Q(x, y)$. x and y are bound variables.

(3) The scope of $(\forall x)$ and $(\forall y)$ is $(P(x, y) \vee Q(y, z))$, where x, y are constrained variables and z are free variables. The scope of $(\exists x)$ is $P(x, y)$, where x is a bound variable and y is a free variable. Then, in the whole formula, x is bound, y is both bound and free, and z is free.

(4) The scope of $(\forall x)$ is $(P(x) \vee (\exists x)Q(x, z) \rightarrow (\exists y)R(x, y))$, where x and y are bound variables and z is free variables, but x in $Q(x, z)$ is bounded by $(\exists x)$, not by $(\forall x)$. x, y in $Q(x, y)$ are free variables.

Through the understanding of bound variables, we can know that $P(x_1, x_2, ..., x_n)$ is an n-ary predicate, which has n independent free variables. If k variables are bounded, it is called n-k-ary predicate. Therefore, if there are no free variables in the predicate formula, the formula becomes a proposition. For example, $(\forall x)Q(x, y, z)$ is a binary predicate. $(\forall x)(\exists y)Q(x, y, z)$ is a unary predicate.

It can be seen from Example 2.3.1 that in a formula, the same variable can be both bounded and free, which is easy to be confused. Therefore, we can change the name of the bound variable so that a variable appears in a form in a formula, that is, it appears freely or restrictively. Because $(\forall x)P(x)$ and $(\forall y)P(y)$, $(\exists x)P(x)$ and $(\exists y)P(y)$ have the same meaning, the bound argument is independent of the symbol representing the argument. According to this feature, bound arguments can be renamed. In order to make the variables in the formula after name changing be either bounded or free, we propose the following rules for renaming.

(1) The scope of the variable name is the guide variable of the quantifier and all the variables in the scope of the quantifier. The rest of the formula remains unchanged.

(2) When changing the name, you must change it to the variable name that does not appear in the scope, preferably the variable name not in the formula.

Example 2.3.2 Change the name of bound variable y in $(\exists x)(\forall y)(P(x, y) \vee Q(y, z)) \rightarrow (\forall x) R(x, y)$.

Solution By replacing the bounded variable y with w, we can obtain the following results:
$$(\exists x)(\forall w)(P(x, w) \vee Q(w, z)) \rightarrow (\forall x) R(x, y)$$
Cannot be replaced by:
$$(\exists x)(\forall w)(P(x, w) \vee Q(w, z)) \rightarrow (\forall x) R(x, w)$$
Because y in $R(x, y)$ is not a bounded variable.

It also can't be replaced by:
$$(\exists x)(\forall w)(P(x, w) \vee Q(y, z)) \rightarrow (\forall x)R(x, y)$$
Because y in $Q(y, z)$ is also a bound variable, and it is also a variable in the scope of $(\forall y)$.

The free variable in the formula can also be changed to solve the same name problem of bound variable and free variable in the formula. This kind of change is called substitution, and the substitution rules are as follows.

(1) The free variable in the predicate formula can be substituted, and the substitution should be carried out at every place where the variable appears freely.

(2) The name of the substituted variable cannot be the same as other variables in the original formula.

Example 2.3.3　Substitute the free variable y in $(\exists x)(P(y) \land R(x, y)) \rightarrow (\forall y)Q(y)$.

Solution　By replacing the free variable y with u, we can obtain the following result:

$$(\exists x)(P(u) \land R(x, u)) \rightarrow (\forall y)Q(y)$$

It can't be replaced by:

$$(\exists x)(P(x) \land R(x, x)) \rightarrow (\forall y)Q(y)$$

Because the substituted variable cannot have the same name as other arguments in the original formula.

It can't be replaced by:

$$(\exists x)(P(u) \land R(x, y)) \rightarrow (\forall y)Q(y)$$

It should be pointed out that when the elements of the individual domain are limited, all possible substitutions of the individual variables are enumerable.

Let the individual domain elements be a_1, a_2, ..., a_n.

So we can get the following equation:

$$(\forall x)A(x) \Leftrightarrow A(a_1) \land A(a_2) \land ... \land A(a_n)$$
$$(\exists x)A(x) \Leftrightarrow A(a_1) \lor A(a_2) \lor ... \lor A(a_n)$$

At the same time, the constraints of quantifiers on variables are often related to the order of quantifiers.

For example $(\forall y)(\forall x)(x < (y-2))$ means that any y has x such that $x < y - 2$. $(\exists y)(\exists x)(x < (y-2))$ indicates that there exists y with x such that $x < y - 2$.

The quantifiers in these propositions are read out in the order from left to right. It should be noted that the order of quantifiers should not be reversed, otherwise it will be inconsistent with the original proposition.

2.4　Equivalence and Implication of Predicate Calculus

The predicate formula often contains the individual variable and the propositional variable. When the individual variable is replaced by the definite individual and the propositional variable is replaced by the definite proposition, it is called the assignment of the predicate formula. After a predicate formula is assigned, it becomes a proposition with a definite truth value T or F.

Definition 2.4.1　Let A and B be any two predicate formulas, and they have the same individual field. If for any assignment of A and B, A and B have the same truth value, then

A and B are said to be equivalent, which is called $A \Leftrightarrow B$.

Definition 2.4.2　Let A be a predicate formula and its individual field is E. If A is true for any assignment to A, then A is said to be tautology or tautological.

Definition 2.4.3　Let A be a predicate formula. If any assignment to A is false, then A is said to be unsatisfiable or contradictory.

Definition 2.4.4　Let A be a predicate formula. If at least one set of assignments makes A true, then A is said to be satisfiable.

With the concepts of equivalence and tautology of predicate formulas, we can discuss some equivalent expressions and implication expressions of predicate calculus.

1. Generalization of Equivalence in Propositional Logic

In propositional calculus, any tautology in which the same proposition variable is replaced by the same formula will also result in tautology. We can extend this situation to the predicate formula. When the formula in predicate calculus replaces the variable of perpetual tautology in propositional calculus, the predicate formula obtained is an effective formula. Therefore, the equivalent formula table and implication formula in propositional calculus Tables can be extended to predicate calculus. For example:

$$(\forall x)(P(x) \to Q(x)) \Leftrightarrow (\forall x)(\neg P(x) \vee Q(x))$$
$$(\forall x)P(x) \vee (\exists y)R(x,y) \Leftrightarrow \neg(\neg(\forall x)P(x) \wedge \neg(\exists y)R(x,y))$$
$$(\exists x)H(x,y) \wedge \neg(\exists x)H(x,y) \Leftrightarrow F$$

2. Quantifier Negative Equivalent

To better illustrate this problem, we will discuss it with examples.

Example 2.4.1　Let $P(x)$ denote that x is late for class today, $\neg P(x)$ means that x is not late for class today.

So "some people are not late for class today" is the same as "not everyone is late for class today", that is, $\neg(\forall x)P(x) \Leftrightarrow (\exists x)\neg P(x)$. Moreover, "there are not some people who are late for class today" is the same as "everyone is not late for class today", that is, $\neg(\exists x)P(x) \Leftrightarrow (\forall x)\neg P(x)$.

The following formula can be obtained:

$$\neg(\forall x)P(x) \Leftrightarrow (\exists x)\neg P(x)$$
$$\neg(\exists x)P(x) \Leftrightarrow (\forall x)\neg P(x)$$

It is agreed that the negative conjunction before the quantifier does not negate the quantifier, but negates the quantifier and its scope.

These two equivalent expressions can be proved on the finite individual field. Let the individual field be $\{a_1, a_2, \ldots, a_n\}$.

$$\neg(\forall x)A(x) \Leftrightarrow \neg(A(a_1) \wedge A(a_2) \wedge \ldots \wedge A(a_n))$$
$$\Leftrightarrow \neg A(a_1) \vee \neg A(a_2) \vee \ldots \vee \neg A(a_n)$$
$$\Leftrightarrow (\exists x)\neg A(x)$$

$$\neg(\exists x)A(x)\Leftrightarrow\neg(A(a_1)\lor A(a_2)\lor...\lor A(a_n))$$
$$\Leftrightarrow\neg A(a_1)\land\neg A(a_2)\land...\land\neg A(a_n)$$
$$\Leftrightarrow(\forall x)\neg A(x)$$

When the individual field is an infinite set, the equivalence is also true.

3. Expansion and Contraction of Quantifier Scope

The scope of the quantifier often contains conjunctive and disjunctive items. If one of them is a proposition or a predicate formula without any variables, the proposition can be moved out of the scope of the quantifier. For example:

$$(\forall x)(A(x)\lor B)\Leftrightarrow(\forall x)A(x)\lor B$$
$$(\forall x)(A(x)\land B)\Leftrightarrow(\forall x)A(x)\land B$$
$$(\exists x)(A(x)\lor B)\Leftrightarrow(\exists x)A(x)\lor B$$
$$(\exists x)(A(x)\land B)\Leftrightarrow(\exists x)A(x)\land B$$

Because there is no constraint variable x in B, it belongs to or does not belong to the scope of the quantifier.

Using the above four formulas, we can deduce the following four formulas:

$$(\forall x)(A(x)\to B)\Leftrightarrow(\exists x)A(x)\to B$$
$$(\exists x)(A(x)\to B)\Leftrightarrow(\forall x)A(x)\to B$$
$$(\forall x)(B\to A(x))\Leftrightarrow B\to(\forall x)A(x)$$
$$(\exists x)(B\to A(x))\Leftrightarrow B\to(\exists x)A(x)$$

When the variable of the predicate is different from the guiding variable of the quantifier, there can be the similar formulas. For example:

$$(\forall x)(P(x)\lor Q(y))\Leftrightarrow(\forall x)P(x)\lor Q(y)$$
$$(\forall x)((\forall y)P(x,y)\land Q(z))\Leftrightarrow(\forall x)(\forall y)P(x,y)\land Q(z)$$

4. Quantifier Distribution Equivalence

Let $A(x)$ and $B(x)$ be any predicate formula with free argument x:

$$(\forall x)(A(x)\land B(x))\Leftrightarrow(\forall x)A(x)\land(\forall x)B(x)$$
$$(\exists x)(A(x)\lor B(x))\Leftrightarrow(\exists x)A(x)\lor(\exists x)B(x)$$

The former can be understood as "All x have property A and property B" and "all x have property A and all x have property B". The latter can be proved by the former.

Example 2.4.2　It is proved that　$(\exists x)(A(x)\lor B(x))\Leftrightarrow(\exists x)A(x)\lor(\exists x)B(x)$.

Proof　Because:

$$(\forall x)(\neg A(x)\land\neg B(x))\Leftrightarrow(\forall x)\neg A(x)\land(\forall x)\neg B(x)$$

And because:

$$(\forall x)(\neg A(x)\land\neg B(x))\Leftrightarrow(\forall x)\neg(A(x)\lor B(x))$$
$$\Leftrightarrow\neg(\exists x)(A(x)\lor B(x))$$
$$(\forall x)\neg A(x)\land(\forall x)\neg B(x)\Leftrightarrow\neg(\exists x)A(x)\land\neg(\exists x)B(x)$$
$$\Leftrightarrow\neg((\exists x)A(x)\lor(\exists x)B(x))$$

So:

$$\neg(\exists x)(A(x)\lor B(x))\Leftrightarrow\neg((\exists x)A(x)\lor(\exists x)B(x))$$

And so：
$$(\exists x)(A(x) \vee B(x)) \Leftrightarrow (\exists x)A(x) \vee (\exists x)B(x)$$

5. The Implication of Quantifiers and Conjunctions

Let $A(x)$ and $B(x)$ be any predicate formula with free variable x.
$$(\forall x)A(x) \vee (\forall x)B(x) \Rightarrow (\forall x)(A(x) \vee B(x))$$
$$(\exists x)(A(x) \wedge B(x)) \Rightarrow (\exists x)A(x) \wedge (\exists x)B(x)$$
$$(\forall x)(A(x) \rightarrow B(x)) \Rightarrow (\forall x)A(x) \rightarrow (\forall x)B(x)$$
$$(\forall x)(A(x) \leftrightarrow B(x)) \Rightarrow (\forall x)A(x) \leftrightarrow (\forall x)B(x)$$

The first formula is explained as follows：these efforts can be explained to students as follows. But these students are smart or hardworking，but they can't push out that these students are smart or these students are working hard.

The second formula can be derived from the first one.

Example 2.4.3 It is proved that $(\exists x)(A(x) \wedge B(x)) \Rightarrow (\exists x)A(x) \wedge (\exists x)B(x)$.

Proof It can be obtained from the first formula：
$$(\forall x)\neg A(x) \vee (\forall x)\neg B(x) \Rightarrow (\forall x)(\neg A(x) \vee \neg B(x))$$
Meanwhile：
$$(\forall x)\neg A(x) \vee (\forall x)\neg B(x) \Leftrightarrow \neg(\exists x)A(x) \vee \neg(\exists x)B(x)$$
$$\Leftrightarrow \neg((\exists x)A(x) \wedge (\exists x) B(x))$$
$$(\forall x)(\neg A(x) \vee \neg B(x)) \Leftrightarrow (\forall x)\neg(A(x) \wedge B(x))$$
$$\Leftrightarrow \neg(\exists x)(A(x) \wedge B(x))$$
So：
$$\neg((\exists x)A(x) \wedge (\exists x)B(x)) \Rightarrow \neg(\exists x)(A(x) \wedge B(x))$$
Hypothetical translocation：
$$(\exists x)(A(x) \wedge B(x)) \Rightarrow (\exists x)A(x) \wedge (\exists x)B(x)$$

6. The Use of Multiple Quantifiers

For convenience，only two quantifiers are listed here. The usage of more quantifiers is similar to theirs. For binary predicates，if free variables are not considered，the following eight combinations can be generated.

$$(\forall x)(\forall y)P(x,y) \qquad (\forall y)(\forall x)P(x,y)$$
$$(\exists x)(\exists y)P(x,y) \qquad (\exists y)(\exists x)P(x,y)$$
$$(\forall x)(\exists y)P(x,y) \qquad (\exists y)(\forall x)P(x,y)$$
$$(\exists x)(\forall y)P(x,y) \qquad (\forall y)(\exists x)P(x,y)$$

For example，let $P(x,y)$ denote that x and y are of the same gender，and the individual domain for x is class 1 of a middle school，and y is class 2 of a middle school.

$(\forall x)(\forall y)P(x,y)$：Everyone in class 1 and class 2 is of the same gender.

$(\forall y)(\forall x)P(x,y)$：Everyone in class 2 and class 1 is of the same gender.

Obviously，the meaning of the above two statements is the same. So：
$$(\forall x)(\forall y)P(x,y) \Leftrightarrow (\forall y)(\forall x)P(x,y)$$
Similarly：

$(\exists x)(\exists y)P(x,y)$: There are people of the same gender in class 1 and class 2.

$(\exists y)(\exists x)P(x,y)$: There are people of the same gender in class 2 and class 1.

The two statements have the same meaning. So

$$(\exists x)(\exists y)P(x,y)\Leftrightarrow(\exists y)(\exists x)P(x,y)$$

However, $(\forall x)(\exists y)P(x,y)$ means that for all the people in class 1, there are people in class 2 who are of the same gender.

$(\exists y)(\forall x)P(x,y)$ means that there is a person in class 2, and a person in class 1 is of the same gender as him.

$(\forall y)(\exists x)P(x,y)$ means that for all the people in class 2, there are people of the same gender in class 1.

$(\exists x)(\forall y)P(x,y)$ means that there is a person in class 1, and all the people in class 2 are of the same gender.

In the above four sentences, the expressions are different, so the order in which the full name quantifier and the existential quantifier appear in the formula cannot be changed at will. The predicate formula with two quantifiers has the following implication relationship:

$$(\forall x)(\forall y)A(x,y)\Rightarrow(\exists y)(\forall x)A(x,y)$$
$$(\forall y)(\forall x)A(x,y)\Rightarrow(\exists x)(\forall y)A(x,y)$$
$$(\exists y)(\forall x)A(x,y)\Rightarrow(\forall x)(\exists y)A(x,y)$$
$$(\exists x)(\forall y)A(x,y)\Rightarrow(\forall y)(\exists x)A(x,y)$$
$$(\forall x)(\exists y)A(x,y)\Rightarrow(\exists y)(\exists x)A(x,y)$$
$$(\forall y)(\exists x)A(x,y)\Rightarrow(\exists x)(\exists y)A(x,y)$$

2.5　Prenex Normal Forms

In the process of propositional calculus, formulas are often transformed into the normal form, which is easier to identify and more beautiful. For predicate calculus, there are similar situations. A predicate calculus formula can be transformed into its equivalent normal form.

Definition 2.5.1　A formula whose quantifiers are at the beginning of the whole formula and whose scope extends to the end of the whole formula is called the prenex normal forms.

According to this definition, the prenex normal forms can be expressed as follows:

$$(\square v_1)(\square v_2)...(\square v_n)A$$

Among them: $\square$ is $\exists$ or $\forall$; v_i is the individual variable, $i=1$, ..., n. A is the predicate formula of non-content words.

For example:

$$(\forall x)(\forall y)(A(x)\wedge B(y)\rightarrow P(x,y))$$
$$(\forall y)(\forall x)(\exists z)(\neg P(x,y)\wedge Q(x)\vee B(x,z))$$

are all the prenex normal forms. But

$$(\forall x)P(x)\vee(\forall y)(Q(y)\wedge A(x,y))$$
$$(\forall y)(\forall x)(\neg H(x,y)\vee F(x))\wedge(\exists z)L(x,z)$$

aren't the prenex normal forms.

Theorem 2.5.1 Any prenex normal forms can be transformed into its equivalent bundle normal form.

Proof Firstly, by using the quantifier transformation formula, the negation is deepened to the front of the propositional variable and predicate filling, and then the negative is used $(\forall x)(A \lor B(x)) \Leftrightarrow A \lor (\forall x)B(x)$ and $(\exists x)(A \land B(x)) \Leftrightarrow A \land (\exists x)B(x)$ are used to move the quantifier to the front of the whole expression, thus obtaining the prenex normal forms.

Example 2.5.1 Find the prenex normal forms of the formula $(\forall x)F(x) \to (\exists x)G(x)$.

Solution
$$(\forall x)F(x) \to (\exists x)G(x) \Leftrightarrow \neg(\forall x)F(x) \lor (\exists x)G(x)$$
$$\Leftrightarrow (\exists x)\neg F(x) \lor (\exists x)G(x)$$
$$\Leftrightarrow (\exists x)(\neg F(x) \lor G(x))$$

Example 2.5.2 Find the prenex normal forms of the formula $(\forall y)A(x, y) \to (\exists x)B(x, y)$.

Solution
$$(\forall y)A(x, y) \to (\exists x)B(x, y) \Leftrightarrow (\forall t)A(x, t) \to (\exists s)B(s, y)$$
$$\Leftrightarrow (\exists t)(\exists s)(A(x, t) \to B(s, y))$$

Definition 2.5.2 A predicate formula A, if it has the following form, is called the prenex conjunctive normal form.

$(\Box v_1)(\Box v_2)...(\Box v_n)((A_{11} \lor A_{12} \lor ... \lor A_{1n}) \land (A_{21} \lor A_{22} \lor ... \lor A_{2n}) \land ... \land (A_{n1} \lor A_{n2} \lor ... \lor A_{nn}))$

Among them: $\Box$ is $\exists$ or $\forall$; v_i is the individual variable, $i=1, ..., n$; A_{ij} is the atomic formula or its negation. For example: $(\forall x)(\exists z)(\forall y)\{[P \lor (x=a) \lor (z=d)] \land [Q(y) \lor (a=b)]\}$ is the atomic formula or its negation.

Theorem 2.5.2 Each predicate formula can be translated into its equivalent form of the prenex conjunctive normal form. (proof omitted).

Example 2.5.3 Find the prenex conjunctive normal form of the formula wff D: $(\forall x)[(\forall y)P(x) \lor (\forall z)Q(z, y) \to \neg(\forall y)R(x, y)]$.

Solution The first step is to cancel the multi-residue words:
$$D \Leftrightarrow (\forall x)[P(x) \lor (\forall z)Q(z, y) \to \neg(\forall y)R(x, y)]$$
The second step is to change the name:
$$D \Leftrightarrow (\forall x)[P(x) \lor (\forall z)Q(z, y) \to \neg(\forall w)R(x, w)]$$
The third step is to remove the conditional connectives:
$$D \Leftrightarrow (\forall x)[\neg(P(x) \lor (\forall z)Q(z, y)) \lor \neg(\forall w)R(x, w)]$$
The fourth step is to substitute $\neg$:

$$D \Leftrightarrow (\forall x)[(\neg P(x) \land (\exists z) \neg Q(z, y)) \lor (\exists w) \neg R(x, w)]$$

The fifth step is to push the quantifier to the left:

$$D \Leftrightarrow (\forall x)(\exists z)(\exists w)[(\neg P(x) \land \neg Q(z, y)) \lor \neg R(x, w)]$$
$$\Leftrightarrow (\forall x)(\exists z)(\exists w)[(\neg P(x) \lor \neg R(x, w)) \land (\neg Q(z, y) \lor \neg R(x, w))]$$

Example 2.5.4　Find the prenex conjunctive normal forms of the formula $((\exists x) F(x) \lor (\exists x) G(x)) \to (\exists x)(F(x) \lor G(x))$.

Solution

$$((\exists x) F(x) \lor (\exists x) G(x)) \to (\exists x)(S(x) \lor T(x))$$
$$\Leftrightarrow (\exists x)(F(x) \lor G(x)) \to (\exists y)(S(y) \lor T(y))$$
$$\Leftrightarrow (\forall x)(\exists y)((F(x) \lor G(x)) \to (S(y) \lor T(y)))$$
$$\Leftrightarrow (\forall x)(\exists y)(\neg (F(x) \lor G(x)) \lor (S(y) \lor T(y)))$$
$$\Leftrightarrow (\forall x)(\exists y)((\neg F(x) \land \neg G(x)) \lor (S(y) \lor T(y)))$$
$$\Leftrightarrow (\forall x)(\exists y)((\neg F(x) \lor S(y) \lor T(y)) \land (\neg G(x) \lor S(y) \lor T(y)))$$

Definition 2.5.3　A predicate formula A is called a prenex disjunctive normal form if it has the following form.

$$(\square v_1)(\square v_2)...(\square v_n)[(A_{11} \land A_{12} \land ... \land A_{1n}) \lor (A_{21} \land A_{22} \land ... \land A_{2n}) \lor ... \lor (A_{n1} \land A_{n2} \land ... \land A_{nn})].$$

Among them: $\square$ is $\exists$ or $\forall$; v_i is the individual variable, $i = 1, ..., n$; A_{ij} is the atomic formula or its negation.

Theorem 2.5.3　Each predicate formula can be translated into its equivalent form of the prenex disjunctive normal form. (proof omitted).

The procedure of converting any predicate formula A into an equivalent prenex disjunctive normal form is similar to that in example 2.5.3.

2.6　Inference Theory of Predicate Calculus

In predicate calculus, C is an effective conclusion of a set of premises A_1, A_2, ..., A_n, and still defined as $A_1 \land A_2 \land ... \land A_n \Rightarrow C$. The rule P, rule T, substitution rule, all equivalence and implication in propositional calculus can be used in predicate reasoning. In addition, the equivalence and implication in predicate calculus introduced in Section 2.3 can also be used in predicate reasoning. In addition, there are the following rules.

(1) Universal Specify, which is expressed as US. That is

$$(\forall x)P(x) \Rightarrow P(c)$$

At this time, the condition is that: ① c is any individual in the domain of individuals. ② When x in $A(x)$ is replaced by c, it must be replaced in all places where x appears.

Rule description: If all individuals in the individual domain satisfy predicate A, then any individual c in the individual domain satisfies predicate A. Using this rule, we can deduce a special conclusion without full quantifiers from the premise with full quantifiers. It embodies the derivation method from general to special in logic reasoning.

(2) Universal Generalize, which is expressed as UG. That is

$$P(y) \Rightarrow (\forall x)P(x)$$

At this time, the condition is that: ① y is any individual in the domain of individuals and is true to y in $P(y)$. ② x is an individual variable that does not appear in $A(y)$.

For example, if the individual field is a set of real numbers **R**, $G(x, y)$ denotes $x > y$. Let $P(y) \Leftrightarrow (\exists x)G(x, y)$, it is obvious that $P(y)$ satisfies the condition ①, it is certain that $(\forall z)P(z) \Leftrightarrow (\forall z)(\exists x)G(x, z) \Leftrightarrow (\forall z)(\exists x)(x > z)$, $(\forall z)(\exists x)(x > z)$ is a true proposition. If it is deduced into $(\forall x)A(x) \Leftrightarrow (\forall x)(\exists x)G(x, x) \Leftrightarrow (\forall x)(\exists x)(x > x)$, it will produce an error, because $(\forall x)(\exists x)(x > x)$ is a false proposition. The reason for the mistake is that the condition has been violated condition ②.

(3) Existential Specify, which is expressed as ES. That is

$$(\exists x)P(x) \Rightarrow P(c)$$

At this time, the condition is that: ① c is a certain individual in the individual domain, not an individual variable. ② c is an individual that does not appear in $P(x)$.

Rule description: If some individuals in the individual domain satisfy predicate A, then at least one determined individual c satisfies predicate A.

For example, let the individual field be the set of integers **Z**, $A(x)$ denotes that x is odd, and $B(x)$ denotes that x is even.

$(\exists x)A(x) \Rightarrow A(c)$, It means that if some integers are odd and let c be 3, then c is odd. This reasoning is right.

$(\exists x)B(x) \Rightarrow B(d)$, It means that if some integers are even, let d be 4, then d is even. This reasoning is also true. Therefore, the following reasoning holds:

$$(\exists x)A(x) \wedge (\exists x)B(x) \Rightarrow A(c) \wedge B(d)$$

However, the following reasoning is wrong:

$$(\exists x)A(x) \wedge (\exists x)B(x) \Rightarrow A(c) \wedge B(c)$$
$$(\exists x)A(x) \wedge (\exists x)B(x) \Rightarrow A(d) \wedge B(d)$$

Because 3 cannot be both odd and even; similarly, 4 cannot be both odd and even. The reason for the mistake is that the condition ② has been violated.

(4) Existential Generalize, which is expressed as EG. That is

$$P(c) \Rightarrow (\exists x)P(x)$$

At this time, the condition is that: ① c is the determined individual in the individual domain. ② x cannot be an individual variable that appears in $P(c)$.

Rule description: For an individual c in the individual domain satisfying the predicate P, there is $(\exists x)P(x)$.

Example 2.6.1   To prove Socrates: everybody's going to die. Socrates is a man, Socrates is going to die.

Let $P(x)$: x be human. $Q(x)$: x is dying. s: Socrates

This problem is equivalent to proving implication formula: $(\forall x)(P(x) \rightarrow Q(x)) \wedge P(s) \Rightarrow Q(s)$.

Proof

(1) $(\forall x)(H(x)\to M(x))$ P

(2) $H(s)\to M(s)$ US(1)

(3) $H(s)$ P

(4) $M(s)$ T(2)(3)Hypothetical reasoning

Example 2.6.2　Prove implication formula

$$(\forall x)(H(x)\to M(x))\quad(\exists x)H(x)\Rightarrow(\exists x)M(x)$$

Proof

(1) $(\exists x)H(x)$ P

(2) $H(c)$ ES(1)

(3) $(\forall x)(H(x)\to M(x))$ P

(4) $H(c)\to M(c)$ US(3)

(5) $M(c)$ T(2)(4)Hypothetical reasoning

(6) $(\exists x)M(x)$ EG(5)

If (1), (2)are written after(3), (4), the following reasoning is obtained:

(1) $(\forall x)(H(x)\to M(x))$ P

(2) $H(c)\to M(c)$ US(1)

(3) $(\exists x)H(x)$ P

(4) $H(c)$ ES(3)

(5) $M(c)$ T(2)(4)Hypothetical reasoning

(6) $(\exists x)M(x)$ EG(5)

This reasoning is logically wrong. Because c in (2) is an individual in the domain of individuals, we can't select c in (2) from (3) to (4) with the ES rule. Because the individual it chooses may not be the same individual as the individual c in (2), the reasoning is wrong.

Example 2.6.3　Prove the following formula with the CP rule.

$$(\forall x)(F(x)\vee G(x))\Rightarrow(\forall x)F(x)\vee(\exists x)G(x)$$

The original formula can be equivalent to the following formula

$$(\forall x)(F(x)\vee G(x))\Rightarrow\neg(\forall x)F(x)\to(\exists x)G(x)$$

Proof

(1) $\neg(\forall x)F(x)$ P(Additional premises)

(2) $(\exists x)\neg F(x)$ T(1)(Negative equivalent of quantifier)

(3) $\neg F(c)$ ES(2)

(4) $(\forall x)(F(x)\vee G(x))$ P

(5) $F(c)\vee G(c)$ US(4)

(6) $G(c)$ T(3)(5)(Disjunctive syllogism)

(7) $(\exists x)G(x)$ EG(6)

(8) $\neg(\forall x)F(x)\to(\exists x)G(x)$ CP

Example 2.6.4　Let the individual domain be the total individual domain. Prove to reason: the members of the academic society are workers and experts. Some of the members are young people. So some members are young experts.

Firstly, the proposition is symbolized. $F(x)$: x is a member of the academic society. $G(x)$: x is an expert. $H(x)$: x is the worker. $R(x)$: x is a young man.

So the question is to prove: $(\forall x)(F(x) \to G(x) \wedge H(x))$, $(\exists x)(F(x) \wedge R(x)) \Rightarrow (\exists x)(F(x) \wedge R(x) \wedge G(x))$.

Proof

(1) $(\exists x)(F(x) \wedge R(x))$	P
(2) $F(c) \wedge R(c)$	ES(1)
(3) $F(c)$	T(2)(Law of reduction)
(4) $(\forall x)(F(x) \to G(x) \wedge H(x))$	P
(5) $F(c) \to G(c) \wedge H(c)$	US(4)
(6) $G(c) \wedge H(c)$	T(3)(5)(Hypothetical Reasoning)
(7) $R(c)$	T(2)(Law of reduction)
(8) $G(c)$	T(6)(Law of reduction)
(9) $F(c) \wedge R(c) \wedge G(c)$	T(2)(7)(8)(Conjunctive Introduction)
(10) $(\exists x)(F(x) \wedge R(x) \wedge G(x))$	EG(9)

2.7 Application of Predicate Logic

We can apply the knowledge of predicate logic to our daily life, especially to solve reasoning problems in real life.

Some examples of comprehensive applications are given below.

Example 2.7.1 A family went to Shanghai Disneyland for a visit. They went to the Thunder Mountain rafting, the creation Aurora, the Pirates of the Caribbean, and finally went to the store to buy souvenirs. Known:

(1) Someone bought Mickey earmuffs.

(2) Someone didn't buy Mickey earmuffs.

(3) Mom and dad bought Mickey earmuffs.

If one of the above three sentences is true, who bought Mickey earmuffs?

Solution

If sentence (3) is true. Then father and mother bought Mickey earmuffs, so the first sentence is true. This does not meet the condition, so (3) is false.

If sentence (3) is false, which indicates that at least one of the parents has not bought Mickey earmuffs, then sentence (2) is true, and then sentence (1) is false.

So we can know that "no one buys Mickey earmuffs" is true. So, mom and dad didn't buy Mickey earmuffs.

Example 2.7.2 Symbolize and judge the validity of the following conclusions.

The premise: (1) Each master's degree student is either the promotion exemption student or the unified examination student.

(2) Not every master's degree student is the promotion exempt student.

Conclusion: (3) Some graduate students are students of the national examination.

Solution Suppose $R(x)$: x is a master's student. $S(x)$: x is the promotion exempt student. $C(x)$: x is he unified examination student.

The above problem is equivalent to proving the following formula:

$$(\forall x)(R(x)\rightarrow(S(x)\lor C(x))),\quad \neg(\forall x)(R(x)\rightarrow S(x))\Rightarrow(\exists x)(R(x)\land C(x))$$

Proof

(1) $\neg(\forall x)(R(x)\rightarrow S(x))$	P
(2) $(\exists x)\neg(R(x)\rightarrow S(x))$	T(1)E
(3) $(\exists x)\neg(\neg R(x)\lor S(x))$	T(2)E
(4) $(\exists x)(R(x)\land\neg S(x))$	T(3)E
(5) $R(a)\land\neg S(a)$	T(4)ES
(6) $R(a)$	T(5)I
(7) $\neg S(a)$	T(5)I
(8) $(\forall x)(R(x)\rightarrow(S(x)\lor C(x)))$	P
(9) $R(a)\rightarrow(S(a)\lor C(a))$	T(8)US
(10) $S(a)\lor C(a)$	T(6)(9)I
(11) $C(a)$	T(7)(10)I
(12) $R(a)\land C(a)$	T(6)(11)I
(13) $(\exists x)(R(x)\land C(x))$	T(12)EG

Example 2.7.3   Let the individual domain be all students of the software engineering department, symbolize the following propositions and predicates, and prove the validity of the conclusion.

R: The weather is fine today. S: The award ceremony was held on time. $A(x)$: x was on time.

(1) As long as we encounter bad weather today, some students of the software engineering department will not be able to arrive on time.

(2) If and only if all the students are on time, the award ceremony can be held on time.

Conclusion: If the award ceremony is held on time, the weather will be fine today.

Solution The proposition can be symbolized as:

$$\neg R\rightarrow(\exists x)\neg A(x),\quad (\forall x)A(x)\leftrightarrow S\Rightarrow S\rightarrow R$$

(1) $\neg R\rightarrow(\exists x)\neg A(x)$	P
(2) $\neg R\rightarrow\neg(\forall x)A(x)$	T(1)E
(3) $(\forall x)A(x)\rightarrow R$	T(2)E
(4) $(\forall x)A(x)\leftrightarrow S$	P
(5) $((\forall x)A(x)\rightarrow S)\land(S\rightarrow(\forall x)A(x))$	T(4)E
(6) $S\rightarrow(\forall x)A(x)$	T(5)I
(7) $S\rightarrow R$	T(3)(6)I

1. Write the following propositions with predicate expressions.

(1) Xiao Wang is not a college student.

(2) Zhang Yun studies hard and does well.

(3) Xiao Gang has a sister or a brother.

(4) If a is odd, then $2a$ is not odd.

(5) All rational numbers are real numbers.

(6) Some real numbers are rational.

(7) Not all real numbers are rational.

(8) All athletes admire certain coaches.

(9) No female comrades is both a civil servant and a national athlete.

(10) Some male comrade are both national athletes and coaches.

2. Try symbolizing the following statements into predicate expressions.

(1) Xiao Zhang has studied Spanish and French.

(2) a is greater than b only if a is greater than c.

(3) 4 is not an odd number.

(4) 2 or 5 are prime numbers.

(5) Unless Wang Jianguo is from Hainan, he must be afraid of heat.

3. Use predicate formulas to translate the following propositions.

(1) There are rational numbers a, b and c that make the difference between a and b greater than the sum of a and c.

(2) For any real number a, there exists a real number b smaller than a.

(3) If the product of a finite number is non-zero, then none of the factors is zero.

4. Let the predicate $P(x, y)$ denote "x is greater than y", and the individual fields of individual variables x and y are $E=\{4, 5, 6\}$. Find the truth value of the following formula.

(1) $(\exists x)P(x, 4)$ (2) $(\forall x)P(x, 1)$ (3) $(\forall x)(\forall y)P(x, y)$

(4) $(\exists x)(\exists y)P(x, y)$ (5) $(\exists x)(\forall y)P(x, y)$ (6) $(\forall x)(\exists y)P(x, y)$

5. Use the predicate formula to characterize the following proposition.

A hardworking female college student in a hat is reading a big and thick English book.

6. Indicate the bounded variable and free variable of the following formulas.

(1) $(\forall x)(P(x) \rightarrow Q(x, y, z))$

(2) $(\forall x)P(x, y, z) \wedge (\exists y)Q(x, y, z)$

(3) $(\forall x)(\exists y)(P(x, y) \vee Q(x, z)) \rightarrow (\exists x)R(x, y)$

(4) $(\exists x)(\forall y)(P(x, y) \vee Q(z))$

(5) $(\forall x)(P(x) \vee Q(x)) \rightarrow (\exists x)R(x)$

7. Indicate bounded variable, free variable, guiding variable and the scope of quantifier of the following formulas.

(1) $(\forall x)(P(x) \vee Q(x)) \rightarrow (\exists x)(P(x) \vee R(x))$

(2) $(\forall x)(P(x) \wedge Q(z)) \vee (\forall x)(P(x) \to R(x))$

(3) $(\forall x)(P(x) \leftrightarrow Q(x)) \wedge (\exists x)P(x) \vee R(x)$

8. Suppose the individual field is a set $\{1, 2, 3\}$. Try to eliminate the quantifiers in the following formula.

(1) $(\forall x)R(x)$ (2) $(\forall x)(R(x) \to S(x))$

(3) $(\forall x)P(x) \vee (\forall x)Q(x)$ (4) $(\forall x)P(x) \wedge (\forall x)\neg Q(x)$

9. Rename the bound variable in the following predicate formulas.

(1) $(\forall x)(\exists y)(R(x, z) \vee S(y)) \to Q(x, y)$

(2) $(\forall x)(R(x, y) \to (\exists z)S(x, z)) \to (\forall y)Q(x, y)$

10. Bring in the free arguments in the following predicate formulas.

(1) $((\exists y)P(x, y) \vee (\forall x)Q(x, z)) \wedge (\forall z)R(x, z)$

(2) $((\forall y)R(x, y) \wedge (\exists z)S(x, z)) \vee (\forall x)Q(x, y)$

11. The following two equivalent expressions are proved:

(1) $(\exists x)(R(x) \to S(x)) \Leftrightarrow (\forall x)R(x) \to (\exists x)S(x)$

(2) $(\forall x)(\forall y)(R(x) \to S(x)) \Leftrightarrow (\exists x)R(x) \to (\forall y)S(y)$

12. Let the individual domain be $E = \{1, 2, 3\}$, and prove the following formula:

$$(\forall x)A(x) \vee (\forall x)B(x) \Rightarrow (\forall x)(A(x) \vee B(x))$$

13. Turn the following into the prenex normal form.

(1) $(\forall x)(R(x) \to (\exists z)S(x, z))$

(2) $(\exists x)(\neg((\exists z)R(x, z)) \to ((\exists y)S(y) \to Q(x)))$

(3) $(\forall x)(\forall y)(((\exists z)R(x, y, z) \wedge (\exists w)S(x, w)) \to (\exists u)Q(y, u))$

14. Find the prenex conjunctive normal form and the prenex disjunctive normal form of the following formula.

(1) $((\exists x)R(x) \vee (\exists x)S(x)) \to (\exists x)(R(x) \vee S(x))$

(2) $(\forall x)R(x) \to (\exists x)((\forall u)S(x, z) \vee (\forall u)Q(x, y, u))$

(3) $(\forall x)(R(x) \to (\forall y)((\forall z)S(x, y) \to \neg(\forall z)Q(y, x)))$

(4) $(\forall x)(R(x) \to S(x, y)) \to ((\exists y)R(y) \wedge (\exists z)S(y, z))$

15. Prove the following formula.

(1) $(\forall x)(\neg R(x) \to S(x)), (\forall x)\neg S(x) \Rightarrow (\exists x)R(x)$

(2) $(\forall x)(R(x) \vee S(x)), (\forall x)(S(x) \to \neg Q(x)), (\forall x)Q(x) \Rightarrow (\exists x)R(x)$

(3) $(\exists x)R(x) \to (\forall x)S(x) \Rightarrow (\forall x)(R(x) \to S(x))$

(4) $(\forall x)(R(x) \to S(x)), (\forall x)(Q(x) \to \neg S(x)) \Rightarrow (\forall x)(Q(x) \to \neg R(x))$

16. Prove the following formula with the CP rule.

(1) $(\forall x)(R(x) \to S(x)) \Rightarrow (\forall x)R(x) \to (\forall x)S(x)$

(2) $(\forall x)(R(x) \vee S(x)) \Rightarrow (\forall x)R(x) \vee (\exists x)S(x)$

第 2 章　谓词逻辑

在第 1 章的命题逻辑中,主要研究的是命题与命题演算,而原子命题作为其不可再分解的组成单位对研究命题间的关系起着非常大的作用。

但是,实际上还可以对原子命题做进一步的分析,尤其是不同的原子命题之间往往存在一些共性。为刻画命题内部的逻辑结构,我们需要研究谓词逻辑。

另外,命题逻辑并不是对所有的推断都能迎刃而解,比如苏格拉底三段论。这些都要求我们对命题的内部关系做深入的研究。

2.1　谓词与量词

命题是反映判断的句子,不反映判断的句子不是命题。一般来说,反映判断的句子是由主语和谓语两部分组成的。例如,手机是通信工具。其中"手机"是主语,"是通信工具"是谓语。主语一般都是个体,个体是指所研究对象中可以独立存在的、具体的或者抽象的客体。它可以是独立存在的人或者物,也可以是抽象的概念,如"马列主义""资本主义""小王""老师"等。客体常用小写英文字母或者小写英文字母带下标表示,叫作个体标识符。

表示具体或者特定个体的标识符称为个体常元,一般用小写英文字母 a、b、c、⋯ 或这些英文字母带下标表示。例如,

a:李林　　　b:小高　　　c:张云　　　d:小花

a,b,c,d 都是个体常元。

将表示任意个体或者泛指某类个体的标识符称为个体变元,常用 x、y、z 等或者这些英文字母带下标表示。

用于刻画个体的性质或者关系的就是谓词。例如,张三是一个工人,李明是一个工人,这两个命题可以用不同的符号 P、Q 表示,但是 P 和 Q 的谓语有同样的属性"是一个工人"。因此,可以引入一个符号表示"是一个工人",再引入一种方法表示个体的名称,这样就把"××是一个工人"这个客体的本质属性刻画出来了。又例如,

(a) 他是学生。　　　(b) 11 是质数。　　　(c) 小张比小刘高。

在上述语句中,"是学生""是质数""⋯⋯比⋯⋯高"都是谓词。前两个是体现个体性质的谓词,最后一个是指明两个个体之间关系的谓词。

我们通常用大写字母表示谓词,用小写字母或者带下标的小写字母表示个体。例如 A 表示"是一个老师",c 表示小张,d 表示小白,则 $A(c)$,$A(d)$ 分别表示"小张是一个老师""小白是一个老师"。

用谓词表达命题时,必须包括个体和谓词两部分。例如,下面三个命题的谓词表示,

$F(a)$:张三是优秀共青团员。

$G(b,c)$：小佩比小莉高。

$H(d,e,f)$：小山坐在小宁和小明之间。

我们把 $F(a)$ 称为一元谓词，$G(b,c)$ 称为二元谓词，$H(d,e,f)$ 称为三元谓词，以此类推。

将谓词后面填上相关联的个体常元所得的式子叫作谓词填式。$F(a),G(b,c),H(d,e,f)$ 都是谓词填式。谓词填式表示的是命题。

一般来说，n 元谓词需要将 n 个个体名称插入固定的位置，如果 A 为 n 元谓词，a_1，$a_2,\cdots,a_n$ 是个体的名称，则 $A(a_1,a_2,\cdots,a_n)$ 就成为一个命题。

通常而言，一元谓词表达个体的性质，而多元谓词表达个体之间的关系。

为说明命题函数的概念，下面先解释命题与谓词的关系。

设 T 是谓词"能够完成任务"，a 表示个体名称张三，b 表示个体名称李四，c 表示小车，那么 $T(a),T(b),T(c)$ 分别表示 3 个不同的命题，但它们都有一个共同的形式，即 $T(x)$。当 x 分别取 a,b,c 时就分别表示为"张三能完成任务""李四能完成任务""小车能完成任务"。

同理，如果 $H(x,y)$ 表示"x 大于 y"，那么 $H(3,2)$ 就表示一个真命题"3 大于 2"，而 $H(4,5)$ 表示一个假命题"4 大于 5"。

从上述可以看出，$T(x),H(x,y)$ 本身不是一个命题，因为其真值是无法确定的，只有当变元 x,y 等取特定的个体时，才确定了一个命题。

定义 2.1.1 ▶ 由一个谓词和一些个体变元组成的表达式称为简单命题函数，简称命题函数。

根据这个定义可知，n 元谓词就是有 n 个个体变元的命题函数 $P(x_1,x_2,\cdots,x_n)$，叫作 n 元命题函数($n\geqslant1$)。例如，$T(x),H(x,y),P(x_1,x_2,\cdots,x_n)$ 分别叫作一元命题函数、二元命题函数、n 元命题函数($n\geqslant1$)。特别注意，当 $n=0$ 时，称为零元命题函数，那么它本身就是一个命题，故命题是 n 元谓词中 $n=0$ 时的情况。$F(a),G(b,c),H(d,e,f)$ 都是零元命题函数，它们都是命题。因此，命题逻辑中的命题均可以表示为谓词逻辑中的零元命题函数(谓词填式)，命题成为命题函数的特例。

定义 2.1.2 ▶ 由一个或者 n 个简单命题函数及联结词($\neg$、$\wedge$、$\vee$、$\rightarrow$、$\leftrightarrow$)组合而成的表达式称为复合命题函数。

例 2.1.1 设 $A(x)$ 表示"x 上课认真"，用 $B(x)$ 表示"x 学习成绩好"，则 $\neg A(x)$ 表示"x 上课不认真"。$A(x)\wedge B(x)$ 表示"x 上课认真，学习成绩也好"。$A(x)\rightarrow B(x)$ 表示"如果 x 上课认真，那么 x 学习成绩好"。

例 2.1.2 设 $H(x,y)$ 表示"x 比 y 轻"。当 x,y 指人或者物时，它是一个命题，但是当 x,y 指实数时，$H(x,y)$ 就不是一个命题。

命题函数不是一个命题，只有当个体变元取特定名称时，才能成为一个命题。但是个体变元在哪个范围内取特定的值，对是否成为命题及命题的真值有极大的影响。

例 2.1.3 $Q(x)$ 表示"x 是一位中学老师"，如果 x 的讨论范围是某中学老师，那么 $Q(x)$ 就是永真式。

如果 x 的讨论范围是一位大学老师，那么 $Q(x)$ 就是永假式。

如果 x 的讨论范围是某电影院的观众，观众中可能有中学老师也可能没有，那么对一些观众，$Q(x)$ 为真；对其他观众，$Q(x)$ 为假。

例 2.1.4 $(R(x,y)\wedge R(y,z))\rightarrow R(x,z)$

如果将 $R(x,y)$ 解释为 "x 大于 y"，当 x,y,z 都是在实数范围内取值时，这个式子表示为 "如果 x 大于 y，y 大于 z，那么 x 大于 z"。这就是一个永真式。

如果将 $R(x,y)$ 解释为 "x 是 y 的父亲"，当 x,y,z 的取值范围都是人时，那么这个式子解释为 "如果 x 是 y 的父亲，y 是 z 的父亲，那么 x 是 z 的父亲"。很明显，这是一个永假式。

如果 $R(x,y)$ 解释为 "x 距离 y 有 10 米"，若 x,y,z 表示地面上的房子，那么 "x 距离 y 有 10 米且 y 距离 z 有 10 米，则 x 距离 z 有 10 米"。这个命题的真值将由 x，y，z 的具体位置而定，它可能为 T，也可能为 F。

由上面几个例子可以看出，命题函数确定为命题实际上与个体变元的论述范围有关。这个范围称为个体域或者论域。个体域可以是有穷集合，也可以是无穷集合，包含任意个体域的个体域称为全总个体域，它是由宇宙间一切对象组成的集合。在本书中，如无特别说明，所采用的都是全总个体域。

使用上面所讲的一些概念还不能完全用符号很好地表示日常生活中的各种命题。例如，$R(x)$ 表示 "x 已经结婚"，当个体域的取值范围是某中学老师时，$R(x)$ 既可以表示这所中学的所有老师都已经结婚，也可以表示这所中学中的一些老师已经结婚。这有可能产生混乱，为避免这种混乱，应引入量词的概念用于刻画 "所有的" 和 "一些" 的不同概念。

根据 "所有的" 和 "存在一些" 可知，量词可以分为两种，分别是全称量词和存在量词。

日常生活和数学中常用的 "所有的""每一个""任意的" 等词统称为全称量词，可将它们符号化为 "$\forall$"，并用 $(\forall x)$，$(\forall y)$ 等表示个体域中所有的个体，而用 $(\forall x)F(x)$ 和 $(\forall y)G(y)$ 等分别表示个体域中的所有个体都有性质 F 和性质 G。

日常生活和数学中常用的 "存在""有一个""有些""至少有一个" 等词统称为存在量词，可将它们符号化为 "$\exists$"，并用 $(\exists x)$，$(\exists y)$ 等表示个体域里的有些个体，而用 $(\exists x)F(x)$ 和 $(\exists y)G(y)$ 等分别表示在个体域中存在个体具有性质 F 和存在个体具有性质 G。

例 2.1.5 将下列命题符号化，并在①、②、③三个个体域中考察命题的真值。

命题：(1) 所有的数都大于 2。

(2) 至少有一个数大于 2。

个体域：① $\{2,1,3,-2,4\}$。② $\{20,30,3\}$。③ $\{2,-3,1\}$。

解 设 $R(x)$：x 大于 5。

(1) "所有的数都大于 2。" 符号化为 $(\forall x)R(x)$。

在个体域①，②，③中，它们的真值分别为假、真、假。

(2) "至少有一个数大于 2。" 符号化为 $(\exists x)R(x)$。

在个体域①，②，③中，它们的真值分别为真、真、假。

在个体变元被量化以后，命题函数变成命题，其真值又与个体域的选定有关，这给命题函数的研究带来了一定的困难，同时由于在讨论带有量词的命题函数时必须确定其个体域，所以为了统一，下面统一使用全总个体域。在全总个体域中，对每一个个体变元的变化范围，用特性谓词加以限制。特性谓词加入的方法：

(1) 对全称量词，特性谓词作为条件命题的前件加入。

(2) 对存在量词，特性谓词作为合取项加入。

例 2.1.6 将下列命题在①，②两个个体域中进行符号化。

命题：(1) 所有的学生都要考试。

（2）存在一个学生要考试。

个体域：① 所有学生组成的集合。② 全总个体域。

解 设 $R(x)$：x 要考试。

当个体域为①时，可以符号化如下。

（1）符号化为 $(\forall x)R(x)$。

（2）符号化为 $(\exists x)R(x)$。

当个体域为②时，设特性谓词 $A(x)$：x 是学生。

（1）符号化为 $(\forall x)(A(x)\to R(x))$。

（2）符号化为 $(\exists x)(A(x)\wedge R(x))$。

2.2　谓词公式与翻译

通过前面的学习我们了解到，简单的命题函数可以通过与命题联结词组合形成一些谓词表达式。有了量词和谓词的概念，谓词表达式就可以表示日常生活的命题。但是，什么样的谓词公式才能成为谓词公式并进行谓词演算呢？下面就介绍谓词公式的相关概念。

我们把 $P(x_1,x_2,\cdots,x_n)$ 称为谓词演算的原子公式，其中 $x_1,x_2,\cdots,x_n$ 称为个体变元，因此原子公式包括下列形式中的各种特殊例子，如 $A(x),A(x,y),A(f(x),y),A(x,y,z)$，$A(a,y)$ 等。总体而言，我们把命题、命题变元、谓词填式和命题函数都叫作谓词演算的原子公式。

定义 2.2.1 ▶ 按下列规则构成的表达式称为谓词演算的合式公式，简称谓词公式。

（1）谓词演算的原子公式是合式公式。

（2）若 A 是合式公式，则 $\neg A$ 是合式公式。

（3）若 A 和 B 是合式公式，则 $(A\wedge B),(A\vee B),(A\to B)$ 和 $(A\leftrightarrow B)$ 是合式公式。

（4）若 A 是合式公式，x 是 A 中出现的任意个体变元，则 $(\forall x)A,(\exists x)A$ 是合式公式。

（5）只有有限次地应用（1）、（2）、（3）、（4）所得的公式才是合式公式。

下面举例说明如何用谓词公式表达自然语言中的一些有关命题。

例 2.2.1 并非每个实数都是有理数。

解 设 $R(x)$：x 是实数，$Q(x)$：x 是有理数。

该命题符号化为 $\neg(\forall x)(R(x)\to Q(x))$。

例 2.2.2 没有不犯错误的人。

解 设 $M(x)$：x 是人，$F(x)$：x 犯错误。

此命题可以理解为任何人都会犯错。此时，符号化为 $(\forall x)(M(x)\to F(x))$。

此命题也可以理解为并非存在一些人不犯错误。此时，符号化为 $\neg(\exists x)(M(x)\wedge\neg F(x))$。

例 2.2.3 并不是所有的兔子都比所有的乌龟跑得快。

解 设 $F(x)$：x 是兔子，$G(x)$：x 是乌龟，$H(x,y)$：x 比 y 跑得快。

该命题符号化为 $\neg(\forall x)(\forall y)(F(x)\wedge G(y)\to H(x,y))$。

2.3 变元的约束

一个谓词公式中的所有个体变元被量化以后便成为命题。要分析一个谓词公式且它是否成为命题,必须看它被量化的情况。

定义 2.3.1 ▶ 如果 A 是谓词公式 B 的一部分且是谓词公式,则称 A 是 B 的子公式。

定义 2.3.2 ▶ 紧接量词以后的最小子公式叫作该量词的辖域或者作用域。

定义 2.3.3 ▶ 量词$(\forall x)$和$(\exists x)$中的 x 叫作该量词的指导变元或作用变元。

定义 2.3.4 ▶ 量词$(\forall x)$和$(\exists x)$的辖域内 x 的一切出现叫作约束出现,x 叫作约束变元,约束变元以外的其他变元的出现叫作自由出现,自由出现的变元叫作自由变元。

例 2.3.1 说明以下各式的作用域与变元的约束情况。

(1) $(\forall x)P(x) \rightarrow Q(y)$

(2) $(\forall x)(P(x) \vee (\exists y)Q(x,y))$

(3) $(\forall x)(\forall y)(P(x,y) \vee Q(y,z)) \wedge (\exists x)P(x,y)$

(4) $(\forall x)(P(x) \vee (\exists x)Q(x,z) \rightarrow (\exists y)R(x,y)) \vee Q(x,y)$

解 (1) $(\forall x)$的作用域为 $P(x)$,x 为约束变元,y 为自由变元。

(2) $(\forall x)$的作用域为$(P(x) \vee (\exists y)Q(x,y))$,$(\exists y)$的作用域为 $Q(x,y)$,x,y 都是约束变元。

(3) $(\forall x)$和$(\forall y)$的作用域为$(P(x,y) \vee Q(y,z))$,其中 x,y 为约束变元,z 为自由变元。$(\exists x)$的作用域为 $P(x,y)$,其中 x 为约束变元,y 为自由变元。那么在整个公式中,x 都是约束出现,y 既是约束出现又是自由出现,z 是自由出现。

(4) $(\forall x)$的作用域为$(P(x) \vee (\exists x)Q(x,z) \rightarrow (\exists y)R(x,y))$,其中 x,y 都是约束变元,z 为自由变元,但是 $Q(x,z)$中的 x 是受$(\exists x)$的约束,而不是受$(\forall x)$约束。$Q(x,y)$中的 x,y 是自由变元。

通过对约束变元的理解可以知道,$P(x_1, x_2, \cdots, x_n)$是 n 元谓词,它有 n 个相互独立的自由变元,若对其中 k 个变元进行约束则称为 $n-k$ 元谓词,因此,谓词公式中如果没有自由变元出现,该式就成为一个命题。例如,$(\forall x)Q(x,y,z)$是二元谓词。$(\forall x)(\exists y)Q(x,y,z)$是一元谓词。

由例 2.3.1 可以看出,在一个公式中,同一个变元既可以是约束的,也可以是自由的,容易混淆。故可以对约束变元进行换名,使一个变元在一个公式中只呈现一种形式,即呈自由出现或者呈约束出现。因为$(\forall x)P(x)$与$(\forall y)P(y)$,$(\exists x)P(x)$与$(\exists y)P(y)$都具有相同的意义,所以约束变元与表示该变元的符号无关。根据这个特点,可以将约束变元换名。为使换名后的公式中出现的变元要么是约束的,要么是自由的,我们提出以下换名规则。

(1) 对约束变元可以换名,其更改的变元名称范围是量词的指导变元,以及该量词辖域中的所有该变元,公式的其余部分不变。

(2) 换名时一定要更改成辖域中没有出现的变元名,最好是公式中没有的变量名。

例 2.3.2 将$(\exists x)(\forall y)(P(x,y) \vee Q(y,z)) \rightarrow (\forall x)R(x,y)$中的约束变元 y 换名。

解 用 w 置换约束变元 y 可得到:

$$(\exists x)(\forall w)(P(x,w)\vee Q(w,z))\rightarrow(\forall x)\,R(x,y)$$

不能换成：

$$(\exists x)(\forall w)(P(x,w)\vee Q(w,z))\rightarrow(\forall x)\,R(x,w)$$

因为 $R(x,y)$ 中的 y 不是约束变元。

也不能换成：

$$(\exists x)(\forall w)(P(x,w)\vee Q(y,z))\rightarrow(\forall x)\,R(x,y)$$

因为 $Q(y,z)$ 中的 y 也是约束变元，同时也是 $(\forall y)$ 辖域中的变元。

对公式中的自由变元也可以进行更改，用于解决公式中约束变元与自由变元的同名问题，这种更改叫作代入，代入规则如下。

（1）谓词公式中的自由变元可以代入，需对公式中该变元自由出现的每一处进行代入。

（2）代入的变元与原公式中其他变元的名称不能相同。

例 2.3.3　对 $(\exists x)(P(y)\wedge R(x,y))\rightarrow(\forall y)Q(y)$ 中的自由变元 y 进行代入。

解　用 u 置换自由变元 y 可得：

$$(\exists x)(P(u)\wedge R(x,u))\rightarrow(\forall y)Q(y)$$

不能换成：

$$(\exists x)(P(x)\wedge R(x,x))\rightarrow(\forall y)Q(y)$$

因为代入的变元不能与原公式中其他变元的名称相同。故不能换成：

$$(\exists x)(P(u)\wedge R(x,y))\rightarrow(\forall y)Q(y)$$

需要指出，当个体域的元素有限时，个体变元的所有可能的取代是可枚举的。

设个体域元素为 $a_1,a_2,\cdots,a_n$。则：

$$(\forall x)A(x)\Leftrightarrow A(a_1)\wedge A(a_2)\wedge\cdots\wedge A(a_n)$$
$$(\exists x)A(x)\Leftrightarrow A(a_1)\vee A(a_2)\vee\cdots\vee A(a_n)$$

同时，量词对变元的约束往往与量词的顺序有关。

例如，$(\forall y)(\forall x)(x<(y-2))$ 表示任何 y 均有 x，使 $x<y-2$。$(\exists y)(\exists x)(x<(y-2))$ 表示存在 y 有 x，使 $x<y-2$。

对这些命题中的多个量词，我们约定按从左到右的顺序读出。需要注意的是，量词顺序不能颠倒，否则将与原命题不符。

2.4　谓词演算的等价式与蕴含式

在谓词公式中常包含命题变元和个体变元，当个体变元由确定的个体取代，命题变元由确定的命题取代时，就称为对谓词公式赋值。一个谓词公式经过赋值以后，就成为具有确定真值 T 或者 F 的命题。

定义 2.4.1　设 A、B 是任意两个谓词公式，并且它们有共同的个体域，对 A、B 的任何赋值，若其真值相同，则称 A 与 B 是等价的，记作 $A\Leftrightarrow B$。

定义 2.4.2　设 A 是谓词公式，其个体域为 E。如果对 A 的任何赋值，A 都为真，则称 A 是有效的或永真的。

定义 2.4.3　设 A 是谓词公式，如果对 A 的任何赋值，A 都为假，则称 A 是不可满足的或永假的。

定义 2.4.4 ▶ 设 A 是谓词公式,如果至少有一组赋值使 A 为真,则称 A 是可满足的。

有了谓词公式的等价和永真等概念,就可以讨论谓词演算的一些等价式和蕴含式了。

1. 命题逻辑中的等价式的推广

在命题演算中,任一永真公式,其中同一命题变元用同一公式取代时,其结果也是永真公式。我们可以把这个情况推广到谓词公式中。当谓词演算中的公式代替命题演算中永真公式的变元时,所得的谓词公式即为有效公式,故命题演算中的等价公式表和蕴含式表都可推广到谓词演算中使用。例如,

$$(\forall x)(P(x) \to Q(x)) \Leftrightarrow (\forall x)(\neg P(x) \lor Q(x))$$

$$(\forall x)P(x) \lor (\exists y)R(x, y) \Leftrightarrow \neg(\neg(\forall x)P(x) \land \neg(\exists y)R(x, y))$$

$$(\exists x)H(x, y) \land \neg(\exists x)H(x, y) \Leftrightarrow F$$

2. 量词否定等价式

为更好地说明这个问题,我们先举例讨论。

例 2.4.1 设 $P(x)$ 表示 x 今天上课却迟到了,$\neg P(x)$ 表示 x 今天上课没有迟到。

那么"存在一些人今天上课没有迟到"和"不是所有人今天上课都迟到"在意义上相同,即 $\neg(\forall x)P(x) \Leftrightarrow (\exists x)\neg P(x)$。并且,"不是存在一些人今天上课迟到"与"所有人今天上课都没有迟到"意义上相同,即 $\neg(\exists x)P(x) \Leftrightarrow (\forall x)\neg P(x)$。

于是可以得到以下公式:

$$\neg(\forall x)P(x) \Leftrightarrow (\exists x)\neg P(x)$$

$$\neg(\exists x)P(x) \Leftrightarrow (\forall x)\neg P(x)$$

现约定:量词之前的否定联结词不是否定该量词,而是否定该量词及其辖域。

这两个等价式可在有限个体域上证明。设个体域为 $\{a_1, a_2, \cdots, a_n\}$。

$$\neg(\forall x)A(x) \Leftrightarrow \neg(A(a_1) \land A(a_2) \land \cdots \land A(a_n))$$

$$\Leftrightarrow \neg A(a_1) \lor \neg A(a_2) \lor \cdots \lor \neg A(a_n)$$

$$\Leftrightarrow (\exists x)\neg A(x)$$

$$\neg(\exists x)A(x) \Leftrightarrow \neg(A(a_1) \lor A(a_2) \lor \cdots \lor A(a_n))$$

$$\Leftrightarrow \neg A(a_1) \land \neg A(a_2) \land \cdots \land \neg A(a_n)$$

$$\Leftrightarrow (\forall x)\neg A(x)$$

当个体域为无限集时,等价式也是成立的。

3. 量词作用域的扩张和收缩

在量词的作用域中常常含有合取和析取项,如果其中一个是命题或者不含任何变元的谓词公式,则可将该命题移至量词作用域之外。例如,

$$(\forall x)(A(x) \lor B) \Leftrightarrow (\forall x)A(x) \lor B$$

$$(\forall x)(A(x) \land B) \Leftrightarrow (\forall x)A(x) \land B$$

$$(\exists x)(A(x) \lor B) \Leftrightarrow (\exists x)A(x) \lor B$$

$$(\exists x)(A(x) \land B) \Leftrightarrow (\exists x)A(x) \land B$$

主要是因为 B 中不出现任何约束变元 x,故它属于或者不属于量词的作用域均有同等意义。

利用上述四个式子可以推出以下四个式子：

$$(\forall x)(A(x) \rightarrow B) \Leftrightarrow (\exists x)A(x) \rightarrow B$$
$$(\exists x)(A(x) \rightarrow B) \Leftrightarrow (\forall x)A(x) \rightarrow B$$
$$(\forall x)(B \rightarrow A(x)) \Leftrightarrow B \rightarrow (\forall x)A(x)$$
$$(\exists x)(B \rightarrow A(x)) \Leftrightarrow B \rightarrow (\exists x)A(x)$$

当谓词的变元与量词的指导变元不同时，也能有类似上述的公式。例如，

$$(\forall x)(P(x) \vee Q(y)) \Leftrightarrow (\forall x)P(x) \vee Q(y)$$
$$(\forall x)((\forall y)P(x,y) \wedge Q(z)) \Leftrightarrow (\forall x)(\forall y)P(x,y) \wedge Q(z)$$

4. 量词分配等价式

设 $A(x)$ 和 $B(x)$ 是含自由变元 x 的任意谓词公式，则

$$(\forall x)(A(x) \wedge B(x)) \Leftrightarrow (\forall x)A(x) \wedge (\forall x)B(x)$$
$$(\exists x)(A(x) \vee B(x)) \Leftrightarrow (\exists x)A(x) \vee (\exists x)B(x)$$

前者可以理解为"所有的 x 有性质 A 和性质 B"和"所有的 x 有性质 A 且所有的 x 有性质 B"是等同的。后者可以利用前者来证明。

例 2.4.2 证明 $(\exists x)(A(x) \vee B(x)) \Leftrightarrow (\exists x)A(x) \vee (\exists x)B(x)$。

证明 因为：

$$(\forall x)(\neg A(x) \wedge \neg B(x)) \Leftrightarrow (\forall x)\neg A(x) \wedge (\forall x)\neg B(x)$$

又因为：

$$(\forall x)(\neg A(x) \wedge \neg B(x)) \Leftrightarrow (\forall x)\neg(A(x) \vee B(x))$$
$$\Leftrightarrow \neg(\exists x)(A(x) \vee B(x))$$
$$(\forall x)\neg A(x) \wedge (\forall x)\neg B(x) \Leftrightarrow \neg(\exists x)A(x) \wedge \neg(\exists x)B(x)$$
$$\Leftrightarrow \neg((\exists x)A(x) \vee (\exists x)B(x))$$

所以：

$$\neg(\exists x)(A(x) \vee B(x)) \Leftrightarrow \neg((\exists x)A(x) \vee (\exists x)B(x))$$

于是：

$$(\exists x)(A(x) \vee B(x)) \Leftrightarrow (\exists x)A(x) \vee (\exists x)B(x)$$

5. 量词与联结词的蕴含式

设 $A(x)$ 和 $B(x)$ 是含自由变元 x 的任意谓词公式。

$$(\forall x)A(x) \vee (\forall x)B(x) \Rightarrow (\forall x)(A(x) \vee B(x))$$
$$(\exists x)(A(x) \wedge B(x)) \Rightarrow (\exists x)A(x) \wedge (\exists x)B(x)$$
$$(\forall x)(A(x) \rightarrow B(x)) \Rightarrow (\forall x)A(x) \rightarrow (\forall x)B(x)$$
$$(\forall x)(A(x) \leftrightarrow B(x)) \Rightarrow (\forall x)A(x) \leftrightarrow (\forall x)B(x)$$

对第一式做如下说明：这些学生都聪明或这些学生都努力，可以推出这些学生都聪明或努力。但是这些学生都聪明或努力却不能推出这些学生都聪明或这些学生都努力。

第二式可用第一式推出。

例 2.4.3 证明 $(\exists x)(A(x) \wedge B(x)) \Rightarrow (\exists x)A(x) \wedge (\exists x)B(x)$。

证明 由第一式可得：

$$(\forall x)\neg A(x) \vee (\forall x)\neg B(x) \Rightarrow (\forall x)(\neg A(x) \vee \neg B(x))$$

而：

$$(\forall x) \neg A(x) \vee (\forall x) \neg B(x) \Leftrightarrow \neg(\exists x)A(x) \vee \neg(\exists x)B(x)$$

$$\Leftrightarrow \neg((\exists x)A(x) \wedge (\exists x)B(x))$$

$$(\forall x)(\neg A(x) \vee \neg B(x)) \Leftrightarrow (\forall x) \neg (A(x) \wedge B(x))$$

$$\Leftrightarrow \neg(\exists x)(A(x) \wedge B(x))$$

故有：

$$\neg((\exists x)A(x) \wedge (\exists x)B(x)) \Rightarrow \neg(\exists x)(A(x) \wedge B(x))$$

由假言易位式(逆反式)有：

$$(\exists x)(A(x) \wedge B(x)) \Rightarrow (\exists x)A(x) \wedge (\exists x)B(x)$$

6. 多个量词的使用

为方便起见,这里只列举两个量词的情况,更多量词的使用方法是类似的。对于二元谓词,如果不考虑自由变元,那么就可以产生以下八种组合情况。

$$(\forall x)(\forall y)P(x,y) \qquad (\forall y)(\forall x)P(x,y)$$

$$(\exists x)(\exists y)P(x,y) \qquad (\exists y)(\exists x)P(x,y)$$

$$(\forall x)(\exists y)P(x,y) \qquad (\exists y)(\forall x)P(x,y)$$

$$(\exists x)(\forall y)P(x,y) \qquad (\forall y)(\exists x)P(x,y)$$

例如,设 $P(x,y)$ 表示 x 和 y 同性别,个体域 x 是某中学一班, y 是某中学二班。

$(\forall x)(\forall y)P(x,y)$:一班和二班所有人都同性别。

$(\forall y)(\forall x)P(x,y)$:二班和一班所有人都同性别。

显然上述两个语句的含义是相同的。故：

$$(\forall x)(\forall y)P(x,y) \Leftrightarrow (\forall y)(\forall x)P(x,y)$$

同理：

$(\exists x)(\exists y)P(x,y)$:一班与二班有人同性别。

$(\exists y)(\exists x)P(x,y)$:二班与一班有人同性别。

这两个语句的含义也相同。故：

$$(\exists x)(\exists y)P(x,y) \Leftrightarrow (\exists y)(\exists x)P(x,y)$$

但是, $(\forall x)(\exists y)P(x,y)$ 表示对于一班所有人,二班都有人和他同性别。

$(\exists y)(\forall x)P(x,y)$ 表示存在一个二班的人,一班的所有人和他同性别。

$(\forall y)(\exists x)P(x,y)$ 表示对于二班所有的人,一班都有人与他同性别。

$(\exists x)(\forall y)P(x,y)$ 表示存在一个一班的人,二班所有人和他同性别。

上面四个语句表达的情况各不相同,所以全称量词与存在量词在公式中出现的顺序不能随意更换。具有两个量词的谓词公式,却有以下的蕴含关系：

$$(\forall x)(\forall y)A(x,y) \Rightarrow (\exists y)(\forall x)A(x,y)$$

$$(\forall y)(\forall x)A(x,y) \Rightarrow (\exists x)(\forall y)A(x,y)$$

$$(\exists x)(\forall y)A(x,y) \Rightarrow (\forall y)(\exists x)A(x,y)$$

$$(\exists x)(\forall y)A(x,y) \Rightarrow (\forall y)(\exists x)A(x,y)$$

$$(\forall x)(\exists y)A(x,y) \Rightarrow (\exists y)(\exists x)A(x,y)$$

$$(\forall y)(\exists x)A(x,y) \Rightarrow (\exists x)(\exists y)A(x,y)$$

2.5　前束范式

在命题演算的过程中,常常会将公式转化成规范的形式,这样会更易识别,也更美观。谓词演算也有类似情况,一个谓词演算公式可以转化成与它等价的范式。

定义 2.5.1 ▶一个公式,如果量词均在全式的开头,它们的作用域延伸到整个公式的末尾,则称其为前束范式。

根据这个定义,前束范式可表示为以下的形式:

$$(\square v_1)(\square v_2)\cdots(\square v_n)A$$

其中:$\square$ 是 $\exists$ 或 $\forall$;v_i 是个体变元,$i=1,\cdots,n$;A 是不含量词的谓词公式。

例如,

$$(\forall x)(\forall y)(A(x)\wedge B(y)\rightarrow P(x,y))$$

$$(\forall y)(\forall x)(\exists z)(\neg P(x,y)\wedge Q(x)\vee B(x,z))$$

都是前束范式。而

$$(\forall x)P(x)\vee(\forall y)(Q(y)\wedge A(x,y))$$

$$(\forall y)(\forall x)(\neg H(x,y)\vee F(x))\wedge(\exists z)L(x,z)$$

都不是前束范式。

定理 2.5.1 ▶任何谓词公式都可以转化成与其等价的前束范式。

证明　首先利用量词转化公式,把否定深入命题变元和谓词填式的前面。其次利用 $(\forall x)(A\vee B(x))\Leftrightarrow A\vee(\forall x)B(x)$ 和 $(\exists x)(A\wedge B(x))\Leftrightarrow A\wedge(\exists x)B(x)$ 把量词移到全式的最前面,这样即可得到前束范式。

例 2.5.1　求公式 $(\forall x)F(x)\rightarrow(\exists x)G(x)$ 的前束范式。

解
$$(\forall x)F(x)\rightarrow(\exists x)G(x)\Leftrightarrow\neg(\forall x)F(x)\vee(\exists x)G(x)$$
$$\Leftrightarrow(\exists x)\neg F(x)\vee(\exists x)G(x)$$
$$\Leftrightarrow(\exists x)(\neg F(x)\vee G(x))$$

例 2.5.2　将公式 $(\forall x)(\forall y)((\exists x)(P(x,s)\vee P(y,z)))$ 转化为等价的前束范式。

解
$$(\forall y)A(x,y)\rightarrow(\exists x)B(x,y)\Leftrightarrow(\forall t)A(x,t)\rightarrow(\exists s)B(s,y)$$
$$\Leftrightarrow(\exists t)(\exists s)(A(x,t)\rightarrow B(s,y))$$

定义 2.5.2 ▶一个谓词公式 A 如果具有以下形式,则称其为前束合取范式。

$(\square v_1)(\square v_2)\cdots(\square v_n)((A_{11}\vee A_{12}\vee\cdots\vee A_{1n})\wedge(A_{21}\vee A_{22}\vee\cdots\vee A_{2n})\wedge\cdots\wedge(A_{n1}\vee A_{n2}\vee\cdots\vee A_{nn}))$

其中:$\square$ 是 $\exists$ 或 $\forall$;v_i 是个体变元,$i=1,\cdots,n$;A_{ij} 是原子公式或其否定。例如,$(\forall x)(\exists z)(\forall y)\{[P\vee(x=a)\vee(z=d)]\wedge[Q(y)\vee(a=b)]\}$ 是前束合取范式。

定理 2.5.2 ▶每个谓词公式都可转化为与其等价的前束合取范式。(证明略)

例 2.5.3　将 wff $D:(\forall x)[(\forall y)P(x)\vee(\forall z)Q(z,y)\rightarrow\neg(\forall y)R(x,y)]$ 转化为与之等价的前束合取范式。

解　第一步是取消多余量词:

$$D\Leftrightarrow(\forall x)[P(x)\vee(\forall z)Q(z,y)\rightarrow\neg(\forall y)R(x,y)]$$

第二步是换名：

$$D \Leftrightarrow (\forall x)[P(x) \lor (\forall z)Q(z,y) \to \neg(\forall w)R(x,w)]$$

第三步是消去条件联结词：

$$D \Leftrightarrow (\forall x)[\neg(P(x) \lor (\forall z)Q(z,y)) \lor \neg(\forall w)R(x,w)]$$

第四步是将¬深入：

$$D \Leftrightarrow (\forall x)[(\neg P(x) \land (\exists z)\neg Q(z,y)) \lor (\exists w)\neg R(x,w)]$$

第五步是将量词推到左边：

$$D \Leftrightarrow (\forall x)(\exists z)(\exists w)[(\neg P(x) \land \neg Q(z,y)) \lor \neg R(x,w)]$$
$$\Leftrightarrow (\forall x)(\exists z)(\exists w)[(\neg P(x) \lor \neg R(x,w)) \land (\neg Q(z,y) \lor \neg R(x,w))]$$

例 2.5.4 将 $((\exists x)F(x) \lor (\exists x)G(x)) \to (\exists x)(F(x) \lor G(x))$ 转化为与其等价的前束合取范式。

解 $((\exists x)F(x) \lor (\exists x)G(x)) \to (\exists x)(S(x) \lor T(x))$
$$\Leftrightarrow (\exists x)(F(x) \lor G(x)) \to (\exists y)(S(y) \lor T(y))$$
$$\Leftrightarrow (\forall x)(\exists y)((F(x) \lor G(x)) \to (S(y) \lor T(y)))$$
$$\Leftrightarrow (\forall x)(\exists y)(\neg(F(x) \lor G(x)) \lor (S(y) \lor T(y)))$$
$$\Leftrightarrow (\forall x)(\exists y)((\neg F(x) \land \neg G(x)) \lor (S(y) \lor T(y)))$$
$$\Leftrightarrow (\forall x)(\exists y)((\neg F(x) \lor S(y) \lor T(y)) \land (\neg G(x) \lor S(y) \lor T(y)))$$

定义 2.5.3 一个谓词公式 A 如果具有以下形式,则称其为前束析取范式。

$(\Box v_1)(\Box v_2) \cdots (\Box v_n)[(A_{11} \land A_{12} \land \cdots \land A_{1n}) \lor (A_{21} \land A_{22} \land \cdots \land A_{2n}) \lor \cdots \lor (A_{n1} \land A_{n2} \land \cdots \land A_{nn})]$

其中：$\Box$ 是 $\exists$ 或 $\forall$；v_i 是个体变元,$i = 1, \cdots, n$；A_{ij} 是原子公式或其否定。

定理 2.5.3 每个谓词公式都可转化为与其等价的前束析取范式。(证明略)

任何一个谓词公式 A 转化为等价的前束析取范式的步骤与例 2.5.3 类似。

2.6 谓词演算的推理理论

在谓词演算中,C 是一组前提 $A_1, A_2, \cdots, A_n$ 的有效结论,仍然定义为 $A_1 \land A_2 \land \cdots \land A_n \Rightarrow C$。命题演算推理中的 P 规则、T 规则、置换规则、所有的等价式和蕴含式在谓词推理中都可以使用。另外,2.3 节中介绍的谓词演算中的等价式与蕴含式也可以在谓词推理中使用。除此之外,还有以下的规则。

(1) 全称指定规则,表示为 US。即

$$(\forall x)P(x) \Rightarrow P(c)$$

此时成立的条件是：① c 是个体域中任一个体；② 用 c 取代 $A(x)$ 中 x 时,一定在 x 出现的所有地方进行取代。

全称指定规则说明：若个体域中的所有个体都满足谓词 A,则个体域中任一个体 c 也满足谓词 A。利用这个规则,可以从带有全称量词的前提中,推导出不带全称量词的特殊结论。它体现了在逻辑推理中由一般到特殊的推导方法。

(2) 全称推广规则,表示为 UG。即

$$P(y) \Rightarrow (\forall x)P(x)$$

此时成立的条件是:① y 是个体域中任一个体且对 y、$P(y)$ 为真;② x 是不出现在 $A(y)$ 中的个体变元。

例如,个体域为实数集合 **R**,$G(x,y)$ 表示 $x>y$,设 $P(y)\Leftrightarrow(\exists x)G(x,y)$,显然 $P(y)$ 满足条件①,一定能推出 $(\forall z)P(z)\Leftrightarrow(\forall z)(\exists x)G(x,z)\Leftrightarrow(\forall z)(\exists x)(x>z)$,$(\forall z)(\exists x)(x>z)$ 是一个真命题。若有 $(\forall x)A(x)\Leftrightarrow(\forall x)(\exists x)G(x,x)\Leftrightarrow(\forall x)(\exists x)(x>x)$,就产生了错误,因为 $(\forall x)(\exists x)(x>x)$ 是一个假命题。错误的原因是违背了条件②。

(3) 存在指定规则,表示为 ES。即
$$(\exists x)P(x)\Rightarrow P(c)$$

此时成立的条件是:① c 是个体域中的某个特定的个体,而不是个体变元;② c 是不出现在 $P(x)$ 中的个体。

存在指定规则说明:若个体域中存在一些个体满足谓词 A,则至少有某个特定的个体 c 满足谓词 A。

例如,设个体域为整数集合 **Z**,$A(x)$ 表示 x 是奇数,$B(x)$ 表示 x 是偶数。

$(\exists x)A(x)\Rightarrow A(c)$ 表示:若存在一些整数是奇数,令 c 为 3,则 c 是奇数。这个推理是对的。

$(\exists x)B(x)\Rightarrow B(d)$ 表示:若存在一些整数是偶数,令 d 为 4,则 d 是偶数。这个推理也是对的。因此有下列推理成立:
$$(\exists x)A(x)\wedge(\exists x)B(x)\Rightarrow A(c)\wedge B(d)$$

而下列推理是错误的:
$$(\exists x)A(x)\wedge(\exists x)B(x)\Rightarrow A(c)\wedge B(c)$$
$$(\exists x)A(x)\wedge(\exists x)B(x)\Rightarrow A(d)\wedge B(d)$$

因为 3 不能既是奇数又是偶数;同样,4 也不能既是奇数又是偶数。错误的原因是违背了条件②。

(4) 存在推广规则,表示为 EG。即
$$P(c)\Rightarrow(\exists x)P(x)$$

此时成立的条件是:① c 是个体域中确定的个体;② x 不能是出现在 $P(c)$ 中的个体变元。

存在推广规则说明:对于个体域中的某个个体 c 满足谓词 P,当然有 $(\exists x)P(x)$。

例 2.6.1　证明苏格拉底论证:所有人都要死。苏格拉底是人,苏格拉底要死。

设　$P(x)$:x 是人,$Q(x)$:x 要死,s:苏格拉底。

本题要证明的蕴含式是 $(\forall x)(P(x)\rightarrow Q(x))\wedge P(s)\Rightarrow Q(s)$。

证明

(1) $(\forall x)(H(x)\rightarrow M(x))$　　　　　　　　P

(2) $H(s)\rightarrow M(s)$　　　　　　　　US(1)

(3) $H(s)$　　　　　　　　P

(4) $M(s)$　　　　　　　　T(2)(3)(假言推理)

例 2.6.2　证明 $(\forall x)(H(x)\rightarrow M(x))$,$(\exists x)H(x)\Rightarrow(\exists x)M(x)$

证明

(1) $(\exists x)H(x)$　　　　　　　　P

(2) $H(c)$　　　　　　　　ES(1)

$$(3) \quad (\forall x)(H(x) \to M(x)) \qquad\qquad P$$

$$(4) \quad H(c) \to M(c) \qquad\qquad US(3)$$

$$(5) \quad M(c) \qquad\qquad T(2)(4)(假言推理)$$

$$(6) \quad (\exists x)M(x) \qquad\qquad EG(5)$$

若把(1),(2)写在(3),(4)的后面,得到以下推理:

$$(1) \quad (\forall x)(H(x) \to M(x)) \qquad\qquad P$$

$$(2) \quad H(c) \to M(c) \qquad\qquad US(1)$$

$$(3) \quad (\exists x)H(x) \qquad\qquad P$$

$$(4) \quad H(c) \qquad\qquad ES(3)$$

$$(5) \quad M(c) \qquad\qquad T(2)(4)(假言推理)$$

$$(6) \quad (\exists x)M(x) \qquad\qquad EG(5)$$

这个推理在逻辑上是错误的。因为(2)中的 c 为个体域中的一个个体,用 ES 规则由(3)推到(4)不能选择(2)中的 c,因为它要选的个体和(2)中的个体 c 不一定是同一个个体,故推理是错误的。

例 2.6.3 用 CP 规则证明:

$$(\forall x)(F(x) \lor G(x)) \Rightarrow (\forall x)F(x) \lor (\exists x)G(x)$$

原题可改写成:

$$(\forall x)(F(x) \lor G(x)) \Rightarrow \neg(\forall x)F(x) \to (\exists x)G(x)$$

证明

$$(1) \quad \neg(\forall x)F(x) \qquad\qquad P(附加前提)$$

$$(2) \quad (\exists x)\neg F(x) \qquad\qquad T(1)(量词否定等价式)$$

$$(3) \quad \neg F(c) \qquad\qquad ES(2)$$

$$(4) \quad (\forall x)(F(x) \lor G(x)) \qquad\qquad P$$

$$(5) \quad F(c) \lor G(c) \qquad\qquad US(4)$$

$$(6) \quad G(c) \qquad\qquad T(3)(5)(析取三段论)$$

$$(7) \quad (\exists x)G(x) \qquad\qquad EG(6)$$

$$(8) \quad \neg(\forall x)F(x) \to (\exists x)G(x) \qquad\qquad CP$$

例 2.6.4 设个体域为全总个体域。证明推理:学术会的成员都是工人并且是专家。有些成员是青年人。所以有的成员都是青年专家。

首先将命题符号化。$F(x)$:x 是学术会成员,$G(x)$:x 是专家,$H(x)$:x 是工人,$R(x)$:x 是青年人。

本题要证明:$(\forall x)(F(x) \to G(x) \land H(x))$,$(\exists x)(F(x) \land R(x)) \Rightarrow (\exists x)(F(x) \land R(x) \land G(x))$。

证明

$$(1) \quad (\exists x)(F(x) \land R(x)) \qquad\qquad P$$

$$(2) \quad F(c) \land R(c) \qquad\qquad ES(1)$$

$$(3) \quad F(c) \qquad\qquad T(2)(化简律)$$

$$(4) \quad (\forall x)(F(x) \to G(x) \land H(x)) \qquad\qquad P$$

$$(5) \quad F(c) \to G(c) \land H(c) \qquad\qquad US(4)$$

(6) $G(c) \wedge H(c)$　　　　　　　　　　T(3)(5)(假言推理)

(7) $R(c)$　　　　　　　　　　　　　　　T(2)(化简律)

(8) $G(c)$　　　　　　　　　　　　　　　T(6)(化简律)

(9) $F(c) \wedge R(c) \wedge G(c)$　　　　　　T(2)(7)(8)(合取引入)

(10) $(\exists x)(F(x) \wedge R(x) \wedge G(x))$　　EG(9)

2.7　谓词逻辑的应用

我们可以将学过的谓词逻辑知识应用于日常生活中,特别是用于解决现实生活中的推理问题。

下面给出一些综合应用例题。

例 2.7.1　某家人去上海迪士尼游玩,他们游玩了雷鸣山漂流、创极速光轮、加勒比海盗,最后去商店购买纪念品。已知:

(1) 有人买了米奇耳套。

(2) 有人没有买米奇耳套。

(3) 爸爸和妈妈都买了米奇耳套。

如果以上三句话中有一句话为真,请问爸爸和妈妈谁买了米奇耳套?

解　如果(3)为真,则爸爸和妈妈都买了米奇耳套,那么(1)也为真。这与条件不符合,所以(3)为假。

如果(3)为假,表明爸爸和妈妈至少一人没有买米奇耳套,那么(2)为真,进而(1)为假。

所以可以知道,"有人没有买米奇耳套"为真。那么,爸爸和妈妈都没有买米奇耳套。

例 2.7.2　符号化并推证下列结论的有效性。

前提:(1) 每个硕士生或是推免生或者是统考生。

(2) 并非每个硕士生都是推免生。

结论:(3) 有些研究生是统考生。

解　设 $R(x)$:x 是硕士生,$S(x)$:x 是推免生,$C(x)$:x 是统考生。

上面问题等价于证明以下的蕴含式:

$(\forall x)(R(x) \rightarrow (S(x) \vee C(x)))$,　$\neg(\forall x)(R(x) \rightarrow S(x)) \Rightarrow (\exists x)(R(x) \wedge C(x))$

证明

(1) $\neg(\forall x)(R(x) \rightarrow S(x))$　　　　　　　　P

(2) $(\exists x)\neg(R(x) \rightarrow S(x))$　　　　　　　T(1)E

(3) $(\exists x)\neg(\neg R(x) \vee S(x))$　　　　　　　T(2)E

(4) $(\exists x)(R(x) \wedge \neg S(x))$　　　　　　　T(3)E

(5) $R(a) \wedge \neg S(a)$　　　　　　　　　　　T(4)ES

(6) $R(a)$　　　　　　　　　　　　　　　　T(5)I

(7) $\neg S(a)$　　　　　　　　　　　　　　　T(5)I

(8) $(\forall x)(R(x) \rightarrow (S(x) \vee C(x)))$　　　　P

(9) $R(a) \rightarrow (S(a) \vee C(a))$　　　　　　　T(8)US

(10) $S(a) \vee C(a)$　　　　　　　　　　　　T(6)(9)I

(11) $C(a)$	T(7)(10)I
(12) $R(a) \wedge C(a)$	T(6)(11)I
(13) $(\exists x)(R(x) \wedge C(x))$	T(12)EG

例 2.7.3 设个体域为软件工程系全体学生,用给定的命题和谓词将以下命题符号化,并推证结论的有效性。

R:今天天气很好,S:颁奖典礼准时举行,$A(x)$:x 准时到场。

(1) 只要今天遇到不好的天气,软件工程系的学生中就一定有人不能准时到场。

(2) 当且仅当所有的学生都准时到场时,颁奖典礼才能准时举行。

结论:如果颁奖典礼准时举行,那么今天天气很好。

解 该命题可以符号化为

$$\neg R \rightarrow (\exists x) \neg A(x), \quad (\forall x)A(x) \leftrightarrow S \Rightarrow S \rightarrow R$$

(1) $\neg R \rightarrow (\exists x) \neg A(x)$	P
(2) $\neg R \rightarrow \neg (\forall x)A(x)$	T(1)E
(3) $(\forall x)A(x) \rightarrow R$	T(2)E
(4) $(\forall x)A(x) \leftrightarrow S$	P
(5) $((\forall x)A(x) \rightarrow S) \wedge (S \rightarrow (\forall x)A(x))$	T(4)E
(6) $S \rightarrow (\forall x)A(x)$	T(5)I
(7) $S \rightarrow R$	T(3)(6)I

习 题

1. 用谓词表达式写出下列命题。

(1) 小王不是大学生。

(2) 张云学习努力并且成绩优异。

(3) 小刚有个姐姐或者有个弟弟。

(4) 如果 a 是奇数,那么 $2a$ 就不是奇数。

(5) 所有有理数都是实数。

(6) 存在一些实数是有理数。

(7) 并不是所有的实数都是有理数。

(8) 所有运动员都仰慕某些教练。

(9) 没有一位女同志既是国家公务员又是国家运动员。

(10) 有些男同志既是国家运动员也是教练员。

2. 试将下列语句符号转化为谓词表达式。

(1) 小张学过西班牙语和法语。

(2) 仅当 a 大于 c 时 a 大于 b。

(3) 4 不是奇数。

(4) 2 或 5 是质数。

(5) 除非王建国是海南人,否则他一定怕热。

3. 利用谓词公式翻译下列命题。

(1) 存在有理数 a,b 和 c，能够使 a 与 b 之差大于 a 与 c 之和。

(2) 对于任意一个实数 a，都存在一个比 a 更小的实数 b。

(3) 若有限个数的乘积结果为非 0，那么没有任何一个因子为 0。

4. 令谓词 $P(x,y)$ 表示为"x 大于 y"，个体变元 x 和 y 的个体域都是 $E=\{4,5,6\}$。求下列各式的真值。

(1) $(\exists x)P(x,4)$　　　　(2) $(\forall x)P(x,1)$　　　　(3) $(\forall x)(\forall y)P(x,y)$

(4) $(\exists x)(\exists y)P(x,y)$　　　(5) $(\exists x)(\forall y)P(x,y)$　　　(6) $(\forall x)(\exists y)P(x,y)$

5. 用谓词公式刻画下列命题。

一位戴着帽子的用功的女大学生正在看一本大且厚实的英文著作。

6. 指出下列每个公式的约束变元和自由变元。

(1) $(\forall x)(P(x)\rightarrow Q(x,y,z))$

(2) $(\forall x)P(x,y,z)\wedge(\exists y)Q(x,y,z)$

(3) $(\forall x)(\exists y)(P(x,y)\vee Q(x,z))\rightarrow(\exists x)R(x,y)$

(4) $(\exists x)(\forall y)(P(x,y)\vee Q(z))$

(5) $(\forall x)(P(x)\vee Q(x))\rightarrow(\exists x)R(x)$

7. 指出下列每个公式的指导变元、约束变元、自由变元及量词的辖域。

(1) $(\forall x)(P(x)\vee Q(x))\rightarrow(\exists x)(P(x)\vee R(x))$

(2) $(\forall x)(P(x)\wedge Q(z))\vee(\forall x)(P(x)\rightarrow R(x))$

(3) $(\forall x)(P(x)\leftrightarrow Q(x))\wedge(\exists x)P(x)\vee R(x)$

8. 假设个体域是集合 $\{1,2,3\}$，试着消去下列公式中的量词。

(1) $(\forall x)R(x)$　　　　　　(2) $(\forall x)(R(x)\rightarrow S(x))$

(3) $(\forall x)P(x)\vee(\forall x)Q(x)$　　(4) $(\forall x)P(x)\wedge(\forall x)\neg Q(x)$

9. 将下列谓词公式中的约束变元进行换名。

(1) $(\forall x)(\exists y)(R(x,z)\vee S(y))\rightarrow Q(x,y)$

(2) $(\forall x)(R(x,y)\rightarrow(\exists z)S(x,z))\rightarrow(\forall y)Q(x,y)$

10. 将下列谓词公式中的自由变元进行代入。

(1) $((\exists y)P(x,y)\vee(\forall x)Q(x,z))\wedge(\forall z)R(x,z)$

(2) $((\forall y)R(x,y)\wedge(\exists z)S(x,z))\vee(\forall x)Q(x,y)$

11. 求证以下两个等价式。

(1) $(\exists x)(R(x)\rightarrow S(x))\Leftrightarrow(\forall x)R(x)\rightarrow(\exists x)S(x)$

(2) $(\forall x)(\forall y)(R(x)\rightarrow S(x))\Leftrightarrow(\exists x)R(x)\rightarrow(\forall y)S(y)$

12. 令个体域为 $E=\{1,2,3\}$，求证：

$$(\forall x)A(x)\vee(\forall x)B(x)\Rightarrow(\forall x)(A(x)\vee B(x))$$

13. 把下列各式化为前束范式。

(1) $(\forall x)(R(x)\rightarrow(\exists z)S(x,z))$

(2) $(\exists x)(\neg((\exists z)R(x,z))\rightarrow((\exists y)S(y)\rightarrow Q(x)))$

(3) $(\forall x)(\forall y)(((\exists z)R(x,y,z)\wedge(\exists w)S(x,w))\rightarrow(\exists u)Q(y,u))$

14. 求下列各式的前束合取范式和前束析取范式。

(1) $((\exists x)R(x)\vee(\exists x)S(x))\rightarrow(\exists x)(R(x)\vee S(x))$

(2) $(\forall x)R(x)\rightarrow(\exists x)((\forall u)S(x,z)\vee(\forall u)Q(x,y,u))$

(3) $(\forall x)(R(x)\rightarrow(\forall y)((\forall z)S(x,y)\rightarrow\neg(\forall z)Q(y,x)))$

(4) $(\forall x)(R(x)\rightarrow S(x,y))\rightarrow((\exists y)R(y)\wedge(\exists z)S(y,z))$

15. 证明下列各式。

(1) $(\forall x)(\neg R(x)\rightarrow S(x)),(\forall x)\neg S(x)\Rightarrow(\exists x)R(x)$

(2) $(\forall x)(R(x)\vee S(x)),(\forall x)(S(x)\rightarrow\neg Q(x)),(\forall x)Q(x)\Rightarrow(\exists x)R(x)$

(3) $(\exists x)R(x)\rightarrow(\forall x)S(x)\Rightarrow(\forall x)(R(x)\rightarrow S(x))$

(4) $(\forall x)(R(x)\rightarrow S(x)),(\forall x)(Q(x)\rightarrow\neg S(x))\Rightarrow(\forall x)(Q(x)\rightarrow\neg R(x))$

16. 用 CP 规则证明下列各式。

(1) $(\forall x)(R(x)\rightarrow S(x))\Rightarrow(\forall x)R(x)\rightarrow(\forall x)S(x)$

(2) $(\forall x)(R(x)\vee S(x))\Rightarrow(\forall x)R(x)\vee(\exists x)S(x)$

集 合 论

Set Theory

Chapter 3　Set and Relation

Set is a basic concept that can't be precisely defined. Simply speaking, putting some things together to form a whole is called a set, and these things are the elements or members of this set. The elements of the collection must be deterministic. The so-called "definite" refers to whether any object is an element of a set. It is clear and definite and can not be ambiguous. For example, students in a school, college students in the country, the number of books in a library and so on can form a collection.

3.1　The Concept and Representation of the Set

Definition 3.1.1　The whole set of certain and distinguishable objects is a collection, it is usually expressed in capital letters. The objects that make up a collection are called elements or members of a collection and are often represented in lowercase letters.

The elements of a set are distinguishable, which means that the elements in the set are different from each other. If several elements in a collection are the same, it is counted as one. For example, the set $\{a, b, b, d\}$ and $\{a, b, d\}$ are the same set.

The elements of a collection are arbitrary objects, and objects are concrete or abstract objects that can exist independently. It can be an independent number, letter, person, or other objects, it can also be an abstract concept, and of course, it can also be a set. For example, the elements $\{c\}$ and $\{a\}$ of set $\{1, b, \{c\}, \{a\}\}$ are sets.

The elements of the set are unordered, the $\{a, c, b\}$ and $\{a, b, c\}$ are the same set.

Let S be a set, a is an element of S, denoted as $a \in S$, read as "a belongs to S", also can be read as "a in S" or "S contains a". If a is not an element of S, it is recorded as $a \notin S$, which is read as "a does not belong to S" or "a is not in S".

For example:

(1) 26 English letters form a set, and any English letter is an element of the set.

(2) All the points on the line constitute the real number set **R**, and each real number is an element of the set **R**.

(3) All the students of Jiangsu University form a collection, and each student in the university is an element of this collection.

There are three representations of sets.

The first representation is an enumeration: list the elements of the set in curly brackets "{ }", separated by commas. For example, $I=\{1, 2, 3, 4, 5\}$, $A=\{a, b, 3, \ldots\}$, $C=\{0, 1, -1, \{a\}, \{c, d\}, \ldots\}$, $S=\{T, F\}$.

The second representation is a description: the elements of a set are defined by predicates. For example, $\mathbf{Q}=\{x \mid x \text{ is a rational number}\}$, $\mathbf{R}=\{x \mid x \text{ is a real number}\}$, $\mathbf{C}=\{x \mid x \text{ is a complex number}\}$, $A=\{x \mid x \in I \wedge 0 < x \wedge x < 5\}$.

If $P(x)$ denotes that x is a rational number, then $\mathbf{Q}$ can be expressed as $\mathbf{Q}=\{x \mid P(x)\}$

Generally speaking, a set can be described as: $S=\{x \mid A(x)\}$. Where $A(x)$ is a predicate. Obviously, when $a \in S$ s, then $A(a)$ is true; conversely, when $A(a)$ is true, then $a \in S$. The necessary and sufficient condition of $a \in S$ is that $A(a)$ is true.

In middle school textbooks, natural numbers are defined as $\mathbf{N}=\{1, 2, 3, \ldots\}$.

In discrete mathematics, it is considered that natural numbers begin with 0, $\mathbf{N}=\{0, 1, 2, 3, \ldots\}$.

This set of natural numbers starting from 0 is called an extended set of natural numbers. The extended set of natural numbers is used in discrete mathematics.

Definition 3.1.2　A set with finite elements is called a finite set, otherwise, it is called an infinite set. The number of elements in a finite set is called the cardinality of the set, also known as the potential of the set. The cardinality of the finite set A is denoted as $|A|$.

For example, let $A=\{1, a, \{b, c\}, 5\}$, A is a finite set. The cardinal number of A is $|A|=4$.

An Extended set of natural numbers $\mathbf{N}=\{0, 1, 2, 3, \ldots\}$ is an infinite set. Set of integers $\mathbf{I}$、Set of rational numbers $\mathbf{Q}$, Set of real numbers $\mathbf{R}$ and Set of complex number $\mathbf{C}$ are common infinite sets.

Definition 3.1.3　Let A and B be arbitrary sets, When every element of A is an element of B, then A is said to be a subset of B. It is also called that A is contained in B or B contains A. It is recorded as $A \subseteq B$ or $B \supseteq A$.

When A is not a subset of B, it is denoted as $A \nsubseteq B$. $A \subseteq B$ is expressed by the predicate formula as $A \subseteq B \Leftrightarrow (\forall x)(x \in A \rightarrow x \in B)$.

$A \nsubseteq B$ is expressed by the predicate formula as follows: $A \nsubseteq B \Leftrightarrow (\exists x)(x \in A \wedge x \notin B)$.

For example, let $A=\{1\}$, $B=\{1, 2\}$, $C=\{1, 2, 3\}$, then $A \subseteq A$; $A \subseteq B$, $B \subseteq C$, $A \subseteq C$; $C \nsubseteq B$.

It can be proved that the inclusion of a set has the following properties.

(1) Reflexivity: that is, for any set A, $A \subseteq A$.

(2) Transitivity: that is, for any set A, B, C, when $A \subseteq B$ and $B \subseteq C$, $A \subseteq C$.

Definition 3.1.4　Let A and B be sets, if $A \subseteq B$ and $B \subseteq A$, then A is equal to B, which is written as $A=B$. If A is not equal to B, written as $A \neq B$.

Set equality can also be expressed by predicate formula as follows:
$$A=B \Leftrightarrow A \subseteq B \wedge B \subseteq A$$
$$\Leftrightarrow (\forall x)(x \in A \rightarrow x \in B) \wedge (\forall x)(x \in B \rightarrow x \in A)$$
$$\Leftrightarrow (\forall x)(x \in A \leftrightarrow x \in B)$$

For example, let $A=\{c, e\}$, $B=\{1, \{d\}\}$, $C=\{e, c\}$, then $A=C$, $A \neq B$.

Theorem 3.1.1　From the definition of set equality, we can see that set equality has the following properties.

(1) Reflexivity: that is, for any set A, $A=A$.

(2) Symmetry: that is, for any set A, B, when $A=B$, $B=A$.

(3) Transitivity: that is, for any set A, B, C, when $A=B$ and $B=C$, $A=C$.

Definition 3.1.5　Let A and B be sets, if $A\subseteq B$ and $A\neq B$, then A is called a proper subset of B. It is written as $A\subset B$. If A is not a proper subset of B, which is written as $A\not\subset B$.

The proper subset is expressed by the predicate formula as follows:

$$A\subset B\Leftrightarrow A\subseteq B\land A\neq B$$
$$\Leftrightarrow(\forall x)(x\in A\rightarrow x\in B)\land(\exists x)(x\in B\land x\notin A)$$

For example, let $A=\{a\}$, $B=\{a,b\}$, $C=\{a,b,c\}$, then $A\subset B$, $B\subset C$, $A\subset C$, $A\not\subset A$.

For example, the set of natural numbers is the proper subset of the set of integers, and also the proper subset of the set of rational numbers and the set of real numbers, $\mathbf{N}\subset\mathbf{I}$, $\mathbf{N}\subset\mathbf{Q}$, $\mathbf{N}\subset\mathbf{R}$.

Definition 3.1.6　A set that does not contain any elements is called an empty set. It is recorded as $\varnothing$.

Empty sets can be expressed as follows:

$$\varnothing=\{x\,|\,P(x)\}\land\neg P(x)$$

Where $P(x)$ is any predicate. An empty set $\varnothing$ is a set that does not contain any elements, so, $|\varnothing|=0$.

Theorem 3.1.2　An empty set is a subset of any set.

According to Theorem 3.1.2, an empty set is a subset of any set, that is, $\varnothing\subseteq A$; for any set A, $A\subseteq A$. Generally speaking, any set A has at least two subsets, one of which is an empty set the other is itself A.

Inference　The empty set is unique.

Proof　There are two empty sets $\varnothing_1$ and $\varnothing_2$. According to Theorem 3.1.2, there are $\varnothing_1\subseteq\varnothing_2$ and $\varnothing_2\subseteq\varnothing_1$. According to the definition of set equality, $\varnothing_1=\varnothing_2$.

Definition 3.1.7　Let A be a set, the set of all subsets of A is called the power set of A, written as $P(A)$, then $P(A)=\{S\,|\,S\subseteq A\}$.

Example　3.1.1　Let $A=\{a,b,c\}$, $\varnothing$ is an empty set, try to find $P(A)$, $P(P(Q))$.

Solution

$$P(A)=\{\varnothing,\{a\},\{b\},\{c\},\{a,b\},\{a,c\},\{b,c\},\{a,b,c\}\}$$
$$P(\varnothing)=\{\varnothing\}$$
$$P(P(\varnothing))=\{\varnothing,\{\varnothing\}\}$$

Theorem 3.1.3　Let A is a finite set, then $|P(A)|=2^{|A|}$.

Definition 3.1.8　In a specific problem, if the set involved is a subset of a set, then the set is called a complete set. Written as E.

The complete set is relative, different problems have different complete sets. Even for the same problem, different complete sets can be taken.

Another representation of sets is the Venn diagram. Venn diagrams are often used to describe set operations and their relationships. The Venn diagram of the set is as follows.

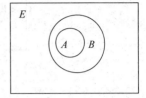

The complete set E is represented by a rectangle. Some circles are drawn in the rectangle to represent other sets. Different circles represent different sets. If not specified, any two circles intersect each other. For example, The Venn diagram of $A \subseteq B$ is shown in Figure 3.1.1.

Figure 3.1.1

3.2 Operation of Set

After understanding the basic concept of a set, we need to further study the operation of the set. As the name implies, the operation of the set is to operate the specified set according to relevant rules. The common set operations include Union, intersection, relative complement, absolute complement and symmetric difference.

Definition 3.2.1 Let A and B be arbitrary sets, The set consisting of elements in A or elements in B is called the union of A and B, and is recorded as $A \cup B$.

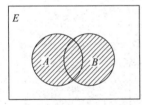

It can be expressed as $A \cup B = \{x \mid x \in A \lor x \in B\}$. The Venn Diagram of the union set is shown in Figure 3.2.1.

From the definition of union, we can get the following conclusions:

$$A \subseteq A \cup B, B \subseteq A \cup B$$

For example, $A = \{1, a\}, B = \{3, c\}$, then $A \cup B = \{1, 3, a, c\}$.

Figure 3.2.1

Definition 3.2.2 Let A and B be sets, the set of common elements of A and B is called the intersection of A and B, which is recorded as $A \cap B$.

It can be expressed as $A \cap B = \{x \mid x \in A \land x \in B\}$. The Venn Diagram of the intersection set is shown in Figure 3.2.2.

From the definition of intersection, we can get the following conclusions:

$$A \cap B \subseteq A, A \cap B \subseteq B$$

If A and B have no common element, $A \cap B = \varnothing$, then A and B are said to be disjoint.

For example, let $A = \{a, e, c\}, B = \{1, 5\}$. then $A \cap B = \varnothing$, A and B are disjoint.

Definition 3.2.3 Let A and B be sets, the set of elements that belong to A but do not belong to B is called the complement set of B to A, also known as the relative complement set of B to A. It is recorded as $A - B$.

It can be expressed as $A - B = \{x \mid x \in A \land x \notin B\}$. The Venn Diagram of relative complement is shown in Figure 3.2.3.

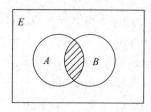
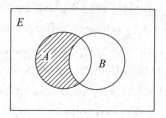

Figure 3.2.2   Figure 3.2.3

For example, let $A=\{\varnothing,\{\varnothing\}\}$, $B=\varnothing$, then $A-B=\{\varnothing,\{\varnothing\}\}-\varnothing=\{\varnothing,\{\varnothing\}\}$.

Let $C=\{a\}$, $D=\{a,b\}$, then $C-D=\{a\}-\{a,b\}=\varnothing$, $C-C=\varnothing$, $D-C=\{b\}$.

Definition 3.2.4   Let A be a set, the relative complement of A to the complete set E is called the absolute complement of A, recorded as $\sim A$. It can be expressed as

$$\sim A=E-A=\{x\mid x\in E\wedge x\notin A\}=\{x\mid x\notin A\}$$

The Venn Diagram of $\sim A$ is shown in Figure 3.2.4.

For example, let $E=\{c,2,d,3\}$, $A=\{2,3\}$, then $\sim A=\{c,2,d,3\}-\{2,3\}=\{c,d\}$.

Theorem 3.2.1 $A-B=A\bigcap(\sim B)$.

Definition 3.2.5   Let A and B be sets, which are composed of elements in A or B, but not the common elements of A and B, which is called the symmetric difference between A and B, recorded as $A\oplus B$.

It can be expressed as $A\oplus B=\{x\mid x\in A\bigcup B\wedge x\notin A\bigcap B\}$.

The Venn Diagram of $A\oplus B$ is shown in Figure 3.2.5.

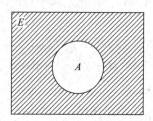
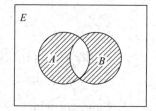

Figure 3.2.4   Figure 3.2.5

For example, let $A=\{1,c,3,e\}$, $B=\{1,2,e,f\}$, then $A\oplus B=A\bigcup B-A\bigcap B=\{1,2,3,c,e,f\}-\{1,e\}=\{2,3,c,f\}$.

Theorem 3.2.2   Let A and B be arbitrary sets, $A\oplus B=(A-B)\bigcup(B-A)=(A\bigcap\sim B)\bigcup(B\bigcap\sim A)$.

The following properties of symmetric difference $A\oplus B$ can be proved by using the above formula.

Let A, B be arbitrary sets.

(1) $A\oplus A=\varnothing$.

Proof $A\oplus A=(A-A)\bigcup(A-A)=\varnothing\oplus\varnothing=\varnothing$.

(2) $A\oplus\varnothing=A$.

Proof $A \oplus \varnothing = (A - \varnothing) \cup (\varnothing - A) = A \cup \varnothing = A$.

(3) $A \oplus E = \sim A$.

Proof $A \oplus E = (A - E) \cup (E - A) = \varnothing \cup \sim A = \sim A$.

To make the expression of the set more concise, the priority of set operation is specified as follows.

The operation level of absolute complement is higher than that of the other four operations. The absolute complement operation is carried out first, and then the other four operations are carried out. The operation order of the other four operations is determined by brackets.

The Venn diagram can not only represent the operation of sets and the relationship between them but also solve the counting problem of finite sets conveniently.

The method to solve the counting problem of finite sets with Venn's graph is as follows.

Each property defines a set and draws a circle to represent the set. If not specified, any two circles are drawn intersecting. The number of elements of a known set is filled in the area representing the set. Usually from the intersection of n sets, according to the calculation results, the number is gradually filled into other blank areas.

If the value of the intersection is unknown, it can be set to x. According to the conditions of the title, list the equations or equations, and get the required results.

Example **3.2.1** 15 students studied English, 5 students studied Japanese, 12 students studied German, 10 students studied French, and 2 of them studied English and Japanese at the same time. In the group of students studying English, German, and French, 5 students are studying any two of these languages. It is known that the students studying Japanese do not know either French or German. Please calculate the number of students who are studying only one language(English, German, French, or Japanese) and the number of students who are studying three languages.

Solution Let A, B, C, D denote the set of people who study English, French, German and Japanese respectively. Suppose that there are x students who study three languages at the same time, and y_1, y_2 and y_3 students who only study English, French or German. According to the information, we can draw a Venn diagram, as shown in Figure 3.2.6.

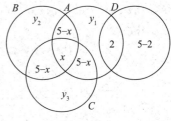

Figure　3.2.6

According to the Venn diagram and known conditions, the system of equations can be listed.

$$\begin{cases} y_1 + 2(5-x) + x + 2 = 17 \\ y_2 + 2(5-x) + x = 10 \\ y_3 + 2(5-x) + x = 12 \\ y_1 + y_2 + y_3 + 3(5-x) + x = 27 - 5 \end{cases}$$

The answer is $x = 2$, $y_1 = 5$, $y_2 = 2$, $y_3 = 4$. In addition, there are only three students who study Japanese.

3.3 Inclusion Exclusion Principle

In counting, the number of all objects contained in a certain content is calculated first without considering the overlap, and then the number of repeated calculations during counting is excluded so that the final result has neither omission nor repetition. This idea is the basic idea of the principle of inclusion and exclusion.

Theorem 3.3.1 Let S be a finite set, P_1, P_2,..., P_m is m property, A_i is a subset of elements with property P_i in S, and $\overline{A}_i$ is the complement set of relative to S, $i=1, 2, ...,$ m. Then S has no property P_1, P_2 The number of elements of P_m is

$$|\overline{A}_1 \cap \overline{A}_2 \cap \cdots \cap \overline{A}_m| = |S| - \sum_{i=1}^{m} |A_i| + \sum_{1 \leqslant i < j \leqslant m} |A_i \cap A_j| -$$

$$\sum_{1 \leqslant i < j < k \leqslant m} |A_i \cap A_j \cap A_k| + \cdots + (-1)^m |A_1 \cap A_2 \cap \cdots \cap \Lambda_m|$$

Inference The number of elements with at least one property in S is

$$|A_1 \cup A_2 \cup \cdots \cup A_m| = \sum_{i=1}^{m} |A_i| - \sum_{1 \leqslant i < j \leqslant m} |A_i \cap A_j| +$$

$$\sum_{1 \leqslant i < j < k \leqslant m} |A_i \cap A_j \cap A_k| - \cdots + (-1)^{m-1} |A_1 \cap A_2 \cap \cdots \cap A_m|$$

Example 3.3.1 Find the number of numbers between 1 and 1500 (including 1 and 1500) that cannot be divisible by 4, 5 or 6.

Solution Let $S = \{x \mid x \in Z \wedge 1 \leqslant x \leqslant 1500\}$, $A = \{x \mid x \in S \wedge x$ divisible by 4$\}$, $B = \{x \mid x \in S \wedge x$ divisible by 5$\}$, $C = \{x \mid x \in S \wedge x$ divisible by 6$\}$.

Let $|Y|$ denote the number of elements in a finite set Y, $\lfloor x \rfloor$ denote the largest integer less than or equal to x, lcm $(x_1, x_2... x_n)$ denotes x_1, x_2 the least common multiple of x_n, then

$|A| = \lfloor 1500/4 \rfloor = 375$

$|B| = \lfloor 1500/5 \rfloor = 300$

$|C| = \lfloor 1500/6 \rfloor = 250$

$|A \cap B| = \lfloor 1500/\text{lcm}(4,5) \rfloor = 75$

$|A \cap C| = \lfloor 1500/\text{lcm}(4,6) \rfloor = 125$

$|B \cap C| = \lfloor 1500/\text{lcm}(5,6) \rfloor = 50$

$|A \cap B \cap C| = \lfloor 1500/\text{lcm}(4,5,6) \rfloor = 25$

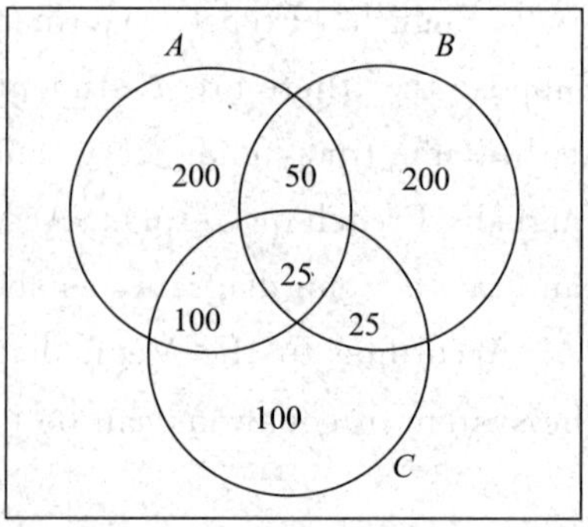

Figure 3.3.1

Fill these numbers in the Venn diagram to get Figure 3.3.1.

Using the inclusion-exclusion principle:

$$|\overline{A} \cap \overline{B} \cap \overline{C}| = |S| - (|A| + |B| + |C|) + (|A \cap B| + |A \cap C| + |B \cap C|) - |A \cap B \cap C|$$

$$= 1500 - (375 + 300 + 250) + (75 + 125 + 50 - 25) = 800$$

800 numbers between 1 and 1500 cannot be divisible by 4, 5 and 6.

3.4　Ordered Pair and Cartesian Product

Ordered pairs are in contact with previous knowledge. For example, the point in a rectangular coordinate system is a sequence pair. If the abscissa and the ordinate are changed before and after, it means another point, so the ordered pairs are ordered, and the Cartesian product is closely related to the ordered pairs.

Definition 3.4.1　The ordered sequence of two individuals x, y is called a doublet also known as ordered pair or tuple. It is recorded as $<x, y>$. x, y is called the first component and the second component of the doublet, respectively.

The so-called ordered sequence means that after changing the position of the first component and the second component, it is different from the original meaning.

Definition 3.4.2　Let $<x, y>$ and $<a, b>$ are two doublets. If $x=a$ and $y=b$, then the doublet $<x, y>$ is equal to $<a, b>$ and denoted as $<x, y>=<a, b>$ respectively.

The doublet $<x, y>$ is equal to $<a, b>$ and is expressed logically as $(<x, y>=<a, b>) \Leftrightarrow (x=a) \wedge (y=b)$.

It can be seen from the definition that when $x \neq y$, $<x, y> \neq <y, x>$.

For example, the point $P_1=<2,1>$ and the point $P_2=<1, 2>$ are two different points in the plane, and they are both doublets.

Definition 3.4.3　The doublet $<<x_1, x_2, ..., x_{n-1}>, x_n>$ is called n-tuple. Recorded as $<x_1, x_2, ..., x_n>$, x_i is called the i th component of the n-tuple, $i=1, ..., n$.

From the definition, it can be seen that: the triple $<x_1, x_2, x_3>$ is defined as $<<x_1, x_2>, x_3>$ in which the first component is the doublet; the quadruple $<x_1, x_2, x_3, x_4>$ is defined as $<<x_1, x_2, x_3>, x_4>$, the first component is the triple; ... .

According to the definition of triple, the first component is doublet, the second component is the ordered pair of an individual, $<<x_1, x_2>, x_3>$ is triple. The first component of the ordered pair $<x_1, <x_2, x_3>>$ is an individual, and the second component is a doublet, so it is not a triple. This conclusion is valid for any n ($n=3$, 4, 5) tuple is also right. For example, $<<x_1, x_2, x_3>, x_4>$ is a quadruple, however, $<x_1, <x_2, x_3, x_4>>$ is not a quadruple; $<<x_1, x_2, x_3, x_4>, x_5>$ is a quintuple, while $<x_1, <x_2, x_3, x_4, x_5>>$ is not a quintuple.

Definition 3.4.4　Let $<x_1, x_2, ..., x_n>$ and $<y_1, y_2, ..., y_n>$ be two n-tuples. If $x_i=y_i$, $i=1, ..., n$, then the two n-tuples are said to be equal, recorded as
$$<x_1, x_2, ..., x_n>=<y_1, y_2, ..., y_n>$$
n-tuple $<x_1, x_2, ..., x_n>$ is equal to $<y_1, y_2, ..., y_n>$ and is represented by logic:
$$(<x_1, x_2, ..., x_n>=<y_1, y_2, ..., y_n>)$$
$$\Leftrightarrow (x_1=y_1) \wedge (x_2=y_2) \wedge ... \wedge (x_n=y_n)$$

Definition 3.4.5 Let A, B be sets, $\{<a,b>|a\in A \wedge b\in B\}$ is called the Cartesian product of A and B, also called the cross product, direct product of A and B. It is recorded as $A\times B$.

If A and B are finite sets, $|A|=n$, $|B|=m$, according to the principle of permutation and combination, $|A\times B|=nm=|A||B|$.

Example 3.4.1 Let $A=\{r,s,t\}$, $B=\{1,2,3\}$.

(1) Find $A\times B$ and $B\times A$.

(2) Verify that $|A\times B|=|A||B|$ and $|B\times A|=|B||A|$.

Solution

(1) Find $A\times B$ and $B\times A$.

$A\times B=\{<r,1>,<r,2>,<r,3>,<s,1>,<s,2>,<s,3>,<t,1>,<t,2>,<t,3>\}$

$B\times A=\{<1,r>,<1,s>,<1,t>,<2,r>,<2,s>,<2,t>,<3,r>,<3,s>,<3,t>\}$

(2) Verify that $|A\times B|=|A||B|$ and $|B\times A|=|B||A|$.

$$|A\times B|=9=3\times 3=|A||B|$$

$$|B\times A|=9=3\times 3=|B||A|$$

Theorem 3.4.1 Cartesian product has the following properties.

(1) Let A be an arbitrary set, then $A\times\varnothing=\varnothing\times A=\varnothing$.

(2) Generally speaking, $\times$ does not satisfy the commutative law: $A\times B\neq B\times A$.

In Example 3.4.1, $A\times B\neq B\times A$.

(3) Generally speaking, $\times$ does not satisfy the law of association $(A\times B)\times C\neq A\times(B\times C)$.

Theorem 3.4.2 Let A, B, C be sets, then

(1) $A\times(B\cup C)=(A\times B)\cup(A\times C)$

(2) $A\times(B\cap C)=(A\times B)\cap(A\times C)$

(3) $(A\cup B)\times C=(A\times C)\cup(B\times C)$

(4) $(A\cap B)\times C=(A\times C)\cap(B\times C)$

Theorem 3.4.3 Let A, B, C be sets, $C\neq\varnothing$, then

(1) The necessary and sufficient condition of $A\subseteq B$ is $A\times C\subseteq B\times C$.

(2) The necessary and sufficient condition of $A\subseteq B$ is $C\times A\subseteq C\times B$.

Theorem 3.4.4 Let A, B, C, D be nonempty sets, then the necessary and sufficient condition of $A\times B\subseteq C\times D$ is $A\subseteq C$ and $B\subseteq D$.

Definition 3.4.6 Cross product $A_1\times A_2\times...\times A_n$ is defined as $(A_1\times A_2\times...\times A_{n-1})\times A_n$, then

$$A_1\times A_2\times...\times A_n=\{<a_1,a_2,\cdots,a_n>|a_1\in A_1 \wedge a_2\in A_2 \wedge...\wedge a_n\in A_n\}$$

It can be seen from the definition:

When $n=3$, $A_1\times A_2\times A_3$ is defined as $(A_1\times A_2)\times A_3$.

$$A_1\times A_2\times A_3=\{<a_1,a_2,a_3>|a_1\in A_1 \wedge a_2\in A_2 \wedge a_3\in A_3\}$$

When $n=4$, $A_1 \times A_2 \times A_3 \times A_4$ is defined as $(A_1 \times A_2 \times A_3) \times A_4$.

$A_1 \times A_2 \times A_3 \times A_4 = \{<a_1,a_2,a_3,a_4> | a_1 \in A_1 \land a_2 \in A_2 \land a_3 \in A_3 \land a_4 \in A_4\}$...

$A_n = A \times A \times ... \times A$ (The cross product of n $A's$)

Example 3.4.2 Let $A=\{1, 2\}$, $B=\{a, b\}$, $C=\{x, y\}$, find: $A \times B \times C$, $A \times (B \times C)$.

Solution $A \times B \times C = (A \times B) \times C$

$$= \{<1,a>,<1,b>,<2,a>,<2,b>\} \times \{x,y\}$$

$$= \{<<1,a>,x>,<<1,b>,x>,<<2,a>,x>,<<2,b>,x>,$$

$$<<1,a>,y>,<<1,b>,y>,<<2,a>,y>,<<2,b>,y>\}$$

$$= \{<1,a,x>,<1,b,x>,<2,a,x>,<2,b,x>,<1,a,y>,$$

$$<1,b,y>,<2,a,y>,<2,b,y>\}$$

$A \times (B \times C) = \{1,2\} \times \{<a,x>,<a,y>,<b,x>,<b,y>\}$

$$= \{<1,<a,x>>,<1,<a,y>>,<1,<b,x>>,<1,<b,y>>,$$

$$<2,<a,x>>,<2,<a,y>>,<2,<b,x>>,<2,<b,y>>\}$$

Obviously, $A \times B \times C \neq A \times (B \times C)$.

3.5　Relation and Its Nature

The relation is a basic concept. In mathematics, the relationship between elements in a set is mainly studied. For example, "0 is less than 1"; "x is not equal to y", etc.

Definition 3.5.1 Let A and B be arbitrary sets. If $R \subseteq A \times B$, then R is a binary relation from A to B. If R is a binary relation from A to A, then R is a binary relation from A.

Let $A=\{1,2,3\}$, $B=\{a,b\}$, $R=\{<1,a>,<2,a>,<3,b>\}$, R is a binary relation from A to B. If $S=\{<3,1>,<2,2>,<2,1>,<1,1>\}$, then R is a binary relation from A.

Definition 3.5.2 Let A, B be arbitrary sets, $R \subseteq A \times B$, if $<x, y> \in R$, then it is said that x and y have R relation, recorded as xRy. If $<x, y> \notin R$, then it is said that x and y don't have an R relation, recorded as $x\overline{R}y$.

If R is a binary relation from A to B, according to Definition 3.5.2, $<x, y> \in R$ and xRy, $<x, y> \notin R$ and $x\overline{R}y$ have the same meaning separately.

Definition 3.5.3 Let A and B be arbitrary sets, and the empty set $\varnothing$ is called the empty relation from A to B, which is still denoted as $\varnothing$. The Cartesian product $A \times B$ of A, B is called the global relation from A to B, denoted as E. The set $\{<a, a> | a \in A\}$ is called the identity relation on A. It is recorded as I_A.

Example 3.5.1 Let $A=\{c,d\}$, $B=\{2,5\}$, find the identity relation I_A on a and the global relation $A \times B$ from A to B.

Solution The identity relation $I_A = \{<c,c>,<d,d>\}$ on A, the global relation from A to B, $E = A \times B = \{<c,2>, <c,5>, <d,2>, <d,5>\}$.

Theorem 3.5.1 Let A be a finite set with n elements, then there are 2^{n^2} kinds of binary relations on A.

Proof Let A be a finite set with n elements, that is $|A| = n$. According to the principle of permutation and combination, $|A \times A| = n^2$. According to Theorem 3.1.3, there is $|P(A \times A)| = 2^{|A \times A|} = 2^{n^2}$, that is, there are 2^{n^2} subsets of $A \times A$. So there are 2^{n^2} kinds of binary relations on the finite set A with n elements.

1. Using enumeration to express binary relation

Global relationship from A to B in Example 3.5.1
$$E = A \times B = \{<c,2>,<c,5>,<d,2>,<d,5>\}$$
Identities on A
$$I_A = \{<c,c>,<d,d>\}$$
It's all in the form of enumeration.

2. Using description to express binary relation

Let $\mathbf{R}$ be a set of real numbers, $L_{\mathbf{R}} = \{<x,y> | x \in \mathbf{R} \wedge y \in \mathbf{R} \wedge x \leqslant y\}$, and $L_{\mathbf{R}}$ is a binary relation on the real number set $\mathbf{R}$.

3. Expressing binary relation by matrix

If A and B are finite sets, $A = \{a_1, a_2, \ldots, a_m\}$, $B = \{b_1, b_2, \ldots, b_n\}$, R is a binary relation from A to B. The relation matrix of R is defined as
$$M_R = (r_{ij})_{m \times n}$$
$$r_{ij} = \begin{cases} 1, <a_i,b_j> \in R \\ 0, <a_i,b_j> \notin R \end{cases} \quad i=1,\cdots,m \quad j=1,\cdots,n$$
The relation matrix is called the binary relation R.

Example 3.5.2 Let $A = \{a_1,a_2,a_3,a_4\}$, $B = \{b_1,b_2,b_3\}$, R is a binary relationship from A to B, which is defined as:
$$R = \{<a_1,b_1>,<a_1,b_3>,<a_2,b_2>,<a_2,b_3>,<a_3,b_1>,<a_4,b_1>,<a_4,b_2>\}$$
Write the relation matrix of R.

Solution The relation matrix of R is
$$M_R = \begin{bmatrix} 1 & 0 & 1 \\ 0 & 1 & 1 \\ 1 & 0 & 0 \\ 1 & 1 & 0 \end{bmatrix}$$

Example 3.5.3 Let $A = \{1,2,3,4\}$, R is a binary relationship from A to A, which is defined as:
$$R = \{<1,1>,<1,2>,<2,1>,<3,2>,<3,1>,<4,3>,<4,2>,<4,1>\}$$
Write the relation matrix of binary relation R on A.

Solution The relation matrix of R is:

$$M_R = \begin{bmatrix} 1 & 1 & 0 & 0 \\ 1 & 0 & 0 & 0 \\ 1 & 1 & 0 & 0 \\ 1 & 1 & 1 & 0 \end{bmatrix}$$

The binary relation R in Example 3.5.3 is a binary relation on A, which can be easily written by using the above definition. The definition of the binary relation on A is the same as that of the relation matrix between A and B.

4. A binary relation is represented by a graph.

If A and B are finite sets and R is a binary relation from A to B, the binary relation R can also be represented by a graph. A graph representing a binary relation R is called a graph of R. The definition of the graph of the binary relationship between A and B is different from that of the binary relationship on A.

They are described as follows.

(1) The graph of A to B binary relation R.

Let $A = \{a_1, a_2, \ldots, a_m\}, B = \{b_1, b_2, \ldots, b_n\}$, R is a binary relationship from A to B. The drawing method of R's relation diagram is as follows: ① Draw m small circles to represent the elements of A, and then draw n small circles to represent the elements of B. These small circles are called the nodes of the graph. ② If $<a_i, b_j> \in R$, draw a line with an arrow from a_i to b_j. These directional lines are called the edges of the graph.

The diagram of binary relation R in Example 3.5.2 is shown in Figure 3.5.1.

(2) The graph of binary relation R on A.

Let $A = \{a_1, a_2, \ldots, a_m\}$, R is a binary relation on A, and its diagram is as follows: ① Draw m small circles to represent the elements of A. ② If $<a_i, a_j> \in R$, draw a line with an arrow from a_i to a_j. The diagram of binary relation R in Example 3.5.3 is shown in Figure 3.5.2.

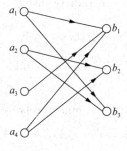

Figure 3.5.1

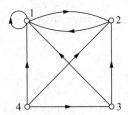

Figure 3.5.2

Definition 3.5.4 Let $R \subseteq X \times X, (\forall x)(x \in X \rightarrow <x, x> \in R)$, then R is said to be reflexive on X.

Let R be a reflexive relation on X. According to Definition 3.5.4, the main diagonals of the relation matrix M_R of R are all 1; there are self-loops at every node in the graph.

Let $X = \{1, 2, 3\}$, the binary relation on X

$$R = \{<1,1>, <2,2>, <3,3>, <1,2>\}$$

R is reflexive, and its relation diagram is shown in Figure 3.5.3, and the relation matrix is as follows:

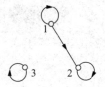

Figure 3.5.3

$$\boldsymbol{M}_R = \begin{bmatrix} 1 & 1 & 0 \\ 0 & 1 & 0 \\ 0 & 0 & 1 \end{bmatrix}$$

Theorem 3.5.2 Let R be a binary relation on X and R be reflexive if and only if $I_X \subseteq R$.

Proof Let R be reflexive on X and prove $I_X \subseteq R$.

$<x,y> \in I_X \Rightarrow x = y \Rightarrow <x,y> \in R$, then $I_X \subseteq R$.

Let $I_X \subseteq R$. Prove that R is reflexive on X.

$\forall x \in X \Rightarrow <x,x> \in I_X \Rightarrow <x,x> \in R$, then R is reflexive on X.

Definition 3.5.5 Let $R \subseteq X \times X, (\forall x)(x \in X \rightarrow <x,x> \notin R)$, R is said to be reflexive on X.

Figure 3.5.4

Let R be an anti-reflexive relation on X, from Definition 3.5.5, we can see that the main diagonals of the relation matrix $\boldsymbol{M}_R$ of R are all 0; there is no self-loop at every node in the graph of R, its relation diagram is shown in Figure 3.5.4, and the relation matrix is as follows:

$$\boldsymbol{M}_R = \begin{bmatrix} 0 & 1 & 0 \\ 0 & 0 & 1 \\ 1 & 0 & 0 \end{bmatrix}$$

Theorem 3.5.3 Let R be a binary relation on X and R be reflexive if and only if $R \cap I_X = \varnothing$.

Proof Let R be reflexive on X, and prove that $R \cap I_X = \varnothing$.

Suppose $R \cap I_X \neq \varnothing$, there must be $<x,y> \in R \cap I_X \Rightarrow <x,y> \in R \wedge <x,y> \in I_X \Rightarrow <x,y> \in R \wedge x = y$, that is $<x, x> \in R$, which contradicts that R is the reflexive relation on X. So $R \cap I_X = \varnothing$.

Let $R \cap I_X = \varnothing$, and prove that R is reflexive on X.

Any $x \in X, <x, x> \in I_X$, because $R \cap I_X = \varnothing$, it is inevitable that $<x, x> \notin R$, that is, R is reflexive on X.

Example 3.5.4 Let $A = \{1, 2, 3\}$ and define the binary relation on A as follows:

$$R = \{<1,1>, <2,2>, <3,3>, <1,3>\}$$
$$S = \{<1,3>\}$$
$$T = \{<1,1>\}$$

Try to explain whether R, S, T are reflexive or reflexive on A.

Solution Because $<1,1>, <2, 2>$ and $<3, 3>$ are all elements of R, R is a reflexive relationship on a, not an anti reflexive relationship. $<1,1>, <2, 2>$ and $<3, 3>$ are not elements of S, so S is an anti reflexive relationship on A, not a reflexive relationship on A.

Because $<2,2>\notin T$, T is not a reflexive relation on A. Because $<1,1>\in T$, T is not a reflexive relation on A.

Definition 3.5.6　Let $R\subseteq X\times X$, $(\forall x)(\forall y)(x\in X\wedge y\in X\wedge<x,y>\in R\rightarrow<y,x>\in R)$, Then R is said to be symmetric on X.

R is a symmetric relation on X. According to Definition 3.5.6, the relation matrix $\boldsymbol{M}_R$ of R is a symmetric matrix. In the graph of R, if there are edges between two different nodes, there must be two edges in opposite directions.

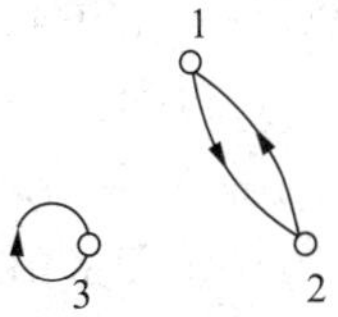

Let $X=\{1,2,3\}$ and the binary relation $R=\{<1,2>,<2,1>,<3,3>\}$, R is symmetric. Its relation diagram is shown in Figure 3.5.5, and the relation matrix is as follows:

Figure　3.5.5

$$\boldsymbol{M}_R=\begin{bmatrix}0&1&0\\1&0&0\\0&0&1\end{bmatrix}$$

Theorem 3.5.4　Let R be a binary relation on X and R be symmetric if and only if $R=R^{c}$.

Proof　Let R be symmetric and $R=R^{c}$.

$<x,y>\in R\Leftrightarrow<y,x>\in R\Leftrightarrow<x,y>\in R^{c}$, so $R=R^{c}$.

Let $R=R^{c}$, and prove that R is symmetric.

$<x,y>\in R\Rightarrow<y,x>\in R^{c}\Rightarrow<y,x>\in R$, so R is symmetric.

Definition 3.5.7　Let $R\subseteq X\times X$,

$$(\forall x)(\forall y)(x\in X\wedge y\in X\wedge<x,y>\in R\wedge<y,x>\in R\rightarrow(x=y))$$

Then R is said to be anti-symmetric on X.

The definition of anti-symmetry can be described as follows:

$$(\forall x)(\forall y)(x\in X\wedge y\in X\wedge<x,y>\in R\wedge(x\neq y)\rightarrow<y,x>\notin R)$$

Let R be an antisymmetric relation on X. according to Definition 3.5.7, the symmetric position with the principal diagonal as the axis in the relation matrix $\boldsymbol{M}_R$ of R cannot be 1 at the same time (except the main diagonal). In the graph of R, every two different nodes cannot have two opposite sides.

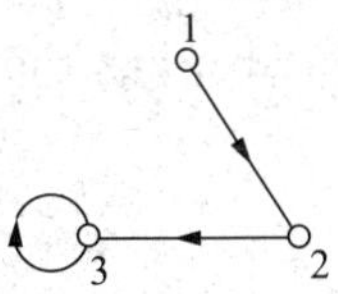

Let $X=\{1,2,3\}$ and the binary relation $R=\{<1,2>,<2,3>,<3,3>\}$, R is antisymmetric. Its relation diagram is shown in Figure 3.5.6, and the relation matrix is as follows:

Figure　3.5.6

$$\boldsymbol{M}_R=\begin{bmatrix}0&1&0\\0&0&1\\0&0&1\end{bmatrix}$$

Example　3.5.5　Let $A=\{1,2,3\}$, the binary relation on definition A is as follows:

$$R=\{<1,1>,<2,2>\}$$
$$S=\{<1,1>,<1,2>,<2,1>\}$$
$$T=\{<1,2>,<1,3>\}$$

$$U=\{<1,3>,<1,2>,<2,1>\}$$

Whether R, S, T, U are symmetric and antisymmetric on A.

Solution R is a symmetric relation on A and an antisymmetric relation on A.

S is a symmetric relation on A. Because $<1,2>$ and $<2,1>$ are S elements, and $1\neq 2$, S is not an antisymmetric relation on A.

Because $<1,2>\in T$, and $<2,1>\notin T$, T is not a symmetric relation on A. T is the antisymmetric relation on A.

U is neither symmetric nor antisymmetric on A.

Theorem 3.5.5 Let R be a binary relation on X, then R is antisymmetric if and only if $R\cap R^c\subseteq I_X$.

Definition 3.5.8 Let $R\subseteq X\times X$

$$(\forall x)(\forall y)(\forall z)(x\in X\wedge y\in X\wedge z\in X\wedge <x,y>\in R\wedge <y,z>\in R\rightarrow <x,z>\in R)$$

R is said to be transitive on X.

Theorem 3.5.6 Let R be a binary relation on X and R be transitive if and only if $R\circ R\subseteq R$.

The above Theorems 3.5.2~3.5.6 can be used as definitions of corresponding concepts to judge and prove the properties of relations. It is very convenient to deal with some problems with these five theorems.

Theorem 3.5.7 Let R, S be a binary relation on X, then

(1) If R and S are reflexive, then $R\cup S$ and $R\cap S$ are reflexive.

(2) If R and S are symmetric, then $R\cup S$ and $R\cap S$ are also symmetric.

(3) If R and S are transitive, then $R\cap S$ is transitive.

3.6　Inverse and Compound Relations

In relation operation, not only union, intersection, complement and other operations can be included, but also the compound operation of relation can be carried out. Because the relation is the set of order pairs, the ordered pair has the order, so the relation also has the inverse operation.

Definition 3.6.1 Let A, B be a set, $R\subseteq A\times B$.

dom $R=\{x|<x,y>\in R\}$ is called the domain of R.

ran $R=\{y|<x,y>\in R\}$ is called the range of R.

FLD $R=$ dom $R\cup$ ran R is called the domain of R.

A is called the pre domain of R; B is called the companion domain of R.

1. Operation of intersection, union, complement and symmetric difference of binary relation

Theorem 3.6.1 Let R, S be a binary relation from X to Y, then $R\cup S$, $R\cap S$, $R-S$, $\sim R$, $R\oplus S$ are also binary relations from X to Y.

Proof Because R and S are binary relations from X to Y, $R\subseteq X\times Y$ and $S\subseteq X\times Y$.

Obviously，$R \cup S \subseteq X \times Y$, that is，$R \cup S$ is a binary relation from X to Y.

$R \cap S \subseteq X \times Y$, that is，$R \cap S$ is a binary relation from X to Y.

$R - S \subseteq X \times Y$, that is，$R - S$ is a binary relation from X to Y.

$R - S \subseteq X \times Y$, that is，$R - S$ is a binary relation from X to Y.

In binary relation operation，the global relation is considered to be a complete set. So$\sim R = X \times Y - R \subseteq X \times Y$, $\sim R$ is a binary relation from X to Y.

From the above conclusion，it can be concluded that：$(R-S) \subseteq X \times Y$ and $(S-R) \subseteq X \times Y$, thus $(R-S) \cup (S-R) \subseteq X \times Y$, so $R \oplus S = (R-S) \cup (S-R) \subseteq X \times Y$, that is，$R \oplus S$ is a binary relationship from X to Y.

Example 3.6.1　Let $X = \{1, 2, 3, 4\}$, the binary relations H and S on X are defined as follows：$H = \{x, y < (x-y)/2$ is an integer$\}$, $S = \{x, y < (x-y)/3$ is a positive integer$\}$. Try to find $H \cup S$, $H \cap S$, $\sim H$, $S - H$.

Solution　H and S are represented by enumeration：

$H = \{<1,1>,<1,3>,<2,2>,<2,4>,<3,3>,<3,1>,<4,4>,<4,2>\}$

$S = \{<4,1>\}$

$H \cup S = \{<1,1>,<1,3>,<2,2>,<2,4>,<3,3>,<3,1>,<4,4>,<4,2>,<4,1>\}$

$H \cap S = \varnothing$

$\sim H = X^2 - H = \{<1,2>,<1,4>,<2,1>,<2,3>,<3,2>,<3,4>,<4,3>,<4,1>\}$

$S - H = \{<4,1>\}$

2. Compound operation of binary relation

Definition 3.6.2　Let X, Y, Z be a set，$R \subseteq X \times Y, S \subseteq Y \times Z$,Set $\{<x,z> | x \in X \wedge z \in Z \wedge (\exists y)(y \in Y \wedge <x,y> \in R \wedge <y,z> \in S)\}$ is called the compound relation of R and S. It is recorded as $R \circ S$, $R \circ S \subseteq X \times Z$, $R \circ S$ is a binary relation from X to Z.

Example 3.6.2　$X = \{1,2,3,4,5\}$, the binary relations R and S on X are defined as follows：$R = \{<1,2>,<3,4>,<2,2>\}$, $S = \{<4,2>,<2,5>,<3,1>,<1,3>\}$, Try to find $R \circ S$, $S \circ R$, $R \circ (S \circ R)$, $(R \circ S) \circ R$, $R \circ R$, $S \circ S$, $R \circ R \circ R$.

Solution　$R \circ S = \{<1,5>,<3,2>,<2,5>\}$

$S \circ R = \{<4,2>,<3,2>,<1,4>\}$

$(R \circ S) \circ R = \{<3,2>\}$

$R \circ (S \circ R) = \{<3,2>\}$

$R \circ R = \{<1,2>,<2,2>\}$

$S \circ S = \{<4,5>,<3,3>,<1,1>\}$

$R \circ R \circ R = \{<1,2>,<2,2>\}$

It can be seen from Example 3.6.2 that $R \circ S \neq S \circ R$, which shows that the compound operation of binary relation does not satisfy the commutative law.

Theorem 3.6.2　Let X, Y, Z, W be sets，$R \subseteq X \times Y, S \subseteq Y \times Z, T \subseteq Z \times W$, then $(R \circ S) \circ T = R \circ (S \circ T)$.

Proof　$<x,w> \in (R \circ S) \circ T \Leftrightarrow (\exists z)(<x,z> \in R \circ S \wedge <z,w> \in T)$

$$\Leftrightarrow(\exists z)((\exists y)(<x,y>\in R \wedge <y,z>\in S) \wedge <z,w>\in T)$$
$$\Leftrightarrow(\exists z)(\exists y)((<x,y>\in R \wedge <y,z>\in S) \wedge <z,w>\in T)$$
$$\Leftrightarrow(\exists y)(\exists z)(<x,y>R \wedge (<y,z>\in S \wedge <z,w>\in T))$$
$$\Leftrightarrow(\exists y)(<x,y>\in R \wedge (\exists z)(<y,z>\in S \wedge <z,w>\in T))$$
$$\Leftrightarrow(\exists y)(<x,y>\in R \wedge <y,w>\in S\circ T)\Leftrightarrow<x,w>\in R\circ$$
$$(S\circ T)$$

So $(R\circ S)\circ T=R\circ(S\circ T)$.

Theorem 3.6.3 Let R be a binary relation on A, $R\circ I_A=I_A\circ R=R$.

Proof Proband $R\circ I_A=R$.

$$<x,y>\in R\circ I_A \Rightarrow (\exists z)(<x,z>\in R \wedge <z,y>\in I_A)$$
$$\Rightarrow(\exists z)(<x,z>\in R \wedge z=y)\Rightarrow<x,y>\in R$$

So $R\circ I_A\subseteq R$.

$$<x,y>\in R\Rightarrow<x,y>\in R \wedge <y,y>\in I_A\rightarrow<x,y>\in R\circ I_A$$

So $R\subseteq R\circ I_A$.

So $R\circ I_A=R$.

Similarly, it can be proved that $I_A\circ R=R$.

Theorem 3.6.4 Let R, S, T be a binary relation on A, then

(1) $R\circ(S\cup T)=R\circ S\cup R\circ T$

(2) $(R\cup S)\circ T=R\circ T\cup S\circ T$

(3) $R\circ(S\cap T)\subseteq R\circ S\cap R\circ T$

(4) $(R\cap S)\circ T\subseteq R\circ T\cap S\circ T$

Proof Only proof (3). (1), (2) and (4) are the same.

$$<x,y>\in R\circ(S\cap T)\Rightarrow(\exists z)(<x,z>\in R \wedge <z,y>\in S\cap T)$$
$$\Rightarrow(\exists z)(<x,z>\in R \wedge (<z,y>\in S \wedge <z,y>\in T))$$
$$\Rightarrow(\exists z)((<x,z>\in R \wedge <z,y>\in S) \wedge (<x,z>\in R \wedge <z,y>\in T))$$
$$\Rightarrow(\exists z)(<x,z>\in R \wedge <z,y>\in S) \wedge (\exists z)(<x,z>\in R \wedge$$
$$<z,y>\in T)$$
$$\Rightarrow<x,y>\in R\circ S \wedge <x,y>\in R\circ T$$
$$\Rightarrow<x,y>\in R\circ S\cap R\circ T$$

So $R\circ(S\cap T)\subseteq R\circ S\cap R\circ T$.

Generally speaking, $R\circ S\cap R\circ T\nsubseteq R\circ(S\cap T)$, the counter examples are as follows: Let $A=\{1,2,3,4,5\}$, $R=\{<4,1>,<4,2>\}\subseteq A\times A$, $S=\{<1,5>,<3,5>\}\subseteq A\times A$, $T=\{<2,5>,<3,5>\}\subseteq A\times A$, then $R\circ S\cap R\circ T=\{<4,5>\}\cap\{<4,5>\}=\{<4,5>\}$, $R\circ(S\cap T)=\{<4,1>,<4,2>\}\circ\{<3,5>\}=\varnothing$, $R\circ S\cap R\circ T\nsubseteq R\circ(S\cap T)$.

Theorem 3.6.4 (1) shows that the compound operation "$\circ$" and the union operation "$\cup$" satisfy the left distribution law. (2) it is shown that the compound operation "$\circ$" and the union operation "$\cup$" satisfy the right distribution law. Therefore, the compound operation "$\circ$" and the union operation "$\cup$" satisfy the distributive law. According to (3) and (4), the compound operation "$\circ$" and the intersection operation "$\cap$" do not satisfy the distributive law.

Definition 3.6.3　Let R be a binary relation on A, n be a natural number, and the n-th power of R is denoted as R^n, then $R^0 = I_A$, $R^{n+1} = R^n \circ R$.

From Definition 3.6.3, it can be seen that any binary relation on A has equal power 0, which is equal to the identity relationship I_A on A. It can also be seen from Definition 3.6.3 that:

$$R^1 = R^0 \circ R = I_A \circ R = R$$
$$R^2 = R^1 \circ R = R \circ R$$
$$R^3 = R^2 \circ R = (R \circ R) \circ R$$

$$\cdots$$

Because the compound operation satisfies the associative law, R^3 can be written as follows:

$$R^3 = R \circ R \circ R$$

In the same way, R^n can be written as:

$$R^n = R \circ R \circ \ldots \circ R \text{ (the composition of } n \text{ } R's)$$

Theorem 3.6.5　Let A be a finite set with n elements and R be a binary relation on A, then there must be natural numbers s and t such that $R^s = R^t$, $0 \leqslant s < t \leqslant 2^{n^2}$.

Proof　R is a binary relation on A, for any natural number k, according to the definition of compound relation, R^k is still a binary relation on A, that is, $R^k \subseteq A \times A$. On the other hand, according to Theorem 3.5.1, there are only 2^{n^2} kinds of binary relations on A. List the powers of R^0, R^1, R^2, ..., $R^{2^{n^2}}$, there are $2^{n^2} + 1$ in total. There must be natural numbers s and t such that $R^s = R^t$, $0 \leqslant s < t \leqslant 2^{n^2}$.

Example　3.6.3　$A = \{1, 2, 3, 4\}$, the binary relation R on A is defined as follows: $R = \{<1,2>, <2,1>, <2,3>, <3,4>\}$. Find the power of binary relation R, and verify Theorem 3.6.5.

Solution　$|A| = 4$

$$R^0 = I_A = \{<1,1>, <2,2>, <3,3>, <4,4>\}$$
$$R^1 = R = \{<1,2>, <2,1>, <2,3>, <3,4>\}$$
$$R^2 = R^1 \circ R = R \circ R = \{<1,1>, <1,3>, <2,2>, <2,4>\}$$
$$R^3 = R^2 \circ R = \{<1,2>, <2,1>, <2,3>, <1,4>\}$$
$$R^4 = R^3 \circ R = \{<1,1>, <1,3>, <2,2>, <2,4>\} = R^2$$
$$0 \leqslant 2 < 4 \leqslant 2^{16} = 65536$$
$$R^5 = R^4 \circ R = R^2 \circ R = R^3$$
$$R^6 = R^5 \circ R = R^3 \circ R = R^4 = R^2$$

$$\cdots$$

$$R^{2n} = R^2$$
$$R^{2n+1} = R^3, \ n = 1, 2, 3, \ldots$$

Theorem 3.6.6　Let R be a binary relation on A, m and n be natural numbers, then $R^m \circ R^n = R^{m+n}$, $(R^m)^n = R^{mn}$.

Definition 3.6.4 Let X, Y be sets, $R \subseteq X \times Y$, the set $\{<y,x> | <x,y> \in R\}$ is called the inverse of R, denoted as R^c, $R^c \subseteq Y \times X$, R^c is a binary relation from Y to X.

It is easy to prove that the relation matrix of R^c, $\boldsymbol{M}_{R^c}$ is the transposition matrix of $\boldsymbol{M}_R$ of R, $\boldsymbol{M}_{R^c} = \boldsymbol{M}_R^T$.

It can be verified that the R^c relation graph can be obtained by reversing the arrow of the arc in the R relation graph.

Example 3.6.4 Let $X = \{1,2,3,4\}$, $Y = \{a,b,c\}$ and the binary relation from X to Y is $R = \{<1,a>,<2,b>,<4,c>\}$.

(1) Try to find R^c, write $\boldsymbol{M}_R$ and $\boldsymbol{M}_{R^c}$, verify that $\boldsymbol{M}_{R^c} = \boldsymbol{M}_R^T$.

(2) Draw the relation graph of R and R^c, and verify that the R^c relation graph can be obtained by reversing the arrow of arc in the R relation graph.

Solution $R^c = \{<a,1>,<b,2>,<c,4>\}$

The relation matrix of R and R^c is as follows:

$$\boldsymbol{M}_R = \begin{bmatrix} 1 & 0 & 0 \\ 0 & 1 & 0 \\ 0 & 0 & 0 \\ 0 & 0 & 1 \end{bmatrix} \qquad \boldsymbol{M}_{R^c} = \begin{bmatrix} 1 & 0 & 0 & 0 \\ 0 & 1 & 0 & 0 \\ 0 & 0 & 0 & 1 \end{bmatrix}$$

Obviously, $\boldsymbol{M}_{R^c} = \boldsymbol{M}_R^T$.

The relationship between R and R^c is Figure 3.6.1 and Figure. 3.6.2 respectively, and the direction of the arc in them is opposite.

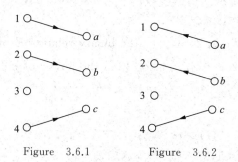

Figure 3.6.1 Figure 3.6.2

Theorem 3.6.7 Let X, Y be sets, $R \subseteq X \times Y$, then $(R^c)^c = R$, dom $R^c = $ ran R, ran $R^c = $ dom R.

Proof

(1) $<x,y> \in R \Leftrightarrow <y,x> \in R^c \Leftrightarrow <x,y> \in (R^c)^c$, so $(R^c)^c = R$.

(2) $x \in$ dom $R^c \Leftrightarrow (\exists y)(<x,y> \in R^c) \Leftrightarrow (\exists y)(<y,x> \in R) \Leftrightarrow x \in$ ran R, so dom $R^c = $ ran R.

Similarly, it can be proved that ran $R^c = $ dom R.

Theorem 3.6.8 Let X, Y be sets, R, S be a binary relation from X to Y, then

(1) $(R \cup S)^c = R^c \cup S^c$

(2) $(R \cap S)^c = R^c \cap S^c$

(3) $(A \times B)^{c} = B \times A$

(4) $(\sim R)^{c} = \sim (R^{c})$

(5) $(R - S)^{c} = R^{c} - S^{c}$

Theorem 3.6.9 Let X, Y, Z be sets, $R \subseteq X \times Y$, $S \subseteq Y \times Z$, then $(R \circ S)^{c} = S^{c} \circ R^{c}$.

Proof $<z,x> \in (R \circ S)^{c} \Leftrightarrow <x,z> \in (R \circ S)$

$\Leftrightarrow (\exists y)(<x,y> \in R \wedge <y,z> \in S)$

$\Leftrightarrow (\exists y)(<y,x> \in R^{c} \wedge <z,y> \in S^{c})$

$\Leftrightarrow (\exists y)(<z,y> \in S^{c} \wedge <y,x> \in R^{c})$

$\Leftrightarrow <z,x> \in S^{c} \circ R^{c}$

So $(R \circ S)^{c} = S^{c} \circ R^{c}$.

3.7 Closure Operations

The composition of relations and the inverse operations mentioned in the previous section can form new relationships. If a new relation with special properties is obtained by extending some order pairs in a given relation, it is a closure operation.

Definition 3.7.1 Let $R \subseteq X \times X$, the binary relation R' on X satisfies:

(1) R' is reflexive (symmetric, transitive).

(2) $R \subseteq R'$.

(3) For any binary relation R'' on X, if $R \subseteq R''$ and R'' is reflexive (symmetric, transitive), there is $R' \subseteq R''$.

Then R' is called the reflexive (symmetric, transitive) closure of R, recorded as $r(R)$ ($s(R)$, $t(R)$).

(1) and (2) of Definition 3.7.1 mean that the reflexive (symmetric, transitive) closure R' is a reflexive (symmetric, transitive) relationship containing R; and (3) of Definition 3.7.1 means that the reflexive (symmetric, transitive) closure R' is the smallest among all the reflexive (symmetric, transitive) relationships involving R. Therefore, the Definition 3.7.1 can be described as the minimum reflexive (symmetric, transitive) relationship involving R is the reflexive (symmetric, transitive) closure of R.

Transitive closure $t(R)$ is sometimes denoted R^{+} and pronounced "R plus". Transitive closure has many important applications in parsing.

When the binary relation R is reflexive (symmetric, transitive), the method for finding the reflexive (symmetric, transitive) closure of R is given by the following theorem.

Theorem 3.7.1 Let $R \subseteq X \times X$, then

(1) R is reflexive if and only if $r(R) = R$.

(2) R is symmetric if and only if $s(R) = R$.

(3) R is transitive if and only if $t(R) = R$.

Proof Only prove (1) and keep the rest for practice.

Let R be reflexive and prove that $r(R)=R$.

Let $R'=R$,

① $R'=R$, R is reflexive, so R' is reflexive.

② Because $R'=R$, $R\subseteq R'$.

③ R'' is an arbitrary reflexive relation on X and $R\subseteq R''$, because $R'=R$, so $R'\subseteq R''$.

So R' is a reflexive closure of R. That is, $r(R)=R$.

Let $r(R)=R$, prove that R is reflexive.

According to the definition of reflexive closure, $r(R)$ is reflexive, so R is reflexive.

The following theorems give general methods for finding reflexive, symmetric and transitive closures of binary relation R.

Theorem 3.7.2 Let $R\subseteq X\times X$, then $r(R)=R\cup I_X$.

Proof Let $R'=R\cup I_X$,

(1) $I_X\subseteq R\cup I_X$, and $R\cup I_X=R'$, so $I_X\subseteq R'$, R' is reflexive.

(2) $R\subseteq R\cup I_X$ and $R\cup I_X=R'$, so $R\subseteq R'$.

(3) R'' is an arbitrary reflexive relation on X and $R\subseteq R''$. Since R' is reflexive, so $I_X\subseteq R''$, thus $R'=R\cup I_X\subseteq R''$.

According to the definition of reflexive closure, $R'=R\cup I_X$ is the reflexive closure of R, that is $r(R)=R\cup I_X$.

Theorem 3.7.3 Let $R\subseteq X\times X$, then $s(R)=R\cup R^c$.

Proof Let $R'=R\cup R^c$, then

(1) $(R')^c=(R\cup R^c)^c=R^c\cup(R^c)^c=R^c\cup R=R\cup R^c=R'$, so R' is symmetric.

(2) $R\subseteq R\cup R^c$ and $R\cup R^c=R'$, so $R\subseteq R'$.

(3) R'' is an arbitrary symmetric relation on X and $R\subseteq R''$.

$<x,y>\in R^c\Rightarrow<y,x>\in R\Rightarrow<y,x>\in R''\Rightarrow<x,y>\in R''$ (R'' is symmetric)

So $R^c\subseteq R''$, thus $R'=R\cup R^c\subseteq R''$. R' is the symmetric closure of R, that is, $s(R)=R\cup R^c$.

Theorem 3.7.4 Let $R\subseteq X\times X$, then $t(R)=R\cup R^2\cup R^3\cup\dots$.

Theorem 3.7.5 Let $|X|=n$, $R\subseteq X\times X$, then $t(R)=R\cup R^2\cup\dots\cup R^n$.

Example 3.7.1 Let $A=\{1,2,3\}$, The binary relation R on A is defined as: $R=\{<1,2>,<2,3>,<3,1>\}$, try to find the transitive closure $t(R)$.

Solution The method of calculating $t(R)$ by relation matrix is as follows:

$$\boldsymbol{M}_R=\begin{bmatrix}0 & 1 & 0\\0 & 0 & 1\\1 & 0 & 0\end{bmatrix}$$

$$\boldsymbol{M}_{R^2}=\boldsymbol{M}_R\circ\boldsymbol{M}_R=\begin{bmatrix}0 & 1 & 0\\0 & 0 & 1\\1 & 0 & 0\end{bmatrix}\circ\begin{bmatrix}0 & 1 & 0\\0 & 0 & 1\\1 & 0 & 0\end{bmatrix}=\begin{bmatrix}0 & 0 & 1\\1 & 0 & 0\\0 & 1 & 0\end{bmatrix}$$

$$\boldsymbol{M}_{R^3}=\boldsymbol{M}_{R^2}\circ\boldsymbol{M}_R=\begin{bmatrix}0&0&1\\1&0&0\\0&1&0\end{bmatrix}\circ\begin{bmatrix}0&1&0\\0&0&1\\1&0&0\end{bmatrix}=\begin{bmatrix}1&0&0\\0&1&0\\0&0&1\end{bmatrix}$$

$$\boldsymbol{M}_{t(R)}=\boldsymbol{M}_R\vee\boldsymbol{M}_{R^2}\vee\boldsymbol{M}_{R^3}=\begin{bmatrix}1&1&1\\1&1&1\\1&1&1\end{bmatrix}$$

Where $\vee$ indicates that the corresponding elements of the matrix are disjunctive.

Theorem 3.7.6　Let $R\subseteq X\times X$, then $(R\cup I_X)^n=(R\cup R^2\cup...\cup R^n)\cup I_X$.

The closure of a binary relation is still a binary relation, and its closure can be found. For example, R is a binary relation on A, $r(R)$ is its reflexive closure, and we can also find the symmetric closure of $r(R)$. The symmetric closure of $r(R)$ is denoted as $s(r(R))$, abbreviated as $sr(R)$, read as the symmetric closure of R's reflexive closure. Similarly, the reflexive closure $r(s(R))$ of symmetric closure of R is abbreviated as $rs(R)$, and the transitive closure $t(s(R))$ of symmetric closure of R is abbreviated as $ts(R)$, ...

R^* is usually used to denote the reflexive closure $rt(R)$ of the transitive closure of R, which is read as "R star". R^* is often used in the study of formal languages and computational models.

Theorem 3.7.7　Let $R\subseteq X\times X$, then
$$rs(R)=sr(R),\ rt(R)=tr(R),\ st(R)\subseteq ts(R)$$

3.8　Equivalence Relation and Compatible Relation

There are two very important relations in binary relation—equivalence relation and compatible relation.

Definition 3.8.1　Let $R\subseteq X\times X$, if R is reflexive, symmetric and transitive, then R is said to be equivalent on X. Let R be an equivalence relation. If $<x,y>\in R$, x is said to be equivalent to y.

Because equivalence is reflexive and symmetric, so the main diagonals of the relation matrix of equivalence relation are all 1 and symmetric. There are self-loops on every node of the graph, and if there are edges between every two nodes, there must be two edges with opposite directions.

Example 3.8.1　Let $A=\{1,2,3,4,5\}$, R is a binary relation on A, $R=\{<1,1>,<1,2>,<2,1>,<2,2>,<3,3>,<3,4>,<4,3>,<4,4>,<5,5>\}$, It is proved that R is an equivalent relation on A.

Proof　Write the relation matrix of R.

$$\boldsymbol{M}_R=\begin{bmatrix}1&1&0&0&0\\1&1&0&0&0\\0&0&1&1&0\\0&0&1&1&0\\0&0&0&0&1\end{bmatrix}$$

The main diagonals of $\boldsymbol{M}_R$ are all 1 and symmetric, so R is reflexive and symmetric. It can also be proved that R is transitive by the definition of transitivity of binary relation. So R is an equivalence relation to A.

We can also use a relation graph to show that R is equivalent. Figure 3.8.1 is a graph of R. In the relation graph of R, every node has its own loop; if there is an edge between every two nodes, there must be two opposite sides. So R is reflexive and symmetric. As before, we can prove that R is transitive by the definition of transitivity of binary relation. So R is an equivalence relation to A.

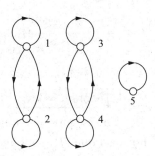

Figure 3.8.1

It is not difficult to see from Figure 3.8.1 that the graph of equivalence relation R is divided into three disconnected parts. The nodes in each part are related, while any two nodes in different parts are not.

Let x and y be two integers, and k be a positive integer. If x, y are divided by the remainder of k, then it is called x and y module k congruence, also known as the equivalence of x and y module k. As $x \equiv y \bmod k$.

Let the quotient of $x(y)$ divided by k be $t_1(t_2)$ and the remainder $a_1(a_2)$, mathematically, $x(y)$ is expressed as:

$$x = k \times t_1 + a_1, \ t_1 \in I, \ a_1 \in I \text{ and } 0 \leqslant a_1 < k$$
$$y = k \times t_2 + a_2, \ t_2 \in I, \ a_2 \in I \text{ and } 0 \leqslant a_2 < k$$

If the remainder of x, y divided by k is equal, $x - y = k \times (t_1 - t_2)$, $t_1 - t_2 \in I$.

That is, $x - y$ can be divisible by k. Therefore, k Congruences of x and y modules can also be described as $x - y$ can be divisible by k.

Example $\boxed{3.8.2}$ Let $R = \{<x, y> | x \in I \wedge y \in I \wedge x \equiv y \bmod k\}$ is a binary relation on the set of integers I. It is proved that R is equivalent.

Proof Let a, b, c be arbitrary integers.

(1) Because $a - a = k \times 0$, so $a \equiv a \bmod k$, $<a, a> \in R$, thus R is reflexive.

(2) If $<a, b> \in R$, then $a \equiv b \bmod k$, $a - b = k \times t$, $t \in I$, $b - a = -(a - b) = k \times (-t)$, $-t \in I$, $b \equiv a \bmod k$, $<b, a> \in R$. So R is symmetric.

(3) If $<a, b> \in R$ and $<b, c> \in R$, $a - b = k \times t_1$, $t_1 \in I$, $b - c = k \times t_2$, $t_2 \in I$, then $a - c = (a - b) + (b - c) = k \times t_1 + k \times t_2 = k \times (t_1 + t_2)$, $t_1 + t_2 \in I$, $<a, c> \in R$ so R is transitive. So R is an equivalence relation on the set of integers I.

Definition 3.8.2 Let R is an equivalence relation to X, $\forall x \in X$, and the set $\{y | y \in X \wedge xRy\}$ is called the R equivalent class formed by x. It is denoted as $[x]_R$.

In Example 3.8.1, the equivalence class of equivalence relation R is: $[1]_R = [2]_R = \{1, 2\}$, $[3]_R = [4]_R = \{3, 4\}$, $[5]_R = \{5\}$. In the graph of R (Figure 3.8.1), there are three disconnected parts, and all nodes in each part form an equivalent class.

The equivalence class of the above module 3 equivalence relation is called module 3 equivalence class. There are three equivalent classes of module 3:

$$[0]_R = \{..., -6, -3, 0, 3, 6, ...\}$$
$$[1]_R = \{..., -5, -2, 1, 4, 7, ...\}$$
$$[2]_R = \{..., -4, -1, 2, 5, 8, ...\}$$

Theorem 3.8.1   Let R be an equivalence relation to X, $\forall x \in X$, then there is $[x]_R \subseteq X$, $x \in [x]_R$.

Theorem 3.8.2   Let R be an equivalence relation to X, for any element a and b of X, the necessary and sufficient condition of aRb is $[a]_R = [b]_R$.

Proof   Let aRb, prove that $[a]_R = [b]_R$.

$\forall c \in [a]_R$, aRc, from the symmetry of R has cRa, from the conditional aRb and the transitivity of R, we can get cRb According to the symmetry of R, there are bRc, $c \in [b]_R$, so $[a]_R \subseteq [b]_R$.

Similarly, it can be proved that $[b]_R \subseteq [a]_R$. This proves that $[a]_R = [b]_R$.

Let $[a]_R = [b]_R$, prove that aRb.

From Theorem 3.8.1, we know $a \in [a]_R$, $[a]_R = [b]_R$, so $a \in [b]_R$, bRa, the symmetry of R has aRb.

Definition 3.8.3   Let R be an equivalence relation to X, the set $\{[x]_R \mid x \in X\}$ is called the quotient set of X concerning R. It is denoted as X/R.

In Example 3.8.1, the equivalence relation R has three equivalence classes: $[1]_R = [2]_R = \{1, 2\}$, $[3]_R = [4]_R = \{3, 4\}$, $[5]_R = \{5\}$, the quotient set of A on R

$$A/R = \{[1]_R, [3]_R, [5]_R\} = \{\{1, 2\}, \{3, 4\}, \{5\}\}$$

The equivalence classes of module 3 equivalence relation R are as follows: $[0]_R$, $[1]_R$ and $[2]_R$, the quotient set I/R of the integer set I concerning R determined by it is $I/R = \{[0]_R, [1]_R, [2]_R\}$.

Definition 3.8.4   Let a nonempty set A, $S = \{A_1, A_2, ..., A_m\}$, where $A_i \subseteq A$ and $A_i \neq \varnothing$, $(i = 1, 2, ..., m)$, if

(1) $A_i \cap A_j = \varnothing$, when $i \neq j$,

(2) $\bigcup_{i=1}^{m} A_i = A$.

Then S is a partition of a set A, each A_i a block of this partition.

Theorem 3.8.3   Let R be an equivalence relation to X, and the quotient set X/R of X concerning R is a partition of X.

Theorem 3.8.4   Let $S = \{S_1, S_2, ..., S_m\}$ is a partition of X, $R = \{<x, y> \mid x \text{ and } y \text{ are in the same partition block}\}$, then R is an equivalence relation to X.

Example 3.8.3   Let $X = \{1, 2, 3, 4\}$, the partition of X is $S = \{\{1\}, \{2, 3\}, \{4\}\}$, try to write the equivalence relation R of S derivation.

Solution $R = \{<1,1>, <2,2>, <2,3>, <3,2>, <3,3>, <4,4>\} = \{1\} \times \{1\} \cup \{2,3\} \times \{2,3\} \cup \{4\} \times \{4\}$

It can be verified that R is an equivalence relation to X.

Theorem 3.8.5 Let R, S set X be equivalent, then $R=S$ if and only if $X/R=X/S$.

Example 3.8.4 Let $X=\{1, 2, 3\}$, write all equivalence relations on set X.

Solution Write out all the partitions on set X, which are:

$$S_1=\{\{1,2,3\}\}, \quad S_2=\{\{1,2\},\{3\}\}$$
$$S_3=\{\{1,3\},\{2\}\}, \quad S_4=\{\{2,3\},\{\{1\}\}$$
$$S_5=\{\{1\},\{2\},\{3\}\}$$

The corresponding equivalence relation is as follows:

$$R_1=\{1,2,3\}\times\{1,2,3\}=X\times X$$
$$R_2=\{\{1,2\}\times\{1,2\}\}\bigcup\{\{3\}\times\{3\}\}$$
$$=\{<1,1>,<1,2>,<2,1>,<2,2>,<3,3>\}$$
$$R_3=\{\{1,3\}\times\{1,3\}\}\bigcup\{\{2\}\times\{2\}\}$$
$$=\{<1,1>,<1,3>,<3,1>,<3,3>,<2,2>\}$$
$$R_4=\{\{2,3\}\times\{2,3\}\}\bigcup\{\{1\}\times\{1\}\}$$
$$=\{<2,2>,<2,3>,<3,2>,<3,3>,<1,1>\}$$
$$R_5=\{\{1\}\times\{1\}\}\bigcup\{\{2\}\times\{2\}\}\bigcup\{\{3\}\times\{3\}\}$$
$$=\{<1,1>,<2,2>,<3,3>\}=I_X$$

Definition 3.8.5 Let $R\subseteq X\times X$, if R is reflexive and symmetric, R is said to be a compatible relation on X.

According to Definition 3.8.5, the compatible relation has the following three properties:

(1) All equivalence relations are compatible.

(2) The main diagonals of the compatible relation matrix are all 1 and symmetric.

(3) In a compatible graph, there are self-loops at every node, and if there are edges between every two nodes, there must be two edges in opposite directions.

Example 3.8.5 Let $A=\{316, 347, 204, 678, 770\}$, the binary relation R on A is defined as: $R=\{<x, y>\mid x\in A\wedge y\in A\wedge x$ and y have the same number$\}$, it is proved that R is a compatible relation on A. Write the relation matrix and diagram. It is proved that R is a compatible relation with A by the relation matrix and graph.

Proof Each number in set A has the same number as itself, so R is reflexive; for any x and y in set A, if x and y have the same number, then y and x also have the same number. So R is symmetric. Thus, R is compatible.

Let $a=316$, $b=347$, $c=204$, $d=678$, $e=770$. R is expressed by enumeration as follows:

$$R=\{<a,a>,<a,b>,<a,d>,<b,a>,<b,b>,<b,c>,<b,d>,<b,e>,$$
$$<c,b>,<c,c>,<c,e>,<d,a>,<d,b>,<d,d>,<d,e>,<e,b>,$$
$$<e,c>,<e,d>,<e,e>\}$$
$$=\{<a,b>,<a,d>,<b,a>,<b,c>,<b,d>,<b,e>,<c,b>,<c,e>,$$
$$<d,a>,<d,b>,<d,e>,<e,b>,<e,c>,<e,d>\}\bigcup I_A$$

The main diagonals of R M_R are all 1 and symmetric, which shows that R is compatible.

The relation graph of R is shown in Figure 3.8.2. In the diagram, every node has its own loop, and if there is an edge between every two nodes, there must be two opposite sides. It is shown that R is compatible.

$$M_R = \begin{bmatrix} 1 & 1 & 0 & 1 & 0 \\ 1 & 1 & 1 & 1 & 1 \\ 0 & 1 & 1 & 0 & 1 \\ 1 & 1 & 0 & 1 & 1 \\ 0 & 1 & 1 & 1 & 1 \end{bmatrix}$$

Because there are self-loops on every node of a compatible graph, and if there are edges between every two nodes, there must be two edges in opposite directions. In this way, the graph with two vertices can be reduced to an edge-free graph. This graph is called a simplified graph of R. A simplified graph of the compatible relation R in Example 3.8.2 is shown in Figure 3.8.3.

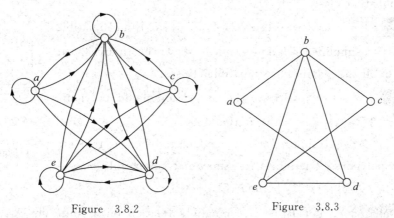

Figure　3.8.2　　　　　　　　　Figure　3.8.3

Definition 3.8.6　　Let R be a compatible relation on X, $C \subseteq X$, if $\forall a$, $b \in C$, there is $<a$, $b> \in R$, then C is a compatible class generated by compatible relation R.

If R is a compatible relation on X, C is a compatible class generated by R. It can be seen from the definition that: ①The compatible class C must be a subset of X. ② $\forall x \in X$, because the compatible relation R is reflexive, $<x$, $x> \in R$, so $\{x\}$ is a compatible class generated by the compatible relation R. In other words, the single element set composed of any element in X is a compatible class generated by compatible relation R.

In Example 3.8.5, $\{a\}$,$\{a,b\}$,$\{b,c\}$,$\{b,d,e\}$ are compatible classes generated by R.

$\{a,b\} \cup \{d\} = \{a,b,d\}$ and $\{b,c\} \cup \{e\} = \{b,c,e\}$ are also compatible classes generated by R.

However, the union of $\{b, d, e\}$ and any nonempty set is no longer a compatible class produced by R, his kind of compatible class is called maximal compatible class.

Definition 3.8.7　　Let R be a compatible relation on X and C be a compatible class generated by R. If it is not a proper subset of any other compatible classes, then C is called a maximal compatible class. It was recorded as C_R.

According to Definition 3.8.7, the maximum compatible class C_R has the following properties:

① Any element x in C_R is compatible with all elements in C_R.

② None of the elements in X-C_R has a compatible relationship with all elements in C_R.

Property ① means that C_R is a compatible class produced by R, that is, the maximal compatible class is the compatible class first. Property ② means that C_R is the maximal compatible class.

The maximum compatible class can be obtained by using the simplified graph of compatibility relation:

① The set of vertices of the largest complete polygon is the maximal compatible class.

② The set of isolated points is the maximal compatible class.

③ If an edge is not an edge of any complete polygon, the set of its two endpoints is the maximum compatible class.

Example 3.8.6 Let $X=\{1, 2, 3, 4, 5, 6\}$, R is a compatible relation on X. its simplified relation graph is shown in Figure 3.8.4. Try to find out all the maximum compatible classes.

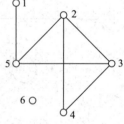

Figure 3.8.4

Solution The set of vertices of the largest complete 3-polygon: $\{2, 3, 5\}$ and $\{2, 3, 4\}$. The set of outliers $\{6\}$.

The set $\{1, 5\}$, is composed of two endpoints of edges that are not complete polygons. Therefore, the maximal compatible classes are: $\{2, 3, 5\}$, $\{2, 3, 4\}$, $\{1, 5\}$ and $\{6\}$.

Theorem 3.8.6 Let R be a compatible relation on a finite set X, and C be a compatible class generated by R, then there must be a maximum compatible class C_R such that $C \subseteq C_R$.

Proof Let $X=\{x_1, x_2, ..., x_n\}$, let $C_0 = C$. The sequence of compatible classes is constructed as follows: $C_0 \subset C_1 \subset C_2 \subset ...$

$$C_{i+1} = C_i \bigcup \{x_j\}$$

Where: j is the minimum subscript of x such that $x_j \notin C_i$ and each element of x_j and C_i have a compatible relation R.

Because $|X| = n$, the process can be ended by at most $n - |C|$ steps. The last set of sequences is the required maximum compatible class C_R.

Definition 3.8.8 Let a nonempty set A, $S = \{A_1, A_2, ..., A_m\}$, where $A_i \subseteq A$ and $A_i \neq \varnothing$, if $\bigcup_{i=1}^{m} A_i = A$, then S is a cover of geometry A.

Theorem 3.8.7 Let X be a finite set, R be a compatible relation on X, and the set composed of all the maximal compatible classes generated by R is the covering of X.

Proof　Let the set of all maximal compatible classes generated by R be$\{C_1, C_2, ..., C_n\}$.

(1) According to the definition of compatible class, any maximal compatible class C_i is a subset of X.

(2) It is proved that $X=C_1\cup C_2\cup...\cup C_n$.

Because of $C_j\subseteq X$, $C_1\cup C_2\cup...\cup C_n\subseteq X$.

$\forall x\in X$, $\{x\}$ are compatible classes. According to Theorem 3.8.6, there must be a maximal compatible class C_i such that $\{x\}\subseteq C_i$, and $C_i\subseteq C_1\cup C_2\cup...\cup C_n$.

So $x\in C_1\cup C_2\cup...\cup C_n$.

This proves that $X\subseteq C_1\cup C_2\cup...\cup C_n$.

So $X=C_1\cup C_2\cup...\cup C_n$.

Definition 3.8.9　Let R be a compatible relation on X, $S=\{R\,|\,\text{Maximal compatible classes}$ generated by $R\}$ is called complete covering of set X, denoted by $C_R(X)$.

For example, in Example 3.8.6, the set formed by the maximum compatible class is $\{\{2,3,5\},\{2,3,4\},\{1,5\},\{6\}\}$, which is a complete cover of the set $X=\{1, 2, 3, 4, 5, 6\}$. According to the definition of compatible relation and compatible class, a single element set composed of any element in X is a compatible class generated by compatible relation R. Therefore, all of $\{1\},\{2\},\{3\},\{4\},\{5\},\{6\}$ are all compatible classes, set $\{\{1\},\{2\},\{3\},\{4\},\{5\},\{6\}\}$ are the covers of X. In other words, the complete cover is the cover of X, and some compatible classes constitute the set of X. Therefore, the coverage of X composed of compatible classes generated by compatible relation R is not unique. But a compatible relation only corresponds to a unique complete covering.

Theorem 3.8.8　Let $S=\{S_1, S_2, ..., S_m\}$ be a covering of X, then $R=(S_1\times S_1)\cup(S_2\times S_2)\cup...\cup(S_m\times S_m)$ is a compatible relation on X.

Example 3.8.7　Let $X=\{1,2,3,4\}$, $S_1=\{\{1,2,3\},\{3,4\}\}$, $S_2=\{\{1,2\},\{2,3\},\{1,3\},\{3,4\}\}$ be two covers of X. Try to write the compatible relation R_1 and R_2 derived from S_1 and S_2.

Solution

$R_1=\{\{1,2,3\}\times\{1,2,3\}\}\cup\{\{3,4\}\times\{3,4\}\}$
$=\{<1,1>,<1,2>,<1,3>,<2,1>,<2,2>,<2,3>,<3,1>,<3,2>,<3,3>,$
$<3,4>,<4,3>,<4,4>\}$
$R_2=\{\{1,2\}\times\{1,2\}\}\cup\{\{2,3\}\times\{2,3\}\}\cup\{\{1,3\}\times\{1,3\}\}\cup\{\{3,4\}\times\{3,4\}\}$
$=\{<1,1>,<1,2>,<2,1>,<2,2>,<2,3>,<3,2>,<3,3>,<1,3>,<3,1>,$
$<3,4>,<4,3>,<4,4>\}$

In example 3.8.7, $S_1\neq S_2$, but $R_1=R_2$. This shows that different covers can derive the same compatible relationship.

Theorem 3.8.9　The compatible relation R on set X is one-to-one corresponding to the complete covering $C_R(X)$ of set X.

3.9 Partial Order Relation

In a set, we often consider the order of elements, one important relation of them is called partial order relation.

Definition 3.9.1 Let $R \subseteq X \times X$ if R is reflexive, antisymmetric and transitive, then R is said to be a partial order relation on X. It is recorded as $\leqslant$. The doublet $<X, \leqslant>$ is called partial order set. If $<x, y> \in \leqslant$, denoted as $x \leqslant y$, and read as x is partially ordered in y.

Example 3.9.1 Let A be a set and $P(A)$ be a power set of A. The inclusion relation $\subseteq$ on $P(A)$ is defined as follows:
$$R_\subseteq = \{<x,y>\} x \in P(A) \wedge y \in P(A) \wedge x \subseteq y\}$$
It is proved that $\subseteq$ is a partial order relation on $P(A)$.

Proof

(1) $\forall x \in P(A), x \subseteq x, <x,x> \in \subseteq$. So $\subseteq$ is reflexive.

(2) Let $<x,y> \in \subseteq \wedge <y,x> \in \subseteq$, defined by the inclusion relation $\subseteq$, $x \subseteq y \wedge y \subseteq x$, according to the definition of set equality, $x = y$. In other words, $\subseteq$ is antisymmetric.

(3) Let $<x,y> \in \subseteq \wedge <y,z> \in \subseteq$, the definition of inclusion relation $\subseteq$ is $x \subseteq y \wedge y \subseteq z$, according to the transitivity of set inclusion, $x \subseteq z$ is defined by inclusion relation $<x, z> \in \subseteq$, $\subseteq$ is transitive, $\subseteq$ is a partial order relation on $P(1)$.

In addition, the identity relation I_A on an arbitrary set A is reflexive, antisymmetric and transitive, I_A is a partial order relation on A; less than or equal to the real number set is also reflexive, antisymmetric and transitive, it is a partial order relation on the set of real numbers.

Definition 3.9.2 Let $<X, \leqslant>$ be a partial order set. For the elements x and y in X, if $x \leqslant y$, $x \neq y$ and no other element z in X makes $x \leqslant z$ and $z \leqslant y$, then we say that y covers x.

Example 3.9.2 Let $A = \{2, 5, 6, 10, 15, 30\}$, the division relation R on A is defined as follows:
$$R = \{<x,y> | x \in A \wedge y \in A \wedge x \text{ divisible by } Y\}$$
Verify that R is a partial order relation on A, and analyze which elements cover other elements and which elements do not cover other elements.

Solution Because any element of A can divide itself, R is reflexive; when x divides y and y divides x, there must be $x = y$, so R is antisymmetric; when x divides y and y divides z, x must divide z, so R is transitive. That is, R is a partial order relation on A.

R is expressed by enumeration:
$$R = \{<2,2>, <2,6>, <2,10>, <2,30>, <5,5>, <5,10>, <5,15>, <5,30>,$$
$$<6,6>, <6,30>, <10,10>, <10,30>, <15,15>, <15,30>, <30,30>\}$$
6 and 10 cover 2; but 30 does not cover 2 because $<2, 6> \in R$ and $<6, 30> \in R$.

10 and 15 cover 5; but 30 doesn't cover 5 because $<5, 10> \in R$ and $<10, 30> \in R$

30 covers 6, 10 and 15.

Definition 3.9.3　Let $<X, \leqslant>$ be a partial order set, and the set $\{<x,y>\mid x \in X \land y \in X \land y \text{ covers } x\}$ is called covering relation on X. Recorded as COV X.

Let COV X be a covering relation on X, if $<x, y> \in$ COV X, according to Definition 3.9.3, y covers x, from Definition 3.9.2, $x \leqslant y$, that is $<x, y> \in \leqslant$. Therefore, COV $X \subseteq \leqslant$. Therefore, any covering relation COV X is a subset of the corresponding partial order relation $\leqslant$.

In Example 3.9.2, the covering relation of a poset $<A, R>$

COV $A = \{<2,6>,<2,10>,<5,10>,<5,15>,<6,30>,<10,30>,<15,30>\}$

Let $<X, \leqslant>$ be a partial order set, and its covering relation COV X is unique. Therefore, we can use the covering relation to construct a graph that represents the partial order set $<X, \leqslant>$, this picture is called Hasse diagram. The Hasse diagram of posets $<X, \leqslant>$ is as follows:

① Use "∘" to represent every element in X.

② If $x \leqslant y$, draw x below y.

③ If $<x, y> \in$ COV X, draw a straight line between x and y. Hasse diagram of Example 3.9.2 is Figure 3.9.1.

Example 3.9.3　Let $A = \{a,b\}, B = P(A) = \{\varnothing, \{a\}, \{b\}, \{a,b\}\}$ be the power sets of A, and the inclusion relations on the power set $P(A)$ be given.

$$R_\subseteq = \{<x,y>\mid x \in P(A) \land y \in P(A) \land x \subseteq y\}$$

Try to write the covering relation COV $P(A)$ and draw its Hasse diagram.

Solution　$R_\subseteq$ is expressed by enumeration

$$R_\subseteq = \{<\varnothing,\{a\}>,<\varnothing,\{b\}>,<\varnothing,\{a,b\}>,<\{a\},\{a,b\}>,<\{b\},\{a,b\}>\} \cup I_B$$

In Example 3.9.1, it has been proved that $R_\subseteq$ is a partial order relation on $P(A)$. The covering relation on $P(A)$ is

COV $P(A) = \{<\varnothing,\{a\}>,<\varnothing,\{b\}>,<\{a\},\{a,b\}>,<\{b\},\{a,b\}>\}$

Hasse diagram is shown in Figure 3.9.2.

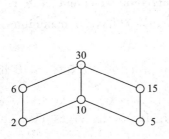

Figure　3.9.1

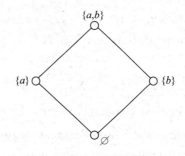

Figure　3.9.2

Definition 3.9.4　Let $<X, \leqslant>$ be a partial order set, $B \subseteq X$, $b \in B$, if no element X in B satisfies $x \neq b$ and $b \leqslant x (x \leqslant b)$, then b is said to be the maximal element (minimal element) of B.

When $B = X$, the maximal element (minimal element) of B is called the maximal element (minimal element) of a partial order set $<X, \leqslant>$.

Example 3.9.4 Let $A = \{a, b, c, d, e, f, g, h\}$, binary relation on A:

$$R = \{<b, d>, <b, e>, <b, f>, <c, d>, <c, e>, <c, f>, <d, f>, <e, f>,$$
$$<g, h>\} \cup I_A$$

Verify that R is a partial order relation on A. Write the covering relation COV A and draw Hastur. Find the maximum and minimum elements of set A.

Proof It is easy to verify that R is reflexive, antisymmetric and transitive, and that R is a partial order relation to A.

COV $A = \{<b, d>, <b, e>, <c, d>, <c, e>, <d, f>, <e, f>, <g, h>\}$

Hasse diagram is shown in Figure 3.9.3.

The maximal elements of set A are a, f, h.

The minimal element of set A are a, b, c, g.

Theorem 3.9.1 From Definition 3.9.4 and Example 3.9.4, the following conclusions can be drawn:

① An isolated point is both a maximal element and a minimal element.

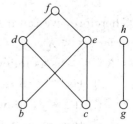

Figure 3.9.3

② The maximum and minimum elements are not unique. The maximal and minimal elements of a finite set B must exist.

③ In the Hasse diagram, if one element of set B does not exist, and other elements of B are connected with it from the upper (lower) side, then the element is the maximum element (minimum element) of B.

Definition 3.9.5 Let $<X, \leqslant>$ be a partial order set, $B \subseteq X$, $b \in B$. If there is $x \leqslant b$ $(b \leqslant x)$ for any $x \in B$, then b is said to be the largest element (minimum element) of B.

When $B = X$, the maximum element (minimum element) of B is called the maximum element (minimum element) of a poset $<X, \leqslant>$. In Example 3.9.4, set A has no maximum and minimum elements.

In Example 3.9.3, let $B = \{\varnothing, \{a\}\}$, the maximum element of B is $\{a\}$, and the minimum element of B is $\varnothing$. Let $B = \{\{a\}, \{b\}\}$, B has no maximum and minimum elements.

Theorem 3.9.2 Let $<X, \leqslant>$ be a partial order set, $B \subseteq X$, if B has the largest element (the smallest element), it must be unique.

Proof Let a and b be the largest element of B. If a is the largest element of B, we can get $b \leqslant a$; if b is the largest element of B, we can get $a \leqslant b$, because $\leqslant$ is antisymmetric, so $a = b$. The largest element of B is unique.

Similarly, it can be proved that the minimum element is unique.

From the above examples and Theorem 3.9.2, the following conclusions can be drawn:

① The maximum and minimum elements do not necessarily exist. If it exists, it must be unique.

② In the Hasse diagram, if an element of set B goes down (up) to all elements of B, then that element is the largest element (smallest element) of B.

Definition 3.9.6 Let $<X,\ \leqslant>$ be a partial order set, $B\subseteq X$, $b\in X$, for any $x\in B$, all $x\leqslant b(b\leqslant x)$, then b is said to be the upper bound (lower bound) of B.

Example 3.9.5 Let $A=\{2,3,6,12,24,36\}$, and the division relation $R=\{<a,b>|$ $a\in A \wedge b\in A \wedge a$ can divide $b\}$, It is a partial order relation on A, try to find the covering relation COV A, draw Hasse diagram, and determine the upper and lower bounds of the following sets.

(1) $B_1=\{2,\ 3,\ 6\}$;

(2) $B_2=\{12,\ 24,\ 36\}$

Solution R is expressed by enumeration as follows:

$R=\{<2,6>,<2,12>,<2,24>,<2,36>,<3,6>,<3,12>,<3,24>,<3,36>,$
$\quad<6,12>,<6,24>,<6,36>,<12,24>,<12,36>\}\cup I_A$

Covering relation COV $A=\{<2,6>,<3,6>,<6,12>,<12,24>,$
$<12,36>\}$

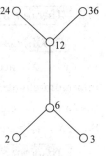

Hasse diagram is shown in Figure 3.9.4.

The upper bound of B_1 is 6, 12, 24, 36. There is no lower bound.

The lower bound of B_2 is 2, 3, 6, 12. There is no upper bound.

It can be seen from this example that:

(1) The upper and lower bounds are not unique.

(2) In the Hasse diagram, if an element of set X goes down (up) to all elements of B, then that element is the upper (lower) bound of B.

Figure 3.9.4

Definition 3.9.7 Let $<X,\ \leqslant>$ be a partial order set, $B\subseteq X$, if there is an upper bound (lower bound) b of B, for any upper bound (lower bound) y of B, there is $b\leqslant y(y\leqslant b)$, then b is said to be the least upper bound (greatest lower bound) of B, also called supremum (infimum), recorded as LUB B(GLB B).

In Example 3.9.5, the minimum upper bound of B_1 is 6; B_1 has no lower bound, of course, no maximum lower bound. The maximum lower bound of B_2 is 12; B_2 has no upper bound, of course, no minimum upper bound.

Definition 3.9.8 Let $<X,\ \leqslant>$ be a partial order set. If $\forall x\in X$, $\forall y\in X$ have $x\leqslant y$ or $y\leqslant x$, then it is called partial order set $<X,\ \leqslant>$ is totally ordered, also known as a linear order set. Partial order relation $\leqslant$ is called total order relation or linear order relation on X.

Example 3.9.6 Let $P=\{\varnothing,\{a\},\{a,b\},\{a,b,c\}\}$, the inclusion relation on P $R_\subseteq=\{<x,y>|x\in P \wedge y\in P \wedge x\subseteq y\}$. Verify that $R_\subseteq$ is a total order relation.

Proof The inclusion relation $R_\subseteq$ on P is expressed by enumeration:

$R_\subseteq=\{<\varnothing,\{a\}>,<\varnothing,\{a,b\}>,<\varnothing,\{a,b,c\}>,<\{a\},\{a,b\}>,<\{a\},\{a,b,c\}>,$
$\quad<\{a,b\},\{a,b,c\}>\}\cup I_P$

It is easy to verify that $R_\subseteq$ is a partial order relation on P.

From the ordered pair set of $R_\subseteq$ we can see that for any two elements x and y in P, there must be $x\subseteq y$ or $y\subseteq x$, so $R_\subseteq$ is the total order relation on P. The covering relationship is as follows:

COV $P = \{<\varnothing,\{a\}>,<\{a\},\{a,b\}>,<\{a,b\},\{a,b,c\}>\}$

Hasse diagram is shown in Figure 3.9.5.

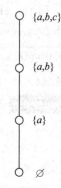

Figure 3.9.5

Definition 3.9.9 Let $<X,\leqslant>$ be a partial order set. If every nonempty subset of X has a minimum element, then the posets $<X,\leqslant>$ are called well-ordered sets. Partial order relation $\leqslant$ is called good order relation on X.

Similar to Example 3.9.1, it can be proved that the less than or equal relationship on the set of natural numbers $\mathbf{N}$ is a partial order relation, and every nonempty subset of the set of natural numbers has a minimum number. Therefore, the less than or equal to relation $<\mathbf{N}, R_\leqslant>$ on the set of natural numbers is a well-ordered set.

Theorem 3.9.3 Every well-ordered set must be a totally ordered set.

Proof Let $<X,\leqslant>$ be a well-ordered set, x and y be any two elements in X. obviously $\{x, y\}\subseteq X$, according to the definition of well ordered set, the set $\{x, y\}$ must have a minimum element, that is, x is the minimum element or y is the minimum element. If x is the minimum element, then there is $x\leqslant y$. If y is the smallest element, then there is $y\leqslant x$. So $<X,\leqslant>$ are totally ordered sets.

Theorem 3.9.4 A finite total ordered set must be a well-ordered set.

3.10 Application of Set and Relation

Set theory is applied in all aspects of computer science research. Set is the basis of constructing discrete structure, and discrete structure is the basic structure of the computer. Set theory has important applications in artificial intelligence, logic and programming language.

Set theory plays a "foundation" role in mathematical theory. It is of great significance in analytic geometry, solution set and matrix operation. For example, the combination of geometric elements, parallel and vertical is a certain relationship between sets; the expansion process from natural number set→rational number set→real number set can be obtained by classifying the former set according to some equivalent relation of a set.

The translation, rotation, reflection, similarity and other geometric transformations of plane geometry are the corresponding relations between the sets in $\mathbf{R}^2$ that meet certain conditions.

1. Applications in Geometry

For example, the circle and open (closed) surface with the origin as the center of the circle and 1 as the radius can be expressed as follows:

$$A = \{(x,y) \mid x^2 + y^2 = 1, (x,y) \in \mathbf{R}^2\}$$
$$A' = \{(x,y) \mid x^2 + y^2 < 1, (x,y) \in \mathbf{R}^2\}$$
$$A'' = \{(x,y) \mid x^2 + y^2 \leqslant 1, (x,y) \in \mathbf{R}^2\}$$

Another example is the description of the "purity" and "completeness" of the curve equation in analytic geometry: "the coordinates of the points on the curve are all the solutions of the equation"; and "the points with the solution of the equation as the coordinates are all the points of the curve". In the language of set theory, it can be simply expressed as $C = \{(x, y) \mid f(x, y) = 0, (x, y) \in \mathbf{R}^2\}$, that is, the point on the curve satisfies the equation $C \subset \{(x,y) \mid f(x,y) = 0, (x,y) \in \mathbf{R}^2\}$, and the point satisfying the equation is $C \supset \{(x,y) \mid f(x,y) = 0, (x,y) \in \mathbf{R}^2\}$.

2. Applications in Operation

(1) Geometric figure property operation:

$$\{x \mid x \text{ is a rectangle}\} \bigcap \{x \mid x \text{ is a diamond}\} = \{x \mid x \text{ is a square}\}$$

(2) Number operation on the number axis:

$$\{x \mid x - 1 > 0\} \bigcap \{x \mid 0 < x < 10\} = \{x \mid 1 < x < 10\}$$

(3) To solve the equations

$$\begin{cases} x - y + 1 = 0 \\ 3x + y - 9 = 0 \end{cases}$$

That is, the intersection coordinates of the two lines: $\{(x, y) \mid x - y + 1 = 0 \text{ and } 3x + y - 9 = 0\}$.

(4) To solve the inequality group:

$$\begin{cases} x^2 + y^2 \geqslant 1 \\ y \leqslant x \end{cases}$$

(5) Operation of difference set and complement set:

$$A - B = \{x \mid x \in A \text{ and } x \notin B\}$$

It is obvious by definition: $A - B \neq B - A$.

Let $A = \{x \mid x \text{ is a freshman in Jiangsu University}\}$, $B = \{x \mid x \text{ is a female freshman in Jiangsu University}\}$, $C = \{x \mid x \text{ is a female student in Jiangsu University}\}$, $D = \{x \mid x \text{ is a student of Jiangsu University}\}$, then there are the following operations:

$A - B = \{x \mid x \text{ is a freshman in Jiangsu University}\}$

$C - B = \{x \mid x \text{ is a female student except for freshman in Jiangsu University}\}$

$D - B = \{x \mid x \text{ is a student of Jiangsu University except for a female freshman}\}$

(6) Inclusive and exclusive problems:

For Example 3.2.1, the number of personnel meeting specific conditions can be calculated by using the inclusion-exclusion principle, and the results can be obtained more clearly and quickly by combining with the Venn diagram.

The relation is a relation between the elements of a set, that is to say, a relation is actually a subset of a set (relation is a subset of Cartesian product). For example, the relationship between employees and their wages.

3. Applications in Database Technology

A binary relation is widely used in database technology.

Database technology is widely used in various fields of society, an relational database (such as MySQL) has become the mainstream database. Cartesian product in set theory is a purely mathematical theory and is an important method to study relational databases, showing an irreplaceable role. It not only provides theoretical and methodological support but also promotes the research and development of database technology. The relational data model is based on strict set algebra, and its logical structure is a two-dimensional table composed of rows and columns to describe the relational data model. Binary relation theory is used to study the possible relationship between domains in an entity set, the determination and design of table structure, the realization of data query and maintenance function of relation operation, lossless connectivity analysis of relation decomposition and connection dependence.

Each element in a relationship is a tuple in the relationship, which is usually represented by t. The number of tuples in the relationship is the cardinality of the relationship.

Since the relation is a subset of the Cartesian product, it can also be regarded as a two-dimensional table, and the relation with the same relation frame is called the same kind of relation.

The nature of the relationship: A normalized set of rows in a two-dimensional table.

(1) The components in each column must come from the same domain and must be the same type of data.

(2) Different columns can come from the same domain. Each column becomes an attribute. Different attributes must have different names.

(3) The order of columns can be exchanged arbitrarily, and the names can be changed at the same time.

(4) The order of tuples in a relation can be arbitrary.

(5) Each component in a relationship must be an indivisible data item.

Exercises

1. Let $A = \{a, \{a\}\}$, which is wrong with the following proposition? (   )

(1) $\{a\} \in P(A)$

(2) $\{a\} \subseteq P(A)$

(3) $\{\{a\}\} \in P(A)$

(4) $\{\{a\}\} \subseteq P(A)$

2. Fill in the correct symbol between 0(   )$\varnothing$.

(1) $=$ (2) $\subseteq$ (3) $\in$ (4) $\notin$

3. If the cardinality of set S, $|S| = 5$, then the cardinality of the power set of S, $|P(s)| =$ (   ).

4. Let $P=\{x\mid (x+1)^2\leqslant 4,\ \text{and}\ x\in R\}$, $Q=\{x\mid 5\leqslant x^2+16\ \text{and}\ x\in R\}$, Which is right with the following proposition? (   )

(1) $Q\subset P$　　　　(2) $Q\subseteq P$　　　　(3) $P\subset Q$　　　　(4) $P=Q$

5. Which of the following sets are equal? (   )

(1) $A_1=\{a,b\}$　　(2) $A_2=\{b,a\}$　　(3) $A_3=\{a,b,a\}$　　(4) $A_4=\{a,b,c\}$

6. If $A-B=\varnothing$, which of the following is not possibly right? (   )

(1) $A=\varnothing$　　　　(2) $B=\varnothing$　　　　(3) $A\subset B$　　　　(4) $B\subset A$

7. Let $A\cap B=A\cap C$, $\overline{A}\cap B=\overline{A}\cap C$, then $B($ $)C$.

8. Which of the following statements is true? (   )

(1) $A-B=B-A\Rightarrow A=B$.

(2) An empty set is a proper subset of any set.

(3) An empty set is a subset of any nonempty set.

(4) If an element of A belongs to B, then $A=B$.

9. Which of the following statements is correct? (   )

(1) All empty sets are not equal.

(2) $\{\varnothing\}\neq\varnothing$.

(3) If A is a nonempty set, then $A\subset A$ holds.

10. Let $A=\{1,2,3,4,5,6\}$, $B=\{1,2,3\}$, the relation from A to B, $R=\{<x,y>\mid x=y^2\}$, find: (1)R; (2)R^{-1}.

11. This paper gives an example of the equivalence relation and partial order relation on set A.

12. What are the three properties of an equivalence relation on set A?

13. Let $S=\{1,2,3,4\}$, the relation on A, $R=\{<1,2>,<2,1>,<2,3>,<3,4>\}$, find:(1)$R\circ R$; (2)$R^{-1}$.

14. Let $A=\{1,2,3,4,5,6\}$, $B=\{1,2,3\}$, the relation from A to B, $R=\{<x,y>\mid x=2y\}$, find: (1)R; (2)R^{-1}.

15. Let $A=\{1,2,3,4,5,6\}$, $B=\{1,2,3\}$, the relation from A to B, $R=\{<x,y>\mid x=y^2\}$, find the relation matrix of R and R^{-1}.

16. The property of relation $R=\{<x,y>\mid x+y=10,\ x,\ y\in A\}$ on set $A=\{1,2,3,\ldots,10\}$ is (   ).

(1) reflexive　(2) symmetrical　(3) transitive, symmetric　(4) transitive

17. If r_1 and r_2 are reflexive relation on a, then there are (   ) reflexive relations in $r_1\cup r_2$, $r_1\cap r_2$, r_1-r_2.

18. Let F, G, R be the binary relation on A, prove that:

(1) $R\circ(F\cup G)=R\circ F\cup R\circ G$

(2) $R\circ(F\cap G)=R\circ F\cap R\circ G$

(3) $R\circ(F\circ G)=R\circ(F\circ G)$

第 3 章　集合与关系

　　集合是不能精确定义的基本概念,简单来说,把一些事物放在一起组成一个整体叫作集合,而这些事物就是这个集合的元素或成员。集合的元素必须是确定的。所谓确定的,是指任何一个对象是否为集合的元素是明确的、确定的,不能模棱两可。例如,一个学校里的学生、全国的大学生、一个图书馆的书本数量等都可以构成一个集合。

3.1　集合的概念与表示

　　定义 3.1.1 ▶一些确定的、能区分的对象的全体是集合,通常用大写的英文字母表示。组成集合的对象叫作集合的元素或成员,常用小写的英文字母表示。

　　集合中的元素是能区分的。能区分是指集合中的元素是互不相同的。如果一个集合中有几个元素相同,应算作一个。例如,集合$\{a,b,b,d\}$和集合$\{a,b,d\}$是同一集合。

　　集合中的元素是任意的对象。对象是可以独立存在的具体的或抽象的客体。它可以是独立存在的数、字母、人或其他物体,也可以是抽象的概念,当然也可以是集合。例如,集合$\{1,b,\{c\},\{a\}\}$的元素$\{c\}$和$\{a\}$就是集合。

　　集合中的元素又是无序的,即集合$\{a,c,b\}$和集合$\{a,b,c\}$是同一集合。

　　设 S 是集合,a 是 S 的一个元素,记为 $a \in S$,读作"a 属于 S",也可读作"a 在 S 中"或"S 包含 a"。如果 a 不是 S 的元素,记为 $a \notin S$,读作"a 不属于 S",也可读作"a 不在 S 中"。

　　例如,

　　(1) 26 个英文字母组成一个集合,任一英文字母是该集合中的元素。

　　(2) 直线上的所有点组成实数集合 **R**,每一个实数是集合 **R** 的元素。

　　(3) 江苏大学全体学生组成一个集合,该校的每一个学生是这个集合中的元素。

　　集合有以下三种表示法。

　　第一种表示法是列举法:在花括号"{ }"中列举出该集合中的元素,元素之间用逗号隔开。例如,$I = \{1,2,3,4,5\}$,$A = \{a,b,3,\cdots\}$,$C = \{0,1,-1,\{a\},\{c,d\},\cdots\}$,$S = \{T,F\}$。

　　第二种表示法是描述法:用谓词界定集合的元素。例如,$\mathbf{Q} = \{x \mid x$ 是有理数$\}$,$\mathbf{R} = \{x \mid x$ 是实数$\}$,$\mathbf{C} = \{x \mid x$ 是复数$\}$,$A = \{x \mid x \in I \wedge 0 < x \wedge x < 5\}$。

　　若用 $P(x)$ 表示 x 是有理数,那么集合 $\mathbf{Q}$ 又可表示为 $\mathbf{Q} = \{x \mid P(x)\}$。

　　一般地说,集合可用描述法表示为 $S = \{x \mid A(x)\}$。其中,$A(x)$ 是谓词。显然,当 $a \in S$ 时,$A(a)$ 为真;反之,当 $A(a)$ 为真时,$a \in S$。即 $a \in S$ 的充分必要条件是 $A(a)$ 为真。

　　在中学的教科书中将自然数定义为 $\mathbf{N} = \{1,2,3,\cdots\}$。

　　在离散数学中,认为自然数是由 0 开始的,即 $\mathbf{N} = \{0,1,2,3,\cdots\}$。

这种由 0 开始的自然数集叫作扩展的自然数集。离散数学中使用扩展的自然数集。本书中的自然数集是指扩展的自然数集。

定义 3.1.2 ▶具有有限个元素的集合叫作有限集，否则叫作无限集。有限集元素的个数称为该集合的基数，也叫作集合的势。有限集 A 的基数记为 $|A|$。

例如，设 $A=\{1,a,\{b,c\},5\}$，A 是有限集，A 的基数 $|A|=4$。

扩展的自然数集 $\mathbf{N}=\{0,1,2,3,\cdots\}$ 是无限集。整数集合 $\mathbf{I}$、有理数集合 $\mathbf{Q}$、实数集合 $\mathbf{R}$ 和复数集合 $\mathbf{C}$ 都是常见的无限集。

定义 3.1.3 ▶设 A,B 是任意的集合，当 A 的每一元素都是 B 的元素时，则称 A 是 B 的子集，也称 A 包含在 B 内或 B 包含 A，记为 $A\subseteq B$ 或 $B\supseteq A$。

当 A 不是 B 的子集时，记为 $A\nsubseteq B$。$A\subseteq B$ 用谓词公式表示为 $A\subseteq B\Leftrightarrow(\forall x)(x\in A\rightarrow x\in B)$
$A\nsubseteq B$ 用谓词公式表示为 $A\nsubseteq B\Leftrightarrow(\exists x)(x\in A\wedge x\notin B)$。

例如，设 $A=\{1\}$，$B=\{1,2\}$，$C=\{1,2,3\}$ 则 $A\subseteq A$；$A\subseteq B,B\subseteq C,A\subseteq C$；$C\nsubseteq B$。

可以证明，集合的包含有下列性质。

(1) 自反性：对任意集合 A，$A\subseteq A$。

(2) 传递性：对任意集合 A,B,C，当 $A\subseteq B$ 且 $B\subseteq C$ 时，$A\subseteq C$。

定义 3.1.4 ▶设 A,B 是集合，如果 $A\subseteq B$ 且 $B\subseteq A$，则称 A 与 B 相等，记为 $A=B$。如果 A 与 B 不相等，记为 $A\neq B$。

集合相等也可用谓词公式表示为
$$A=B\Leftrightarrow A\subseteq B\wedge B\subseteq A$$
$$\Leftrightarrow(\forall x)(x\in A\rightarrow x\in B)\wedge(\forall x)(x\in B\rightarrow x\in A)$$
$$\Leftrightarrow(\forall x)(x\in A\leftrightarrow x\in B)$$

例如，设 $A=\{c,e\}$，$B=\{1,\{d\}\}$，$C=\{e,c\}$ 则 $A=C,A\neq B$。

定理 3.1.1 ▶由集合相等的定义可以看出，集合相等有下列性质。

(1) 自反性：对任意集合 A，$A=A$。

(2) 对称性：对任意集合 A,B，当 $A=B$ 时，$B=A$。

(3) 传递性：对任意集合 A,B,C，当 $A=B$ 且 $B=C$ 时，$A=C$。

定义 3.1.5 ▶设 A,B 是集合，如果 $A\subseteq B$ 且 $A\neq B$，则称 A 是 B 的真子集，记为 $A\subset B$。如果 A 不是 B 的真子集，记为 $A\not\subset B$。

真子集可用谓词公式表示为
$$A\subset B\Leftrightarrow A\subseteq B\wedge A\neq B$$
$$\Leftrightarrow(\forall x)(x\in A\rightarrow x\in B)\wedge(\exists x)(x\in B\wedge x\notin A)$$

例如，设 $A=\{a\}$，$B=\{a,b\}$，$C=\{a,b,c\}$ 则 $A\subset B,B\subset C,A\subset C,A\not\subset A$。

又如，自然数集是整数集的真子集，也是有理数集合和实数集合的真子集，即 $\mathbf{N}\subset\mathbf{I},\mathbf{N}\subset\mathbf{Q},\mathbf{N}\subset\mathbf{R}$。

定义 3.1.6 ▶不包含任何元素的集合叫空集，记为 $\varnothing$。

空集可以表示为
$$\varnothing=\{x\mid P(x)\wedge\neg P(x)\}$$
其中，$P(x)$ 为任意谓词。

空集 $\varnothing$ 是不包含任何元素的集合，所以，$|\varnothing|=0$。

定理 3.1.2 ▶ 空集是任意集合的子集。

根据定理 3.1.2,空集是任意集合的子集,即 $\varnothing \subseteq A$;对任意集合 A 都有 $A \subseteq A$。一般地,任意集合 A 至少有两个子集,一个是空集 $\varnothing$,另一个是它本身 A。

推论 空集是唯一的。

证明 设有两个空集 $\varnothing_1$ 和 $\varnothing_2$,由定理 3.1.2,有 $\varnothing_1 \subseteq \varnothing_2$ 和 $\varnothing_2 \subseteq \varnothing_1$。根据集合相等的定义可知,$\varnothing_1 = \varnothing_2$。

定义 3.1.7 ▶ 设 A 是集合,A 的所有子集构成的集合称为 A 的幂集合,记为 $P(A)$,即 $P(A) = \{S \mid S \subseteq A\}$。

例 3.1.1 设 $A = \{a, b, c\}$,$\varnothing$ 是空集,试求 $P(A)$,$P(P(\varnothing))$。

解
$$P(A) = \{\varnothing, \{a\}, \{b\}, \{c\}, \{a,b\}, \{a,c\}, \{b,c\}, \{a,b,c\}\}$$
$$P(\varnothing) = \{\varnothing\}$$
$$P(P(\varnothing)) = \{\varnothing, \{\varnothing\}\}$$

定理 3.1.3 ▶ 设 A 为有限集合,则 $|P(A)| = 2^{|A|}$。

定义 3.1.8 ▶ 在一个具体问题中,如果所涉及的集合都是某个集合的子集,则称这个集合为全集,记为 E。

全集是相对的,不同的问题有不同的全集。即使是同一问题,也可以取不同的全集。

集合的第三种表示法是文氏图(Venn diagram)。人们常用文氏图描述集合运算和它们之间的关系。集合的文氏图画法如下。

用矩形表示全集 E,在矩形中画一些圆表示其他集合,不同的圆代表不同的集合。如果没有特别说明,任何两个圆彼此相交。例如,$A \subseteq B$ 的文氏图如图 3.1.1 所示。

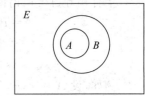

图 3.1.1

3.2 集合的运算

在了解集合的基本概念后,还需要进一步学习集合的运算。顾名思义,集合的运算就是对指定的集合根据相关规则进行运算,常见的集合运算有并、交、相对补、绝对补和对称差。

定义 3.2.1 ▶ 设 A, B 是任意的集合,由 A 中的元素或 B 中的元素组成的集合,称为 A 和 B 的并集,记为 $A \cup B$。

用描述法表示为 $A \cup B = \{x \mid x \in A \vee x \in B\}$。并集的文氏图如图 3.2.1 所示。

从并集的定义中可知:
$$A \subseteq A \cup B, \quad B \subseteq A \cup B$$

例如,$A = \{1, a\}$,$B = \{3, c\}$,则 $A \cup B = \{1, 3, a, c\}$。

定义 3.2.2 ▶ 设 A, B 是集合,由 A 与 B 的公共元素组成的集合,称为 A 和 B 的交集,记为 $A \cap B$。

用描述法表示为 $A \cap B = \{x \mid x \in A \wedge x \in B\}$,交集的文氏图如图 3.2.2 所示。

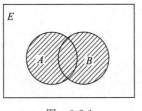
图　3.2.1

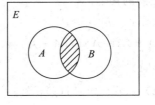
图　3.2.2

从交集的定义中可知：
$$A \cap B \subseteq A, \quad A \cap B \subseteq B$$

如果 A 与 B 无公共元素，即 $A \cap B = \varnothing$，则称 A 和 B 是互不相交的。

例如，令 $A = \{a, e, c\}, B = \{1, 5\}$，则 $A \cap B = \varnothing$，A 和 B 是互不相交的。

定义 3.2.3 ▶ 设 A, B 是集合，由属于 A 而不属于 B 的元素组成的集合，称为 B 对于 A 的补集，也叫 B 对于 A 的相对补集记为 $A - B$。

用描述法表示为 $A - B = \{x \mid x \in A \wedge x \notin B\}$。相对补集的文氏图如图 3.2.3 所示。

例如，令 $A = \{\varnothing, \{\varnothing\}\}, B = \varnothing$，则 $A - B = \{\varnothing, \{\varnothing\}\} - \varnothing = \{\varnothing, \{\varnothing\}\}$。

又如，令 $C = \{a\}, D = \{a, b\}$，则 $C - D = \{a\} - \{a, b\} = \varnothing, C - C = \varnothing, D - C = \{b\}$。

定义 3.2.4 ▶ 设 A 是集合，A 对于全集 E 的相对补集称为 A 的绝对补集，记为 $\sim A$。用描述法表示为
$$\sim A = E - A = \{x \mid x \in E \wedge x \notin A\} = \{x \mid x \notin A\}$$

$\sim A$ 的文氏图如图 3.2.4 所示。

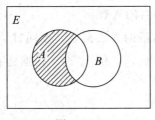
图　3.2.3

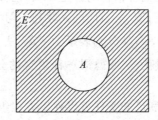
图　3.2.4

例如，令全集 $E = \{c, 2, d, 3\}, A = \{2, 3\}$，则 $\sim A = \{c, 2, d, 3\} - \{2, 3\} = \{c, d\}$。

定理 3.2.1 ▶ $A - B = A \cap (\sim B)$。

定义 3.2.5 ▶ 设 A, B 是集合，由 A 中元素或 B 中元素，但不是 A 与 B 的公共元素组成的集合，称为 A 和 B 的对称差，记为 $A \oplus B$。

用描述法表示为 $A \oplus B = \{x \mid x \in A \cup B \wedge x \notin A \cap B\}$。

$A \oplus B$ 的文氏图如图 3.2.5 所示。

例如，令 $A = \{1, c, 3, e\}, B = \{1, 2, e, f\}$，则 $A \oplus B = A \cup B - A \cap B = \{1, 2, 3, c, e, f\} - \{1, e\} = \{2, 3, c, f\}$。

定理 3.2.2 ▶ 设 A, B 是任意的集合，$A \oplus B = (A - B) \cup (B - A) = (A \cap \sim B) \cup (B \cap \sim A)$。

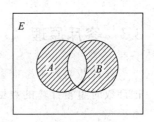
图　3.2.5

利用上述公式可以证明对称差 $A \oplus B$ 的下列性质。

设 A,B 是任意的集合。

(1) $A \oplus A = \varnothing$。

证明 $A \oplus A = (A-A) \bigcup (A-A) = \varnothing \oplus \varnothing = \varnothing$。

(2) $A \oplus \varnothing = A$。

证明 $A \oplus \varnothing = (A-\varnothing) \bigcup (\varnothing-A) = A \bigcup \varnothing = A$。

(3) $A \oplus E = \sim A$。

证明 $A \oplus E = (A-E) \bigcup (E-A) = \varnothing \bigcup \sim A = \sim A$。

为使集合的表达式更加简洁,我们对集合运算的优先顺序做如下规定。

绝对补的运算级别比其他的四种运算高,先进行绝对补运算,再进行其他的四种运算;其他的四种运算的运算顺序由括号决定。

用文氏图不仅可以表示集合的运算和它们之间的关系,还可以很方便地解决有限集合的计数问题。

用文氏图解决有限集合的计数问题的方法如下。

每一条性质定义一个集合,画一个圆表示这个集合。如果没有特别说明,任何两个圆都画成相交的。将已知集合的元素数填入表示该集合的区域内。通常从 n 个集合的交集填起,根据计算的结果逐步将数字填入其他各空白区域内。

如果交集的值是未知的,可以设为 x。根据题目的条件列出方程或方程组求解。

例 3.2.1 对 27 名学习外语的在校生进行学习外语语种情况的调查,结果统计如下:学习英、日、德和法语的人数分别为 15、5、12 和 10 人,其中同时学习英语和日语的有 2 人,学习英、德和法语中任两种语言的都是 5 人,已知学习日语的人既不懂法语也不懂德语,分别求只学习一种语言(英、德、法、日)的人数和学习 3 种语言的人数。

解 令 A,B,C,D 分别表示学习英、法、德、日语的人的集合。设同时学习三种语言的有 x 个学生,只学习英、法或德语一种语言的分别为 y_1,y_2 和 y_3 人。根据题意画出文氏图,如图 3.2.6所示。

根据文氏图和已知条件可以列出方程组。

$$\begin{cases} y_1 + 2(5-x) + x + 2 = 17 \\ y_2 + 2(5-x) + x = 10 \\ y_3 + 2(5-x) + x = 12 \\ y_1 + y_2 + y_3 + 3(5-x) + x = 27-5 \end{cases}$$

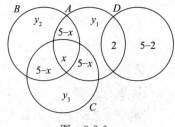

图 3.2.6

解得 $x=2,y_1=5,y_2=2,y_3=4$。此外,学习日语的只有 3 人。

3.3 容斥原理

在计数时先不考虑重叠的情况,把包含于某内容中的所有对象的数目先计算出来,然后再把计数时重复计算的数目排斥出去,使最后的结果既无一遗漏又无重复,这就是容斥原理的基本思想。

定理 3.3.1 ▶ 设 S 为有限集,$P_1,P_2,\cdots,P_m$ 是 m 种性质,A_i 是 S 中具有性质 P_i 的元素

构成的子集，$\overline{A_i}$ 是相对于 S 的补集，$i=1,2,\cdots,m$，则 S 中不具有性质 $P_1,P_2,\cdots,P_m$ 的元素数为

$$|\overline{A_1}\bigcap\overline{A_2}\bigcap\cdots\bigcap\overline{A_m}|=|S|-\sum_{i=1}^{m}|A_i|+\sum_{1\leqslant i<j\leqslant m}|A_i\bigcap A_j|-$$
$$\sum_{1\leqslant i<j<k\leqslant m}|A_i\bigcap A_j\bigcap A_k|+\cdots+(-1)^m|A_1\bigcap A_2\bigcap\cdots\bigcap A_m|$$

推论 S 中至少具有一条性质的元素数

$$|A_1\bigcup A_2\bigcup\cdots\bigcup A_m|=\sum_{i=1}^{m}|A_i|-\sum_{1\leqslant i<j\leqslant m}|A_i\bigcap A_j|+$$
$$\sum_{1\leqslant i<j<k\leqslant m}|A_i\bigcap A_j\bigcap A_k|-\cdots+(-1)^{m-1}|A_1\bigcap A_2\bigcap\cdots\bigcap A_m|$$

例 3.3.1 求 1 到 1500 之间（包含 1 和 1500 在内）既不能被 4 和 5 整除，也不能被 6 整除的数的个数。

解 设 $S=\{x\mid x\in Z\wedge 1\leqslant x\leqslant 1500\}$，$A=\{x\mid x\in S\wedge x$ 可被 4 整除$\}$，$B=\{x\mid x\in S\wedge x$ 可被 5 整除$\}$，$C=\{x\mid x\in S\wedge x$ 可被 6 整除$\}$。

用 $|Y|$ 表示有穷集 Y 中的元素数，$\lfloor x\rfloor$ 表示小于或等于 x 的最大整数，$\text{1cm}(x_1,x_2,\cdots,x_n)$ 表示 $x_1,x_2\cdots,x_n$ 的最小公倍数，则有

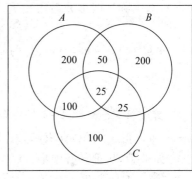

图　3.3.1

$$|A|=\lfloor 1500/4\rfloor=375$$
$$|B|=\lfloor 1500/5\rfloor=300$$
$$|C|=\lfloor 1500/6\rfloor=250$$
$$|A\bigcap B|=\lfloor 1500/\text{1cm}(4,5)\rfloor=75$$
$$|A\bigcap C|=\lfloor 1500/\text{1cm}(4,6)\rfloor=125$$
$$|B\bigcap C|=\lfloor 1500/\text{1cm}(5,6)\rfloor=50$$
$$|A\bigcap B\bigcap C|=\lfloor 1500/\text{1cm}(4,5,6)\rfloor=25$$

将这些数字填入文氏图，得到图 3.3.1。

利用客斥原理：

$$|\overline{A}\bigcap\overline{B}\bigcap\overline{C}|=|S|-(|A|+|B|+|C|)+(|A\bigcap B|+|A\bigcap C|+|B\bigcap C|)-|A\bigcap B\bigcap C|$$
$$=1500-(375+300+250)+(75+125+50-25)=800$$

在 1 到 1500 之间，不能被 4、5、6 整除的数有 800 个。

3.4　序偶与笛卡儿积

序偶其实在以前的知识中就有所接触。例如，直角坐标系中的点就是序偶，如果将横坐标和纵坐标前后调换则表示另一个点，所以序偶是有序的，而笛卡儿积与序偶则是紧密相连的。

定义 3.4.1 ▶ 两个个体 x,y 的有序序列称为二重组，也称为有序对或序偶，记为 $<x,y>$。x,y 分别叫作二重组的第一分量和第二分量。

有序序列是指调换第一分量和第二分量的位置后，就和原来的含义不同了。

定义 3.4.2 ▶ 设 $<x,y>$ 与 $<a,b>$ 是两个二重组，如果 $x=a$ 且 $y=b$，则称二重组 $<x,y>$ 与 $<a,b>$ 相等，记为 $<x,y>=<a,b>$。

二重组$<x,y>$与$<a,b>$相等,用数理逻辑的方法表示为$(<x,y>=<a,b>)\Leftrightarrow(x=a)\wedge(y=b)$。

由定义可以看出,当$x\neq y$时,$<x,y>\neq<y,x>$。

例如,平面上的点$P_1=<2,1>$和点$P_2=<1,2>$是两个不同的点,它们都是二重组。

定义 3.4.3 二重组$<<x_1,x_2,\cdots,x_{n-1}>,x_n>$叫作$n$重组,记为$<x_1,x_2,\cdots,x_n>$,$x_i$叫作该$n$重组的第$i$个分量,$i=1,\cdots,n$。

由定义可以看出:三重组$<x_1,x_2,x_3>$定义为$<<x_1,x_2>,x_3>$,其中第一分量是二重组;四重组$<x_1,x_2,x_3,x_4>$定义为$<<x_1,x_2,x_3>,x_4>$,其中第一分量是三重组……

根据三重组的定义,其第一分量是二重组,第二分量是一个个体的序偶,$<<x_1,x_2>,x_3>$是三重组。而第一分量是一个个体,第二分量是二重组的序偶$<x_1,<x_2,x_3>>$不是三重组。这个结论对任意的$n(n=3,4,5,\cdots)$重组也是正确的。例如,$<<x_1,x_2,x_3>,x_4>$是四重组,而$<x_1,<x_2,x_3,x_4>>$不是四重组;$<<x_1,x_2,x_3,x_4>,x_5>$是五重组,而$<x_1,<x_2,x_3,x_4,x_5>>$不是五重组。

定义 3.4.4 设$<x_1,x_2,\cdots,x_n>$与$<y_1,y_2,\cdots,y_n>$是两个n重组,如果$x_i=y_i$,$i=1,\cdots,n$,则称这两个n重组相等,记为

$$<x_1,x_2,\cdots,x_n>=<y_1,y_2,\cdots,y_n>$$

n重组$<x_1,x_2,\cdots,x_n>$与$<y_1,y_2,\cdots,y_n>$相等,用数理逻辑的方法表示为

$$(<x_1,x_2,\cdots,x_n>=<y_1,y_2,\cdots,y_n>)$$
$$\Leftrightarrow(x_1=y_1)\wedge(x_2=y_2)\wedge\cdots\wedge(x_n=y_n)$$

定义 3.4.5 设A,B是集合,集合$\{<a,b>|a\in A\wedge b\in B\}$叫作$A$和$B$的笛卡儿积,也叫$A$和$B$的叉乘积、直积,记为$A\times B$。

如果A,B都是有限集,$|A|=n$,$|B|=m$,根据排列组合原理,$|A\times B|=nm=|A||B|$。

例 3.4.1 设$A=\{r,s,t\}$,$B=\{1,2,3\}$,(1)试求$A\times B$和$B\times A$,(2)验证$|A\times B|=|A||B|$和$|B\times A|=|B||A|$。

解 (1)求$A\times B$和$B\times A$。

$A\times B=\{<r,1>,<r,2>,<r,3>,<s,1>,<s,2>,<s,3>,<t,1>,<t,2>,<t,3>\}$

$B\times A=\{<1,r>,<1,s>,<1,t>,<2,r>,<2,s>,<2,t>,<3,r>,<3,s>,<3,t>\}$

(2)验证$|A\times B|=|A||B|$和$|B\times A|=|B||A|$。

$$|A\times B|=9=3\times3=|A||B|$$
$$|B\times A|=9=3\times3=|B||A|$$

定理 3.4.1 笛卡儿积运算有以下性质。

(1)设A为任意的集合,则$A\times\varnothing=\varnothing\times A=\varnothing$。

(2)一般来说,$\times$不满足交换律:$A\times B\neq B\times A$。

在例3.4.1中,$A\times B\neq B\times A$。

(3)一般来说,$\times$不满足结合律:$(A\times B)\times C\neq A\times(B\times C)$。

定理 3.4.2 设A,B,C是集合,则

(1)$A\times(B\cup C)=(A\times B)\cup(A\times C)$

(2)$A\times(B\cap C)=(A\times B)\cap(A\times C)$

(3)$(A\cup B)\times C=(A\times C)\cup(B\times C)$

(4) $(A\cap B)\times C=(A\times C)\cap(B\times C)$

定理 3.4.3 ▶ 设 A,B,C 是集合，$C\neq\varnothing$，则

(1) $A\subseteq B$ 的充分必要条件是 $A\times C\subseteq B\times C$。

(2) $A\subseteq B$ 的充分必要条件是 $C\times A\subseteq C\times B$。

定理 3.4.4 ▶ 设 A,B,C,D 是非空集合，则 $A\times B\subseteq C\times D$ 的充分必要条件是 $A\subseteq C$ 且 $B\subseteq D$。

定义 3.4.6 ▶ 叉乘积 $A_1\times A_2\times\cdots\times A_n$ 定义为 $(A_1\times A_2\times\cdots\times A_{n-1})\times A_n$，即 $A_1\times A_2\times\cdots\times A_n=\{<a_1,a_2,\cdots,a_n>|a_1\in A_1\wedge a_2\in A_2\wedge\cdots\wedge a_n\in A_n\}$。

由定义可以看出：

当 $n=3$ 时，$A_1\times A_2\times A_3$ 定义为 $(A_1\times A_2)\times A_3$。
$$A_1\times A_2\times A_3=\{<a_1,a_2,a_3>|a_1\in A_1\wedge a_2\in A_2\wedge a_3\in A_3\}$$

当 $n=4$ 时，$A_1\times A_2\times A_3\times A_4$ 定义为 $(A_1\times A_2\times A_3)\times A_4$。
$A_1\times A_2\times A_3\times A_4=\{<a_1,a_2,a_3,a_4>|a_1\in A_1\wedge a_2\in A_2\wedge a_3\in A_3\wedge a_4\in A_4\}\cdots$
$A_n=A\times A\times\cdots\times A(n$ 个 A 的叉乘积)

例 3.4.2 　设 $A=\{1,2\},B=\{a,b\},C=\{x,y\}$，求：$A\times B\times C,A\times(B\times C)$。

解 $A\times B\times C=(A\times B)\times C$

$\quad=\{<1,a>,<1,b>,<2,a>,<2,b>\}\times\{x,y\}$

$\quad=\{<<1,a>,x>,<<1,b>,x>,<<2,a>,x>,<<2,b>,x>,$
$\quad\ \ <<1,a>,y>,<<1,b>,y>,<<2,a>,y>,<<2,b>,y>\}$

$\quad=\{<1,a,x>,<1,b,x>,<2,a,x>,<2,b,x>,<1,a,y>,<1,b,y>,$
$\quad\ \ <2,a,y>,<2,b,y>\}$

$A\times(B\times C)=\{1,2\}\times\{<a,x>,<a,y>,<b,x>,<b,y>\}$

$\quad=\{<1,<a,x>>,<1,<a,y>>,<1,<b,x>>,<1,<b,y>>,$
$\quad\ \ <2,<a,x>>,<2,<a,y>>,<2,<b,x>>,<2,<b,y>>\}$

显然，$A\times B\times C\neq A\times(B\times C)$。

3.5　关系及其性质

关系是一个基本概念，在数学中主要研究集合中元素间的关系。如"0 小于 1"；"x 不等于 y"等。

定义 3.5.1 ▶ 设 A 和 B 是任意集合，如果 $R\subseteq A\times B$，则称 R 是 A 到 B 的二元关系。如果 R 是 A 到 A 的二元关系，则称 R 是 A 上的二元关系。

设 $A=\{1,2,3\},B=\{a,b\},R=\{<1,a>,<2,a>,<3,b>\}$，则 R 是 A 到 B 的二元关系。若 $S=\{<3,1>,<2,2>,<2,1>,<1,1>\}$，则 S 是 A 上的二元关系。

定义 3.5.2 ▶ 设 A 和 B 是任意集合，$R\subseteq A\times B$，若 $<x,y>\in R$，则称 x 与 y 有 R 关系，记为 xRy。若 $<x,y>\in R$，则称 x 与 y 没有 R 关系，记为 $x\bar{R}y$。

如果 R 是 A 到 B 的二元关系，根据定义 3.5.2，$<x,y>\in R$ 与 xRy，$<x,y>\notin R$ 与 $x\bar{R}y$ 的意义分别相同。

定义 3.5.3 ▶ 设 A 和 B 是任意集合，空集 $\varnothing$ 叫作 A 到 B 的空关系，仍然记为 $\varnothing$。A,B 的

笛卡儿积 $A \times B$ 叫作 A 到 B 的全域关系,记为 E。集合 $\{<a,a>|a \in A\}$ 叫作 A 上的恒等关系,记为 I_A。

例 3.5.1 设 $A=\{c,d\}$, $B=\{2,5\}$, 求 A 上的恒等关系 I_A 和 A 到 B 的全域关系 $A \times B$。

解 A 上的恒等关系 $I_A=\{<c,c>,<d,d>\}$, A 到 B 的全域关系 $E=A \times B=\{<c,2>, <c,5>,<d,2>,<d,5>\}$。

定理 3.5.1 ▶ 设 A 是具有 n 个元素的有限集,则 A 上的二元关系有 2^{n^2} 种。

证明 设 A 为具有 n 个元素的有限集,即 $|A|=n$, 由排列组合原理知 $|A \times A|=n^2$。根据定理 3.1.3,有 $|P(A \times A)|=2^{|A \times A|}=2^{n^2}$, 即 $A \times A$ 的子集有 2^{n^2} 个。所以具有 n 个元素的有限集 A 上有 2^{n^2} 种二元关系。

1. 用列举法表示二元关系

例 3.5.1 中的 A 到 B 的全域关系:

$$E=A \times B=\{<c,2>,<c,5>,<d,2>,<d,5>\}$$

A 上的恒等关系:

$$I_A=\{<c,c>,<d,d>\}$$

这些都是用列举法表示的。

2. 用描述法表示二元关系

设 $\mathbf{R}$ 是实数集, $L_{\mathbf{R}}=\{<x,y>|x \in \mathbf{R} \wedge y \in \mathbf{R} \wedge x \leqslant y\}$, $L_{\mathbf{R}}$ 是实数集 $\mathbf{R}$ 上的二元关系。

3. 用矩阵表示二元关系

如果 A,B 是有限集, $A=\{a_1,a_2,\cdots,a_m\}$, $B=\{b_1,b_2,\cdots,b_n\}$, R 是 A 到 B 的二元关系, R 的关系矩阵定义为

$$\boldsymbol{M}_R=(r_{ij})_{m \times n}$$

$$r_{ij}=\begin{cases} 1, & <a_i,b_j> \in R \\ 0, & <a_i,b_j> \notin R \end{cases} \quad i=1,\cdots,m \quad j=1,\cdots,n$$

称为二元关系 R 的关系矩阵。

例 3.5.2 设 $A=\{a_1,a_2,a_3,a_4\}$, $B=\{b_1,b_2,b_3\}$, R 是 A 到 B 的二元关系,定义为 $R=\{<a_1,b_1>,<a_1,b_3>,<a_2,b_2>,<a_2,b_3>,<a_3,b_1>,<a_4,b_1>,<a_4,b_2>\}$ 写出 R 的关系矩阵。

解 R 的关系矩阵为

$$\boldsymbol{M}_R=\begin{bmatrix} 1 & 0 & 1 \\ 0 & 1 & 1 \\ 1 & 0 & 0 \\ 1 & 1 & 0 \end{bmatrix}$$

例 3.5.3 设 $A=\{1,2,3,4\}$, R 是 A 上的二元关系,定义为

$R=\{<1,1>,<1,2>,<2,1>,<3,2>,<3,1>,<4,3>,<4,2>,<4,1>\}$

写出 A 上的二元关系 R 的关系矩阵。

解 R 的关系矩阵为

$$M_R = \begin{bmatrix} 1 & 1 & 0 & 0 \\ 1 & 0 & 0 & 0 \\ 1 & 1 & 0 & 0 \\ 1 & 1 & 1 & 0 \end{bmatrix}$$

例 3.5.3 中的二元关系 R 是 A 上的二元关系,只需看成 A 到 A 的二元关系,利用上述定义,就可以方便地写出它的关系矩阵。A 上的二元关系和 A 到 B 的二元关系的关系矩阵的定义是相同的。

4. 用图表示二元关系

如果 A 和 B 是有限集,R 是 A 到 B 的二元关系,还可以用图表示二元关系 R。表示二元关系 R 的图叫作 R 的关系图。A 到 B 的二元关系的关系图和 A 上的二元关系的关系图的定义是不一样的。

它们可分别描述如下。

(1) A 到 B 的二元关系 R 的关系图。

设 $A = \{a_1, a_2, \cdots, a_m\}$,$B = \{b_1, b_2, \cdots, b_n\}$,$R$ 是 A 到 B 的二元关系,R 的关系图的绘制方法如下:①画出 m 个小圆圈表示 A 的元素,再画出 n 个小圆圈表示 B 的元素。这些小圆圈叫作关系图的结点(下同)。②如果 $<a_i, b_j> \in R$,则从 a_i 到 b_j 画一根有方向(带箭头)的线。这些有方向(带箭头)的线叫作关系图的边(下同)。

例 3.5.2 中的二元关系 R 的关系图如图 3.5.1 所示。

(2) A 上的二元关系 R 的关系图。

设 $A = \{a_1, a_2, \cdots, a_m\}$,$R$ 是 A 上的二元关系,其关系图的绘制方法如下:①画出 m 个小圆圈表示 A 的元素。②如果 $<a_i, a_j> \in R$,则从 a_i 到 a_j 画一根有方向(带箭头)的线。

例 3.5.3 中的二元关系 R 的关系图如图 3.5.2 所示。

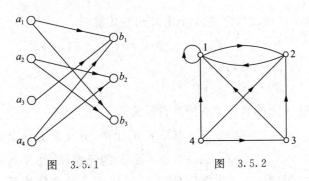

图 3.5.1　　　　　图 3.5.2

定义 3.5.4 ▶设 $R \subseteq X \times X$,$(\forall x)(x \in X \to <x,x> \in R)$,则称 R 在 X 上是自反的。

设 R 是 X 上的自反关系,由定义 3.5.4 可知,R 的关系矩阵 M_R 的主对角线全为 1;在关系图中每一个结点上都有自回路。

设 $X = \{1,2,3\}$,X 上的二元关系

$$R = \{<1,1>, <2,2>, <3,3>, <1,2>\}$$

R 是自反的,它的关系图如图 3.5.3 所示,其关系矩阵如下:

$$\boldsymbol{M}_R = \begin{bmatrix} 1 & 1 & 0 \\ 0 & 1 & 0 \\ 0 & 0 & 1 \end{bmatrix}$$

定理 3.5.2 ▶ 设 R 是 X 上的二元关系,当且仅当 $I_X \subseteq R$ 时 R 是自反的。

图 3.5.3

证明 设 R 在 X 上是自反的,则 $<x,y> \in I_X \Rightarrow x=y \Rightarrow <x,y> \in R$,即 $I_X \subseteq R$。

设 $I_X \subseteq R$,则 $\forall x \in X \Rightarrow <x,x> \in I_X \Rightarrow <x,x> \in R$,即 R 在 X 上是自反的。

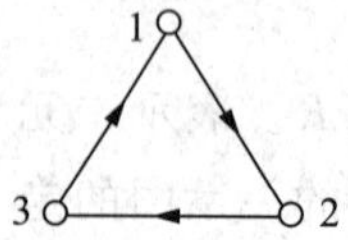

图 3.5.4

定义 3.5.5 ▶ 设 $R \subseteq X \times X$,$(\forall x)(x \in X \to <x,x> \notin R)$,则称 R 在 X 上是反自反的。

设 R 是 X 上的反自反关系,由定义 3.5.5 可知,R 的关系矩阵 $\boldsymbol{M}_R$ 的主对角线全为 0;在 R 的关系图中每一个结点上都没有自回路。它的关系图如图 3.5.4 所示,其关系矩阵如下:

$$\boldsymbol{M}_R = \begin{bmatrix} 0 & 1 & 0 \\ 0 & 0 & 1 \\ 1 & 0 & 0 \end{bmatrix}$$

定理 3.5.3 ▶ 设 R 是 X 上的二元关系,当且仅当 $R \cap I_X = \varnothing$ 时 R 是反自反的。

证明 设 R 在 X 上是反自反的,下证 $R \cap I_X = \varnothing$。

假设 $R \cap I_X \neq \varnothing$,必存在 $<x,y> \in R \cap I_X \Rightarrow <x,y> \in R \wedge <x,y> \in I_X \Rightarrow <x,y> \in R \wedge x=y$,即 $<x,x> \in R$,这与 R 是 X 上的反自反关系相矛盾。所以 $R \cap I_X = \varnothing$。

设 $R \cap I_X = \varnothing$,下证 R 在 X 上是反自反的。

任取 $x \in X$,$<x,x> \in I_X$,由于 $R \cap I_X = \varnothing$,必然有 $<x,x> \notin R$,即 R 在 X 上是反自反的。

例 3.5.4 设 $A = \{1,2,3\}$,定义 A 上的二元关系如下:

$$R = \{<1,1>,<2,2>,<3,3>,<1,3>\}$$
$$S = \{<1,3>\}$$
$$T = \{<1,1>\}$$

试说明 R,S,T 是否是 A 上的自反关系或反自反关系。

解 因为 $<1,1>$、$<2,2>$ 和 $<3,3>$ 都是 R 的元素,所以 R 是 A 上的自反关系,不是反自反关系。因为 $<1,1>$、$<2,2>$ 和 $<3,3>$ 都不是 S 的元素,所以 S 是 A 上的反自反关系,不是 A 上的自反关系。因为 $<2,2> \notin T$,所以 T 不是 A 上的自反关系。因为 $<1,1> \in T$,所以 T 不是 A 上的反自反关系。

定义 3.5.6 ▶ 设 $R \subseteq X \times X$,$(\forall x)(\forall y)(x \in X \wedge y \in X \wedge <x,y> \in R \to <y,x> \in R)$,则称 R 在 X 上是对称的。

R 是 X 上的对称关系,由定义 3.5.6 可知,R 的关系矩阵 $\boldsymbol{M}_R$ 是对称矩阵。在 R 的关系图中,如果两个不同的结点间有边,一定有方向相反的两条边。

设 $X = \{1,2,3\}$,X 上的二元关系 $R = \{<1,2>,<2,1>,<3,3>\}$,$R$ 是对称的。它的

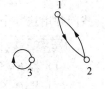

关系图如图 3.5.5 所示,其关系矩阵如下:

$$\boldsymbol{M}_R = \begin{bmatrix} 0 & 1 & 0 \\ 1 & 0 & 0 \\ 0 & 0 & 1 \end{bmatrix}$$

图 3.5.5

定理 3.5.4 ▶ 设 R 是 X 上的二元关系,当且仅当 $R=R^{\mathrm{c}}$ 时 R 是对称的。

证明 设 R 是对称的,下证 $R=R^{\mathrm{c}}$。

$<x,y>\in R \Leftrightarrow <y,x>\in R \Leftrightarrow <x,y>\in R^{\mathrm{c}}$,所以 $R=R^{\mathrm{c}}$。

设 $R=R^{\mathrm{c}}$,下证 R 是对称的。

$<x,y>\in R \Rightarrow <y,x>\in R^{\mathrm{c}} \Rightarrow <y,x>\in R$,所以 R 是对称的。

定义 3.5.7 ▶ 设 $R \subseteq X \times X$,$(\forall x)(\forall y)(x \in X \wedge y \in X \wedge <x,y>\in R \wedge <y,x>\in R \rightarrow (x=y))$,则称 R 在 X 上是反对称的。

反对称的定义又可以等价地描述为

$$(\forall x)(\forall y)(x \in X \wedge y \in X \wedge <x,y>\in R \wedge (x \neq y) \rightarrow <y,x>\notin R)$$

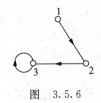

设 R 是 X 上的反对称关系,由定义 3.5.7 可知,在 R 的关系矩阵 $\boldsymbol{M}_R$ 中以主对角线为轴的对称位置上不能同时为 1(主对角线除外)。在 R 的关系图中,每两个不同的结点间不能有方向相反的两条边。

设 $X=\{1,2,3\}$,X 上的二元关系 $R=\{<1,2>,<2,3>,<3,3>\}$,R 是反对称的。它的关系图如图 3.5.6 所示,其关系矩阵如下:

图 3.5.6

$$\boldsymbol{M}_R = \begin{bmatrix} 0 & 1 & 0 \\ 0 & 0 & 1 \\ 0 & 0 & 1 \end{bmatrix}$$

例 3.5.5 设 $A=\{1,2,3\}$,定义 A 上的二元关系如下:

$$R=\{<1,1>,<2,2>\}$$
$$S=\{<1,1>,<1,2>,<2,1>\}$$
$$T=\{<1,2>,<1,3>\}$$
$$U=\{<1,3>,<1,2>,<2,1>\}$$

试说明 R,S,T,U 是否是 A 上的对称关系和反对称关系。

解 R 是 A 上的对称关系,也是 A 上的反对称关系。

S 是 A 上的对称关系。因为 $<1,2>$ 和 $<2,1>$ 都是 S 的元素,而 $1 \neq 2$,所以 S 不是 A 上的反对称关系。

因为 $<1,2>\in T$,而 $<2,1>\notin T$,所以 T 不是 A 上的对称关系。T 是 A 上的反对称关系。

U 不是 A 上的对称关系,也不是 A 上的反对称关系。

定理 3.5.5 ▶ 设 R 是 X 上的二元关系,则当且仅当 $R \cap R^{\mathrm{c}} \subseteq I_X$ 时 R 是反对称的。

定义 3.5.8 ▶ 设 $R \subseteq X \times X$

$(\forall x)(\forall y)(\forall z)(x \in X \wedge y \in X \wedge z \in X \wedge <x,y>\in R \wedge <y,z>\in R \rightarrow <x,z>\in R)$

则称 R 在 X 上是传递的。

定理 3.5.6 ▶ 设 R 是 X 上的二元关系,R 是传递的当且仅当 $R \circ R \subseteq R$。

定理 3.5.2～定理 3.5.6 可以作为相应概念的定义使用,用于判断和证明关系的性质。有些问题用这 5 个定理处理是非常方便的。

定理 3.5.7 ▶ 设 R,S 是 X 上的二元关系,则

(1) 若 R,S 是自反的,则 $R \cup S$ 和 $R \cap S$ 也是自反的。

(2) 若 R,S 是对称的,则 $R \cup S$ 和 $R \cap S$ 也是对称的。

(3) 若 R,S 是传递的,则 $R \cap S$ 也是传递的。

3.6 关系的逆与复合

关系运算中不仅包含并、交、补等,还可以进行关系的复合运算。由于关系是序偶的集合,序偶具有有序性,所以关系还存在逆运算。

定义 3.6.1 ▶ 设 A,B 是集合,$R \subseteq A \times B$。

$\text{dom } R = \{x \mid <x,y> \in R\}$ 叫作 R 的定义域。

$\text{ran } R = \{y \mid <x,y> \in R\}$ 叫作 R 的值域。

$\text{FLD } R = \text{dom } R \cup \text{ran } R$ 叫作 R 的域。

A 叫作 R 的前域;B 叫作 R 的陪域。

1. 二元关系的交、并、补、对称差运算

定理 3.6.1 ▶ 设 R,S 是 X 到 Y 的二元关系,则 $R \cup S,R \cap S,R-S,\sim R,R \oplus S$ 也是 X 到 Y 的二元关系。

证明 因为 R,S 是 X 到 Y 的二元关系,所以,$R \subseteq X \times Y$ 且 $S \subseteq X \times Y$。

显然,$R \cup S \subseteq X \times Y$,即 $R \cup S$ 是 X 到 Y 的二元关系。

$R \cap S \subseteq X \times Y$,即 $R \cap S$ 是 X 到 Y 的二元关系。

$R-S \subseteq X \times Y$,即 $R-S$ 是 X 到 Y 的二元关系。

在二元关系运算中,认为全域关系是全集。所以 $\sim R = X \times Y - R \subseteq X \times Y$,即 $\sim R$ 是 X 到 Y 的二元关系。

由以上结论可以得到:$(R-S) \subseteq X \times Y$ 和 $(S-R) \subseteq X \times Y$,从而 $(R-S) \cup (S-R) \subseteq X \times Y$,所以 $R \oplus S = (R-S) \cup (S-R) \subseteq X \times Y$,即 $R \oplus S$ 是 X 到 Y 的二元关系。

例 3.6.1 设 $X=\{1,2,3,4\}$,X 上的二元关系 H 和 S 的定义如下:$H=\{<x,y> \mid (x-y)/2$ 是整数$\}$,$S=\{<x,y> \mid (x-y)/3$ 是正整数$\}$。试求 $H \cup S,H \cap S,\sim H,S-H$。

解 将 H 和 S 用列举法表示:

$H=\{<1,1>,<1,3>,<2,2>,<2,4>,<3,3>,<3,1>,<4,4>,<4,2>\}$

$S=\{<4,1>\}$

$H \cup S=\{<1,1>,<1,3>,<2,2>,<2,4>,<3,3>,<3,1>,<4,4>,<4,2>,<4,1>\}$

$H \cap S=\varnothing$

$\sim H=X^2-H=\{<1,2>,<1,4>,<2,1>,<2,3>,<3,2>,<3,4>,<4,3>,<4,1>\}$

$S-H=\{<4,1>\}$

2. 二元关系的复合运算

定义 3.6.2 ▶ 设 X,Y,Z 是集合,$R \subseteq X \times Y$,$S \subseteq Y \times Z$,集合 $\{<x,z> \mid x \in X \land z \in Z \land$

$(\exists y)(y\in Y\wedge<x,y>\in R\wedge<y,z>\in S)\}$叫作 R 和 S 的复合关系,记为 $R\circ S,R\circ S\subseteq X\times Z$,即 $R\circ S$ 是 X 到 Z 的二元关系。

例 3.6.2　$X=\{1,2,3,4,5\}$,X 上的二元关系 R 和 S 的定义如下:$R=\{<1,2>,<3,4>,<2,2>\}$,$S=\{<4,2>,<2,5>,<3,1>,<1,3>\}$。试求 $R\circ S,S\circ R,R\circ(S\circ R),(R\circ S)\circ R,R\circ R,S\circ S,R\circ R\circ R$。

解
$$R\circ S=\{<1,5>,<3,2>,<2,5>\}$$
$$S\circ R=\{<4,2>,<3,2>,<1,4>\}$$
$$(R\circ S)\circ R=\{<3,2>\}$$
$$R\circ(S\circ R)=\{<3,2>\}$$
$$R\circ R=\{<1,2>,<2,2>\}$$
$$S\circ S=\{<4,5>,<3,3>,<1,1>\}$$
$$R\circ R\circ R=\{<1,2>,<2,2>\}$$

从例 3.6.2 可以看出,$R\circ S\neq S\circ R$,这说明二元关系的复合运算不满足交换律。

定理 3.6.2　设 X,Y,Z,W 是集合,$R\subseteq X\times Y,S\subseteq Y\times Z,T\subseteq Z\times W$,则 $(R\circ S)\circ T=R\circ(S\circ T)$。

证明　$<x,w>\in(R\circ S)\circ T\Leftrightarrow(\exists z)(<x,z>\in R\circ S\wedge<z,w>\in T)$
$$\Leftrightarrow(\exists z)((\exists y)(<x,y>\in R\wedge<y,z>\in S)\wedge<z,w>\in T)$$
$$\Leftrightarrow(\exists z)(\exists y)((<x,y>\in R\wedge<y,z>\in S)\wedge<z,w>\in T)$$
$$\Leftrightarrow(\exists y)(\exists z)(<x,y>R\wedge(<y,z>\in S\wedge<z,w>\in T))$$
$$\Leftrightarrow(\exists y)(<x,y>\in R\wedge(\exists z)(<y,z>\in S\wedge<z,w>\in T))$$
$$\Leftrightarrow(\exists y)(<x,y>\in R\wedge<y,w>\in S\circ T)\Leftrightarrow<x,w>\in R\circ(S\circ T)$$

所以 $(R\circ S)\circ T=R\circ(S\circ T)$。

定理 3.6.3　设 R 是 A 上的二元关系,$R\circ I_A=I_A\circ R=R$。

证明　先证 $R\circ I_A=R$。
$$<x,y>\in R\circ I_A\Rightarrow(\exists z)(<x,z>\in R\wedge<z,y>\in I_A)\Rightarrow(\exists z)(<x,z>\in R\wedge z=y)\Rightarrow<x,y>\in R$$

所以 $R\circ I_A\subseteq R$。
$$<x,y>\in R\Rightarrow<x,y>\in R\wedge<y,y>\in I_A\Rightarrow<x,y>\in R\circ I_A$$

所以 $R\subseteq R\circ I_A$。故 $R\circ I_A=R$。

类似地,可以证明 $I_A\circ R=R$。

定理 3.6.4　设 R,S,T 是 A 上的二元关系,则

(1) $R\circ(S\cup T)=R\circ S\cup R\circ T$

(2) $(R\cup S)\circ T=R\circ T\cup S\circ T$

(3) $R\circ(S\cap T)\subseteq R\circ S\cap R\circ T$

(4) $(R\cap S)\circ T\subseteq R\circ T\cap S\circ T$

证明　仅证明(3),(1)、(2)和(4)同理。

$$<x,y>\in R\circ(S\cap T)\Rightarrow(\exists z)(<x,z>\in R\wedge<z,y>\in S\cap T)$$
$$\Rightarrow(\exists z)(<x,z>\in R\wedge(<z,y>\in S\wedge<z,y>\in T))$$
$$\Rightarrow(\exists z)((<x,z>\in R\wedge<z,y>\in S)\wedge(<x,z>\in R\wedge$$
$$<z,y>\in T))$$
$$\Rightarrow(\exists z)(<x,z>\in R\wedge<z,y>\in S)\wedge(\exists z)(<x,z>\in R\wedge$$
$$<z,y>\in T)$$
$$\Rightarrow<x,y>\in R\circ S\wedge<x,y>\in R\circ T$$
$$\Rightarrow<x,y>\in R\circ S\cap R\circ T$$

故 $R\circ(S\cap T)\subseteq R\circ S\cap R\circ T$。

一般来说,$R\circ S\cap R\circ T\nsubseteq R\circ(S\cap T)$,举反例如下:

设 $A=\{1,2,3,4,5\}$,$R=\{<4,1>,<4,2>\}\subseteq A\times A$,$S=\{<1,5>,<3,5>\}\subseteq A\times A$,$T=\{<2,5>,<3,5>\}\subseteq A\times A$,则 $R\circ S\cap R\circ T=\{<4,5>\}\cap\{<4,5>\}=\{<4,5>\}$,$R\circ(S\cap T)=\{<4,1>,<4,2>\}\circ\{<3,5>\}=\varnothing$,$R\circ S\cap R\circ T\nsubseteq R\circ(S\cap T)$。

定理 3.6.4 的(1)说明,关系的复合运算"∘"对并运算"∪"满足左分配律。(2)说明,关系的复合运算"∘"对并运算"∪"满足右分配律。所以,复合运算"∘"对并运算"∪"满足分配律。由(3)和(4)可知,复合运算"∘"对交运算"∩"不满足分配律。

定义 3.6.3 ▶ 设 R 是 A 上的二元关系,n 为自然数,R 的 n 次幂记为 R^n,定义为 $R^0=I_A$,$R^{n+1}=R^n\circ R$。

由定义 3.6.3 可以看出,A 上的任何二元关系的 0 次幂都相等,等于 A 上的恒等关系 I_A。由定义 3.6.3 还可以看出:

$$R^1=R^0\circ R=I_A\circ R=R$$
$$R^2=R^1\circ R=R\circ R$$
$$R^3=R^2\circ R=(R\circ R)\circ R$$
$$\cdots$$

因为复合运算满足结合律,所以 R^3 又可以写成:

$$R^3=R\circ R\circ R$$

同理,R^n 也可以写成:

$$R^n=R\circ R\circ\cdots\circ R(n\text{ 个 }R\text{ 的复合})$$

定理 3.6.5 ▶ 设 A 是具有 n 个元素的有限集,R 是 A 上的二元关系,则必存在自然数 s 和 t,使 $R^s=R^t$,$0\leqslant s<t\leqslant 2^{n^2}$。

证明 R 是 A 上的二元关系,对任何自然数 k,由复合关系的定义可知,R^k 仍然是 A 上的二元关系,即 $R^k\subseteq A\times A$。另外,据定理 3.5.1,A 上的二元关系仅有 2^{n^2} 种。列出 R 的各次幂 $R^0,R^1,R^2,\cdots,R^{2^{n^2}}$,共有 $2^{n^2}+1$ 个,必存在自然数 s 和 t,使 $R^s=R^t$,$0\leqslant s<t\leqslant 2^{n^2}$。

例 3.6.3 $A=\{1,2,3,4\}$,A 上的二元关系 R 定义如下:$R=\{<1,2>,<2,1>,<2,3>,<3,4>\}$。求二元关系 R 的各次幂,验证定理 3.6.5。

解 $|A|=4$

$$R^0=I_A=\{<1,1>,<2,2>,<3,3>,<4,4>\}$$
$$R^1=R=\{<1,2>,<2,1>,<2,3>,<3,4>\}$$
$$R^2=R^1\circ R=R\circ R=\{<1,1>,<1,3>,<2,2>,<2,4>\}$$

$$R^3 = R^2 \circ R = \{<1,2>,<2,1>,<2,3>,<1,4>\}$$

$$R^4 = R^3 \circ R = \{<1,1>,<1,3>,<2,2>,<2,4>\} = R^2,$$

$$0 \leqslant 2 < 4 \leqslant 2^{16} = 65536$$

$$R^5 = R^4 \circ R = R^2 \circ R = R^3$$

$$R^6 = R^5 \circ R = R^3 \circ R = R^4 = R^2$$

$$\cdots$$

$$R^{2n} = R^2$$

$$R^{2n+1} = R^3, \; n = 1,2,3,\cdots$$

定理 3.6.6 ▶ 设 R 是 A 上的二元关系，m,n 为自然数，则 $R^m \circ R^n = R^{m+n}$，$(R^m)^n = R^{mn}$。

定义 3.6.4 ▶ 设 X,Y 是集合，$R \subseteq X \times Y$，集合 $\{<y,x> | <x,y> \in R\}$ 则叫作 R 的逆关系，记为 R^C，$R^C \subseteq Y \times X$，R^C 是 Y 到 X 的二元关系。

容易证明，R^C 的关系矩阵 $\boldsymbol{M}_{R^C}$ 是 R 的关系矩阵 $\boldsymbol{M}_R$ 的转置矩阵，即 $\boldsymbol{M}_{R^C} = \boldsymbol{M}_R^{\mathrm{T}}$。

可以验证，将 R 的关系图中的弧线的箭头反置就可以得到 R^C 的关系图。

例 3.6.4 设 $X = \{1,2,3,4\}$，$Y = \{a,b,c\}$，X 到 Y 的二元关系 $R = \{<1,a>,<2,b>,<4,c>\}$。

(1) 试求 R^C，写出 $\boldsymbol{M}_R$ 和 $\boldsymbol{M}_{R^C}$，证 $\boldsymbol{M}_{R^C} = \boldsymbol{M}_R^{\mathrm{T}}$。

(2) 画出 R 和 R^C 的关系图，验证将 R 的关系图中的弧线的箭头反置可得到 R^C 的关系图。

解 $R^C = \{<a,1>,<b,2>,<c,4>\}$

R 和 R^C 的关系矩阵分别是：

$$\boldsymbol{M}_R = \begin{bmatrix} 1 & 0 & 0 \\ 0 & 1 & 0 \\ 0 & 0 & 0 \\ 0 & 0 & 1 \end{bmatrix} \qquad \boldsymbol{M}_{R^C} = \begin{bmatrix} 1 & 0 & 0 & 0 \\ 0 & 1 & 0 & 0 \\ 0 & 0 & 0 & 1 \end{bmatrix}$$

显然，$\boldsymbol{M}_{R^C} = \boldsymbol{M}_R^{\mathrm{T}}$。

R 和 R^C 的关系图分别如图 3.6.1 和图 3.6.2 所示，两张图中的弧线的方向是相反的。

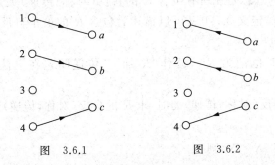

图　3.6.1　　　　　　图　3.6.2

定理 3.6.7 ▶ 设 X,Y 是集合，$R \subseteq X \times Y$，则 $(R^C)^C = R$；$\mathrm{dom}\, R^C = \mathrm{ran}\, R$，$\mathrm{ran}\, R^C = \mathrm{dom}\, R$。

证明 (1) $<x,y> \in R \Leftrightarrow <y,x> \in R^C \Leftrightarrow <x,y> \in (R^C)^C$ 所以 $(R^C)^C = R$。

(2) $x \in \mathrm{dom}\, R^C \Leftrightarrow (\exists y)(<x,y> \in R^C) \Leftrightarrow (\exists y)(<y,x> \in R) \Leftrightarrow x \in \mathrm{ran}\, R$，所以 $\mathrm{dom}\, R^C = \mathrm{ran}\, R$。

类似地,可以证明 $\operatorname{ran} R^c = \operatorname{dom} R$。

定理 3.6.8 ▶ 设 X, Y 是集合,R, S 是 X 到 Y 的二元关系,则

(1) $(R \cup S)^c = R^c \cup S^c$

(2) $(R \cap S)^c = R^c \cap S^c$

(3) $(A \times B)^c = B \times A$

(4) $(\sim R)^c = \sim(R^c)$

(5) $(R - S)^c = R^c - S^c$

定理 3.6.9 ▶ 设 X, Y, Z 是集合,$R \subseteq X \times Y, S \subseteq Y \times Z$,则 $(R \circ S)^c = S^c \circ R^c$。

证明 $<z, x> \in (R \circ S)^c \Leftrightarrow <x, z> \in (R \circ S)$

$$\Leftrightarrow (\exists y)(<x, y> \in R \land <y, z> \in S)$$

$$\Leftrightarrow (\exists y)(<y, x> \in R^c \land <z, y> \in S^c)$$

$$\Leftrightarrow (\exists y)(<z, y> \in S^c \land <y, x> \in R^c)$$

$$\Leftrightarrow <z, x> \in S^c \circ R^c$$

所以 $(R \circ S)^c = S^c \circ R^c$。

3.7 关系的闭包运算

上一节所讲的关系的复合和关系的逆运算都可以构成新的关系。如果在给定的关系中用扩充一些序偶的方法得到具有特殊性质的新关系,就称为闭包运算。

定义 3.7.1 ▶ 设 $R \subseteq X \times X$,X 上的二元关系 R' 满足:

(1) R' 是自反的(对称的、传递的)。

(2) $R \subseteq R'$。

(3) 对 X 上的任意二元关系 R'',如果 $R \subseteq R''$ 且 R'' 是自反的(对称的、传递的),就有 $R' \subseteq R''$。

则称 R' 是 R 的自反(对称、传递)闭包,记为 $r(R)(s(R), t(R))$。

定义 3.7.1 的(1)和(2)的意思是自反(对称、传递)闭包 R' 是包含 R 的自反(对称、传递)关系;定义 3.7.1 的(3)的意思是在所有包含 R 的自反(对称、传递)关系中,自反(对称、传递)闭包 R' 是最小的。所以,定义 3.7.1 又可以描述为包含 R 的最小自反(对称、传递)关系是 R 的自反(对称、传递)闭包。

传递闭包 $t(R)$ 有时也记为 R^+,读作"R 加"。传递闭包在语法分析中有很多重要的应用。

当二元关系 R 是自反(对称,传递)的时,求 R 的自反(对称,传递)闭包的方法由下面的定理给出。

定理 3.7.1 ▶ 设 $R \subseteq X \times X$,则

(1) R 是自反的,当且仅当 $r(R) = R$。

(2) R 是对称的,当且仅当 $s(R) = R$。

(3) R 是传递的,当且仅当 $t(R) = R$。

证明 仅证明(1),其余留做练习。

设 R 是自反的,下证 $r(R) = R$。

令 $R'=R$,

① $R'=R$,R 是自反的,所以 R' 是自反的。

② 因为 $R'=R$,所以 $R\subseteq R'$。

③ R'' 是 X 上的任意自反关系且 $R\subseteq R''$,因为 $R'=R$,所以 $R'\subseteq R''$。

故 R' 是 R 的自反闭包,即 $r(R)=R$。

设 $r(R)=R$,下证 R 是自反的。

由自反闭包的定义可知,$r(R)$ 自反,所以 R 是自反的。

下列几个定理给出了求二元关系 R 的自反闭包、对称闭包和传递闭包的一般方法。

定理 3.7.2 ▶ 设 $R\subseteq X\times X$,则 $r(R)=R\cup I_x$。

证明 令 $R'=R\cup I_x$,

(1) $I_x\subseteq R\cup I_x$,而 $R\cup I_x=R'$,所以 $I_x\subseteq R'$,R' 是自反的。

(2) $R\subseteq R\cup I_x$,而 $R\cup I_x=R'$,所以 $R\subseteq R'$。

(3) R'' 是 X 上的任意自反关系且 $R\subseteq R''$,由于 R'' 是自反的,所以 $I_x\subseteq R''$,从而有 $R'=R\cup I_x\subseteq R''$。

根据自反闭包的定义,$R'=R\cup I_x$ 是 R 的自反闭包,即 $r(R)=R\cup I_x$。

定理 3.7.3 ▶ 设 $R\subseteq X\times X$,则 $s(R)=R\cup R^c$。

证明 令 $R'=R\cup R^c$,

(1) $(R')^c=(R\cup R^c)^c=R^c\cup(R^c)^c=R^c\cup R=R\cup R^c=R'$,所以 R' 是对称的。

(2) $R\subseteq R\cup R^c$,而 $R\cup R^c=R'$,所以 $R\subseteq R'$。

(3) R'' 是 X 上的任意对称关系且 $R\subseteq R''$。

$$<x,y>\in R^c\Rightarrow<y,x>\in R\Rightarrow<y,x>\in R''\Rightarrow<x,y>\in R''\quad(R''是对称的)$$

所以 $R^c\subseteq R''$,从而有 $R'=R\cup R^c\subseteq R''$。$R'$ 是 R 的对称闭包,即 $s(R)=R\cup R^c$。

定理 3.7.4 ▶ 设 $R\subseteq X\times X$,则 $t(R)=R\cup R^2\cup R^3\cup\cdots$。

定理 3.7.5 ▶ 设 $|X|=n$,$R\subseteq X\times X$,则 $t(R)=R\cup R^2\cup\cdots\cup R^n$。

例 3.7.1 设 $A=\{1,2,3\}$,定义 A 上的二元关系 $R=\{<1,2>,<2,3>,<3,1>\}$,试用关系矩阵求传递闭包 $t(R)$。

解 用关系矩阵求 $t(R)$ 的方法如下:

$$M_R=\begin{bmatrix}0&1&0\\0&0&1\\1&0&0\end{bmatrix}$$

$$M_{R^2}=M_R\circ M_R=\begin{bmatrix}0&1&0\\0&0&1\\1&0&0\end{bmatrix}\circ\begin{bmatrix}0&1&0\\0&0&1\\1&0&0\end{bmatrix}=\begin{bmatrix}0&0&1\\1&0&0\\0&1&0\end{bmatrix}$$

$$M_{R^3}=M_{R^2}\circ M_R=\begin{bmatrix}0&0&1\\1&0&0\\0&1&0\end{bmatrix}\circ\begin{bmatrix}0&1&0\\0&0&1\\1&0&0\end{bmatrix}=\begin{bmatrix}1&0&0\\0&1&0\\0&0&1\end{bmatrix}$$

$$M_{t(R)}=M_R\vee M_{R^2}\vee M_{R^2}=\begin{bmatrix}1&1&1\\1&1&1\\1&1&1\end{bmatrix}$$

其中，∨表示矩阵的对应元素进行析取运算。

定理 3.7.6 ▶ 设$R\subseteq X\times X$，则$(R\cup I_X)^n=(R\cup R^2\cup\cdots\cup R^n)\cup I_X$。

二元关系的闭包仍然是二元关系，还可以求它的闭包。例如，R 是 A 上的二元关系，$r(R)$ 是它的自反闭包，还可以求 $r(R)$ 的对称闭包。$r(R)$ 的对称闭包记为 $s(r(R))$，简记为 $sr(R)$，读作 R 的自反闭包的对称闭包。类似地，R 的对称闭包的自反闭包记为 $r(s(R))$，简记为 $rs(R)$；R 的对称闭包的传递闭包记为 $t(s(R))$，简记为 $ts(R)$，$\cdots$

通常用 R^* 表示 R 的传递闭包的自反闭包 $rt(R)$，读作"R 星"。在研究形式语言和计算模型时经常使用 R^*。

定理 3.7.7 ▶ 设$R\subseteq X\times X$，则

$$rs(R)=sr(R),\quad rt(R)=tr(R),\quad st(R)\subseteq ts(R)$$

3.8　等价关系与相容关系

在二元关系中有两个具有非常重要意义的关系——等价关系和相容关系。

定义 3.8.1 ▶ 设$R\subseteq X\times X$，如果 R 是自反的、对称的和传递的，则称 R 是 X 上的等价关系。设 R 是等价关系，若$<x,y>\in R$，称 x 等价于 y。

因为等价关系是自反的和对称的，所以等价关系的关系矩阵的主对角线全为 1，且是对称阵；其关系图每一个结点上都有自回路，且每两个结点间如果有边，一定有方向相反的两条边。

例 3.8.1　设$A=\{1,2,3,4,5\}$，R 是 A 上的二元关系，$R=\{<1,1>,<1,2>,<2,1>,<2,2>,<3,3>,<3,4>,<4,3>,<4,4>,<5,5>\}$，证明 R 是 A 上的等价关系。

证明　写出 R 的关系矩阵 $\boldsymbol{M}_R$。

$$\boldsymbol{M}_R=\begin{bmatrix}1&1&0&0&0\\1&1&0&0&0\\0&0&1&1&0\\0&0&1&1&0\\0&0&0&0&1\end{bmatrix}$$

$\boldsymbol{M}_R$ 的主对角线全为 1 且是对称阵，所以 R 是自反的和对称的；还可以用二元关系传递性的定义证明 R 是传递的。故 R 是 A 上的等价关系。

也可以用关系图说明 R 是等价关系。图 3.8.1 是 R 的关系图。在 R 的关系图中，每一个结点上都有自回路；每两个结点间如果有边，一定有方向相反的两条边。所以 R 是自反的和对称的。与前面一样，也可以用二元关系传递性的定义证明 R 是传递的。故 R 是 A 上的等价关系。

从图 3.8.1 中不难看出，等价关系 R 的关系图被分为三个互不连通的部分。每部分中的结点两两都有关系，不同部分中的任意两个结点则没有关系。

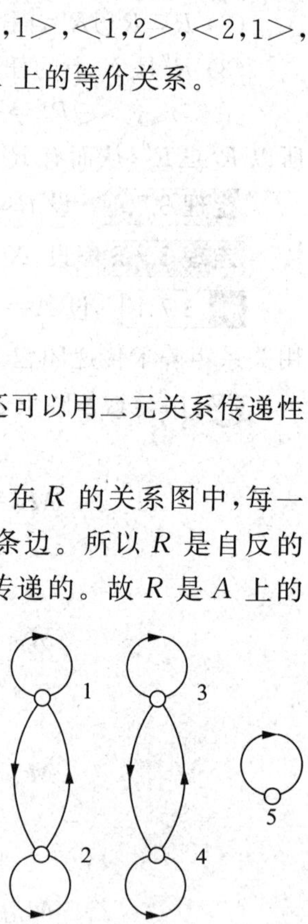

设 x 和 y 是两个整数，k 是一个正整数，若 x,y 用 k 除的余数相等，就称 x 和 y 模 k 同余，也称 x 和 y 模 k 等价，记为 $x\equiv y\bmod k$。

设 $x(y)$ 用 k 除的商为 $t_1(t_2)$，余数为 $a_1(a_2)$，数学上将 $x(y)$ 表

图　3.8.1

示为

$$x=k\times t_1+a_1, t_1\in I, a_1\in I \text{ 且 } 0\leqslant a_1<k$$
$$y=k\times t_2+a_2, t_2\in I, a_2\in I \text{ 且 } 0\leqslant a_2<k$$

若 x,y 用 k 除的余数相等，$x-y=k\times(t_1-t_2),t_1-t_2\in I$。即 $x-y$ 可以被 k 整除。所以，x 和 y 模 k 同余还可以描述为 $x-y$ 可以被 k 整除。

例 3.8.2　设 $R=\{<x,y>|x\in I\land y\in I\land x\equiv y \bmod k\}$ 是整数集合 I 上的二元关系。证明 R 是等价关系。

证明　设 a,b,c 是任意整数。

(1) 因为 $a-a=k\times 0$，所以 $a\equiv a \bmod k$，$<a,a>\in R$。故 R 是自反的。

(2) 若 $<a,b>\in R$，则 $a\equiv b \bmod k$，$a-b=k\times t,t\in I,b-a=-(a-b)=k\times(-t)$，$-t\in I,b\equiv a \bmod k$，$<b,a>\in R$。故 R 是对称的。

(3) 若 $<a,b>\in R$ 且 $<b,c>\in R$，$a-b=k\times t_1$，$t_1\in I,b-c=k\times t_2$，$t_2\in I$，则 $a-c=(a-b)+(b-c)=k\times t_1+k\times t_2=k\times(t_1+t_2),t_1+t_2\in I,<a,c>\in R$，故 R 是传递的。所以 R 是整数集合 I 上的等价关系。

定义 3.8.2　设 R 是 X 上的等价关系，$\forall x\in X$，集合 $\{y|y\in X\land xRy\}$ 叫作 x 形成的 R 等价类，记为 $[x]_R$。

在例 3.8.1 中，等价关系 R 等价类为 $[1]_R=[2]_R=\{1,2\},[3]_R=[4]_R=\{3,4\},[5]_R=\{5\}$。在 R 的关系图(图 3.8.1)中，三个互不连通的部分，每一部分中的所有结点构成一个等价类。

上述模 3 等价关系的等价类叫模 3 等价类，模 3 等价类有以下三个。

$$[0]_R=\{\cdots,-6,-3,0,3,6,\cdots\}$$
$$[1]_R=\{\cdots,-5,-2,1,4,7,\cdots\}$$
$$[2]_R=\{\cdots,-4,-1,2,5,8,\cdots\}$$

定理 3.8.1　设 R 是 X 上的等价关系，$\forall x\in X$，则有 $[x]_R\subseteq X,x\in[x]_R$。

定理 3.8.2　设 R 是 X 上的等价关系，对 X 的任意元素 a 和 b，aRb 的充分必要条件是 $[a]_R=[b]_R$。

证明　设 aRb，下证 $[a]_R=[b]_R$。

$\forall c\in[a]_R,aRc$，由 R 的对称性有 cRa，由条件 aRb 和 R 的传递性得 cRb，再根据 R 的对称性有 $bRc,c\in[b]_R$，故 $[a]_R\subseteq[b]_R$。

类似地，可以证明 $[b]_R\subseteq[a]_R$，从而证明了 $[a]_R=[b]_R$。

设 $[a]_R=[b]_R$，下证 aRb。

由定理 3.8.1 可知 $a\in[a]_R$，因为 $[a]_R=[b]_R$，所以 $a\in[b]_R,bRa$，由 R 的对称性有 aRb。

定义 3.8.3　设 R 是 X 上的等价关系，R 的所有等价类组成的集合 $\{[x]_R|x\in X\}$ 叫作 X 关于 R 的商集，记为 X/R。

在例 3.8.1 中，等价关系 R 有三个等价类：$[1]_R=[2]_R=\{1,2\},[3]_R=[4]_R=\{3,4\}$，$[5]_R=\{5\}$，$A$ 关于 R 的商集

$$A/R=\{[1]_R,[3]_R,[5]_R\}=\{\{1,2\},\{3,4\},\{5\}\}$$

模 3 等价关系 R 的等价类有三个：$[0]_R,[1]_R$ 和 $[2]_R$，由此确定的整数集合 I 关于 R 的

商集 $I/R = \{[0]_R, [1]_R, [2]_R\}$。

定义 3.8.4 ▶ 设有非空集合 A,设有非空集合 A,$S = \{A_1, A_2, \cdots, A_m\}$,其中 $A_i \subseteq A$ 且 $A_i \neq \varnothing$,$(i = 1, 2, \cdots, m)$,若

(1) 当 $i \neq j$ 时,$A_i \cap A_j = \varnothing$,

(2) $\bigcup\limits_{i=1}^{m} A_i = A$。

则称 S 是集合 A 的一个划分,每一个 A_i 称为这个划分的一个分块。

定理 3.8.3 ▶ 设 R 是 X 上的等价关系,X 关于 R 的商集 X/R 是 X 的一个划分。

定理 3.8.4 ▶ 设 $S = \{S_1, S_2, \cdots, S_m\}$ 是 X 的一个划分,$R = \{<x, y> | x$ 和 y 在同一个划分块中$\}$,则 R 是 X 上的等价关系。

例 3.8.3 设 $X = \{1, 2, 3, 4\}$,X 的划分 $S = \{\{1\}, \{2, 3\}, \{4\}\}$,试写出 S 导出的等价关系 R。

解 $R = \{<1,1>, <2,2>, <2,3>, <3,2>, <3,3>, <4,4>\}$
$= \{1\} \times \{1\} \bigcup \{2,3\} \times \{2,3\} \bigcup \{4\} \times \{4\}$

可以验证 R 是 X 上的等价关系。

定理 3.8.5 ▶ 设 R, S 是集合 X 上的等价关系,则当且仅当 $X/R = X/S$ 时 $R = S$。

例 3.8.4 设 $X = \{1, 2, 3\}$,写出集合 X 上的所有等价关系。

解 先写出集合 X 上的所有划分,它们是:

$$S_1 = \{\{1,2,3\}\}, \qquad S_2 = \{\{1,2\}, \{3\}\}$$
$$S_3 = \{\{1,3\}, \{2\}\}, \quad S_4 = \{\{2,3\}, \{\{1\}\}$$
$$S_5 = \{\{1\}, \{2\}, \{3\}\}$$

对应的等价关系为

$$R_1 = \{1,2,3\} \times \{1,2,3\} = X \times X$$
$$R_2 = \{\{1,2\} \times \{1,2\}\} \bigcup \{\{3\} \times \{3\}\}$$
$$= \{<1,1>, <1,2>, <2,1>, <2,2>, <3,3>\}$$
$$R_3 = \{\{1,3\} \times \{1,3\}\} \bigcup \{\{2\} \times \{2\}\}$$
$$= \{<1,1>, <1,3>, <3,1>, <3,3>, <2,2>\}$$
$$R_4 = \{\{2,3\} \times \{2,3\}\} \bigcup \{\{1\} \times \{1\}\}$$
$$= \{<2,2>, <2,3>, <3,2>, <3,3>, <1,1>\}$$
$$R_5 = \{\{1\} \times \{1\}\} \bigcup \{\{2\} \times \{2\}\} \bigcup \{\{3\} \times \{3\}\}$$
$$= \{<1,1>, <2,2>, <3,3>\} = I_X$$

定义 3.8.5 ▶ 设 $R \subseteq X \times X$,如果 R 是自反的和对称的,则称 R 是 X 上的相容关系。

根据定义 3.8.5,相容关系有以下三个性质:

(1) 所有等价关系都是相容关系。

(2) 相容关系的关系矩阵主对角线全为 1 且是对称阵。

(3) 相容关系的关系图每一个结点上都有自回路且每两个结点间如果有边,一定有方向相反的两条边。

例 3.8.5 设 $A = \{316, 347, 204, 678, 770\}$,$A$ 上的二元关系 R 定义为 $R = \{<x, y> |$

$x \in A \wedge y \in A \wedge x$ 和 y 有相同数字},证明 R 是 A 上的相容关系,写出关系矩阵,画出关系图。用关系矩阵和关系图验证 R 是 A 上的相容关系。

证明　集合 A 中的每个数自己和自己有相同数字,故 R 是自反的;对于集合 A 中任意 x 和 y,如果 x 和 y 有相同数字,则 y 和 x 也有相同数字,故 R 是对称的。于是,R 是 A 上的相容关系。

令 $a = 316, b = 347, c = 204, d = 678, e = 770$。

用列举法表示 R:

$$
\begin{aligned}
R = & \{<a,a>,<a,b>,<a,d>,<b,a>,<b,b>,<b,c>,<b,d>,<b,e>,\\
& <c,b>,<c,c>,<c,e>,<d,a>,<d,b>,<d,d>,<d,e>,<e,b>,\\
& <e,c>,<e,d>,<e,e>\}\\
= & \{<a,b>,<a,d>,<b,a>,<b,c>,<b,d>,<b,e>,<c,b>,<c,e>,\\
& <d,a>,<d,b>,<d,e>,<e,b>,<e,c>,<e,d>\} \bigcup I_A
\end{aligned}
$$

R 的关系矩阵 $\boldsymbol{M}_R$ 的主对角线全为 1 且是对称阵,说明 R 是相容关系。

R 的关系图如图 3.8.2 所示。在关系图中,每一个结点上都有自回路且每两个结点间如果有边,一定有方向相反的两条边。这表明 R 是相容关系。

$$
\boldsymbol{M}_R = \begin{bmatrix} 1 & 1 & 0 & 1 & 0 \\ 1 & 1 & 1 & 1 & 1 \\ 0 & 1 & 1 & 0 & 1 \\ 1 & 1 & 0 & 1 & 1 \\ 0 & 1 & 1 & 1 & 1 \end{bmatrix}
$$

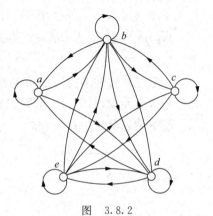

图　3.8.2

因为相容关系的关系图中的每一个结点上都有自回路且每两个结点间如果有边,一定有方向相反的两条边。所以可省去每一个结点上的自回路,将两个结点间方向相反的两条有向边改为一条无向边,得到一个简化图。此图叫作 R 的简化关系图。例 3.8.2 中的相容关系 R 的简化关系图如图 3.8.3 所示。

定义 3.8.6　设 R 是 X 上的相容关系,$C \subseteq X$,如果 $\forall a, b \in C$,有 $<a,b> \in R$,则称 C 是由相容关系 R 产生的相容类。

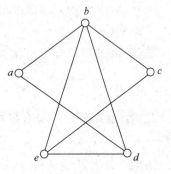

图　3.8.3

如果 R 是 X 上的相容关系,C 是由相容关系 R 产生的相容类。从定义可以看出:①相容类 C 一定是 X 的子集。②$\forall x \in X$,因为相容关系 R 是自反的,$<x,x> \in R$,所以 $\{x\}$ 是由相容关系 R 产生的一个相容类。即 X 中的任何元素组成的单元素集是由相容关系 R 产生的一个相容类。

在例 3.8.5 中,$\{a\}$,$\{a,b\}$,$\{b,c\}$,$\{b,d,e\}$ 都是 R 产生的相容类。

$\{a,b\} \bigcup \{d\} = \{a,b,d\}$ 和 $\{b,c\} \bigcup \{e\} = \{b,c,e\}$ 也是 R 产生的相容类。

但是 $\{b,d,e\}$ 与任何非空集合的并集都不再是 R 产生的相容类,这种相容类叫作最大相容类。

定义 3.8.7 ▶ 设 R 是 X 上的相容关系, C 是 R 产生的相容类。如果它不是其他任何相容类的真子集,则称 C 为最大相容类,记为 C_R。

根据定义 3.8.7,最大相容类 C_R 具有如下的性质:

① C_R 中任意元素 x 与 C_R 中的所有元素都有相容关系 R。

② $X - C_R$ 中没有一个元素与 C_R 中的所有元素都有相容关系 R。

性质①的意思是 C_R 是 R 产生的相容类,即最大相容类首先是相容类。性质②的意思是 C_R 是最大相容类。

利用相容关系的简化关系图可以求最大相容类,方法如下:

① 最大完全多边形的顶点构成的集合是最大相容类。

② 孤立点构成的集合是最大相容类。

③ 如果一条边不是任何完全多边形的边,则它的两个端点构成的集合是最大相容类。

例 3.8.6 设 $X = \{1,2,3,4,5,6\}$, R 是 X 上的相容关系,它的简化关系图如图 3.8.4 所示,试找出所有最大相容类。

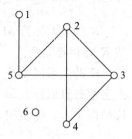

图 3.8.4

解 最大完全 3 边形的顶点构成的集合: $\{2,3,5\}$ 和 $\{2,3,4\}$。孤立点构成的集合: $\{6\}$。

不是任何完全多边形的边的两个端点构成的集合: $\{1,5\}$。所以,最大相容类为 $\{2,3,5\}$、$\{2,3,4\}$、$\{1,5\}$ 和 $\{6\}$。

定理 3.8.6 ▶ 设 R 是有限集合 X 上的相容关系, C 是 R 产生的相容类,那么必存在最大相容类 C_R,使 $C \subseteq C_R$。

证明 设 $X = \{x_1, x_2, \cdots, x_n\}$,令 $C_0 = C$。构造相容类序列: $C_0 \subset C_1 \subset C_2 \subset \cdots$

$$C_{i+1} = C_i \bigcup \{x_j\}$$

其中, j 是使 $x_j \notin C_i$ 且 x_j 与 C_i 的每一个元素都有相容关系 R 的 x 的最小下标。

因为 $|X| = n$,所以至多经过 $n - |C|$ 步就可结束此过程。序列的最后一个集合就是要求最大相容类 C_R。

定义 3.8.8 ▶ 设有非空集合 A, $S = \{A_1, A_2, \cdots, A_m\}$,其中 $A_i \subseteq A$ 且 $A_i \neq \varnothing$ ($i = 1, 2, \cdots, m$),若 $\bigcup_{i=1}^{m} A_i = A$,则称 S 是集合 A 的覆盖。

定理 3.8.7 ▶ 设 X 是有限集合, R 是 X 上的相容关系,由 R 产生的所有最大相容类构成的集合是 X 的覆盖。

证明 设 R 产生的所有最大相容类构成的集合为 $\{C_1, C_2, \cdots, C_n\}$。

(1) 由相容类的定义可知,任何最大相容类 C_i 都是 X 的子集。

(2) 证明 $X = C_1 \bigcup C_2 \bigcup \cdots \bigcup C_n$。

因为 $C_j \subseteq X$,所以 $C_1 \bigcup C_2 \bigcup \cdots \bigcup C_n \subseteq X$。

$\forall x \in X$, $\{x\}$ 是相容类,根据定理 3.8.6,必存在最大相容类 C_i 使 $\{x\} \subseteq C_i$,而 $C_i \subseteq C_1 \bigcup C_2 \bigcup \cdots \bigcup C_n$。

所以 $x \in C_1 \bigcup C_2 \bigcup \cdots \bigcup C_n$。

这就证明 $X \subseteq C_1 \bigcup C_2 \bigcup \cdots \bigcup C_n$。

故 $X = C_1 \bigcup C_2 \bigcup \cdots \bigcup C_n$。

定义 3.8.9 ▶ 设 R 是 X 上的相容关系，$S=\{R\mid R$ 产生的最大相容类$\}$ 叫作集合 X 的完全覆盖，记为 $C_R(X)$。

例如，在例 3.8.6 中最大相容类构成的集合为 $\{\{2,3,5\},\{2,3,4\},\{1,5\},\{6\}\}$，它是集合 $X=\{1,2,3,4,5,6\}$ 的一个完全覆盖。根据相容关系和相容类的定义，X 中的任何元素组成的单元素集是由相容关系 R 产生的一个相容类。所以 $\{1\},\{2\},\{3\},\{4\},\{5\},\{6\}$ 都是相容类，集合 $\{\{1\},\{2\},\{3\},\{4\},\{5\},\{6\}\}$ 是 X 的覆盖。即完全覆盖是 X 的覆盖，某些相容类构成的集合也是 X 的覆盖。所以由相容关系 R 产生的 X 的覆盖并不唯一。但是一个相容关系只对应唯一的一个完全覆盖。

定理 3.8.8 ▶ 设 $S=\{S_1,S_2,\cdots,S_m\}$ 是 X 的一个覆盖，则 $R=(S_1\times S_1)\bigcup(S_2\times S_2)\bigcup\cdots\bigcup(S_m\times S_m)$ 是 X 上的相容关系。

例 3.8.7 设 $X=\{1,2,3,4\}$，$S_1=\{\{1,2,3\},\{3,4\}\}$ 和 $S_2=\{\{1,2\},\{2,3\},\{1,3\},\{3,4\}\}$ 是 X 的两个覆盖。写出 S_1 和 S_2 导出的相容关系 R_1 和 R_2。

解

$R_1=\{\{1,2,3\}\times\{1,2,3\}\}\bigcup\{\{3,4\}\times\{3,4\}\}$
$\quad=\{<1,1>,<1,2>,<1,3>,<2,1>,<2,2>,<2,3>,<3,1>,<3,2>,<3,3>,<3,4>,$
$\qquad<4,3>,<4,4>\}$

$R_2=\{\{1,2\}\times\{1,2\}\}\bigcup\{\{2,3\}\times\{2,3\}\}\bigcup\{\{1,3\}\times\{1,3\}\}\bigcup\{\{3,4\}\times\{3,4\}\}$
$\quad=\{<1,1>,<1,2>,<2,1>,<2,2>,<2,3>,<3,2>,<3,3>,<1,3>,<3,1>,<3,4>,$
$\qquad<4,3>,<4,4>\}$

在例 3.8.7 中，$S_1\neq S_2$，但是 $R_1=R_2$。这说明不同的覆盖可以导出相同的相容关系。

定理 3.8.9 ▶ 集合 X 上的相容关系 R 与集合 X 的完全覆盖 $C_R(X)$ 是一一对应的。

3.9 偏序关系

在集合中，常常要考虑元素的次序关系，其中一种重要关系称作偏序关系。

定义 3.9.1 ▶ 设 $R\subseteq X\times X$，如果 R 是自反的、反对称的和传递的，则称 R 是 X 上的偏序关系，记为 $\leqslant$。二重组 $<X,\leqslant>$ 称为偏序集。如果 $<x,y>\in\leqslant$，记为 $x\leqslant y$，读作 x 偏序于 y。

例 3.9.1 设 A 是集合，$P(A)$ 是 A 的幂集合，$P(A)$ 上的包含关系 $\subseteq$ 定义如下：
$$R_\subseteq=\{<x,y>\}x\in P(A)\wedge y\in P(A)\wedge x\subseteq y\}$$
试证明 $\subseteq$ 是 $P(A)$ 上偏序关系。

证明 (1) $\forall x\in P(A)$，$x\subseteq x$，$<x,x>\in\subseteq$。即 $\subseteq$ 是自反的。

(2) 设 $<x,y>\in\subseteq\wedge<y,x>\in\subseteq$，由包含关系 $\subseteq$ 定义有 $x\subseteq y\wedge y\subseteq x$，由集合相等的定义有 $x=y$。即 $\subseteq$ 是反对称的。

(3) 设 $<x,y>\in\subseteq\wedge<y,z>\in\subseteq$，由包含关系 $\subseteq$ 的定义有 $x\subseteq y\wedge y\subseteq z$，由集合包含的传递性有 $x\subseteq z$，再由包含关系 $\subseteq$ 定义有 $<x,z>\in\subseteq$，即 $\subseteq$ 是传递的。$\subseteq$ 是 $P(A)$ 上的偏序关系。

另外，任意集合 A 上的恒等关系 I_A 是自反的、反对称的和传递的，I_A 是 A 上的偏序关系；

实数集合上的小于或等于关系也是自反的、反对称的和传递的,它是实数集合上的偏序关系。

定义 3.9.2 ▶ 设$<X,\leqslant>$为偏序集,对 X 中的元素 x 和 y,如果 $x\leqslant y,x\neq y$ 且没有 X 中的其他元素 z 使 $x\leqslant z$ 和 $z\leqslant y$,则称 y 盖住了 x。

例 3.9.2 设 $A=\{2,5,6,10,15,30\}$,A 上的整除关系 R 定义如下:
$$R=\{<x,y>|x\in A \wedge y\in A \wedge x\text{ 整除 }y\}$$

验证 R 是 A 上的偏序关系,分析哪些元素盖住了另一些元素,哪些元素没有盖住另一些元素。

解 因为 A 的任何元素都能整除它自己,所以 R 是自反的;当 x 整除 y 且 y 整除 x 时,一定有 $x=y$,所以 R 是反对称的;当 x 整除 y 且 y 整除 z,x 一定整除 z,所以 R 是传递的。即 R 是 A 上的偏序关系。

用列举法表示 R:
$$R=\{<2,2>,<2,6>,<2,10>,<2,30>,<5,5>,<5,10>,<5,15>,<5,30>,$$
$$<6,6>,<6,30>,<10,10>,<10,30>,<15,15>,<15,30>,<30,30>\}$$

6 和 10 盖住了 2;但 30 没有盖住 2,因为$<2,6>\in R$ 和$<6,30>\in R$。

10 和 15 盖住了 5;但 30 没有盖住 5,因为$<5,10>\in R$ 和$<10,30>\in R$。

30 盖住了 6、10 和 15。

定义 3.9.3 ▶ 设$<X,\leqslant>$为偏序集,集合$\{<x,y>|x\in X \wedge y\in X \wedge y\text{ 盖住了 }x\}$叫作 X 上的盖住关系,记为 COV X。

设 COV X 是 X 上的盖住关系,如果$<x,y>\in$COV X,根据定义 3.9.3,y 盖住了 x,由定义 3.9.2 可知,$x\leqslant y$,即$<x,y>\in\leqslant$。故 COV X$\subseteq\leqslant$。所以任何盖住关系 COV X 都是相应偏序关系$\leqslant$的子集。

在例 3.9.2 中,偏序集$<A,R>$的盖住关系
COV $A=\{<2,6>,<2,10>,<5,10>,<5,15>,<6,30>,<10,30>,<15,30>\}$

设$<X,\leqslant>$是偏序集,它的盖住关系 COV X 是唯一的。所以可以利用盖住关系做图,表示该偏序集$<X,\leqslant>$,这个图叫作哈斯图。偏序集$<X,\leqslant>$的哈斯图的画法如下:

① 用"。"表示 X 中的每一个元素。

② 如果 $x\leqslant y$,则将 x 画在 y 的下方。

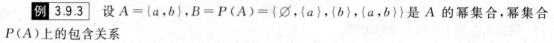
图 3.9.1

③ 若$<x,y>\in$COV X,则在 x 和 y 之间画一条直线。

例 3.9.2 的哈斯图如图 3.9.1 所示。

例 3.9.3 设 $A=\{a,b\}$,$B=P(A)=\{\varnothing,\{a\},\{b\},\{a,b\}\}$是 A 的幂集合,幂集合 $P(A)$上的包含关系
$$R_{\subseteq}=\{<x,y>|x\in P(A) \wedge y\in P(A) \wedge x\subseteq y\}$$

试写出盖住关系 COV $P(A)$,画出它的哈斯图。

解 用列举法表示 $R_{\subseteq}$:
$$R_{\subseteq}=\{<\varnothing,\{a\}>,<\varnothing,\{b\}>,<\varnothing,\{a,b\}>,<\{a\},\{a,b\}>,<\{b\},\{a,b\}>\}\bigcup I_B$$

在例 3.9.1 中,已证 $R_{\subseteq}$ 是 $P(A)$上的偏序关系。$P(A)$上的盖住关系为
COV $P(A)=\{<\varnothing,\{a\}>,<\varnothing,\{b\}>,<\{a\},\{a,b\}>,<\{b\},\{a,b\}>\}$

其哈斯图如图 3.9.2 所示。

定义 3.9.4 ▶ 设$<X,\preccurlyeq>$是偏序集,$B\subseteq X,b\in B$,如果 B 中没有任何元素 x 满足 $x\neq b$ 且 $b\preccurlyeq x(x\preccurlyeq b)$,则称 b 是 B 的极大元(极小元)。

当 $B=X$ 时,B 的极大元(极小元)称为偏序集$<X,\preccurlyeq>$的极大元(极小元)。

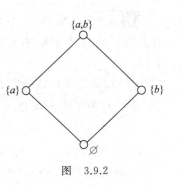

图　3.9.2

例 3.9.4 设 $A=\{a,b,c,d,e,f,g,h\}$,A 上的二元关系

$R=\{<b,d>,<b,e>,<b,f>,<c,d>,<c,e>,<c,f>,$
$<d,f>,<e,f>,<g,h>\}\bigcup I_A$

验证 R 是 A 上的偏序关系。写出盖住关系 COV A,画出哈斯图。找出集合 A 的极大元和极小元。

证明 容易验证 R 是自反的、反对称的和传递的,R 是 A 上的偏序关系。

COV $A=\{<b,d>,<b,e>,<c,d>,<c,e>,<d,f>,$
$\qquad\qquad<e,f>,<g,h>\}$

其哈斯图如图 3.9.3 所示。

集合 A 的极大元是 a,f,h。

集合 A 的极小元是 a,b,c,g。

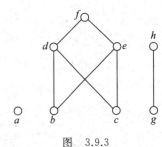

图　3.9.3

定理 3.9.1 ▶ 由定义 3.9.4 和例 3.9.4 可以得出以下结论:

① 孤立点既是极大元又是极小元。

② 极大元和极小元不唯一。有限集合 B 的极大元和极小元一定存在。

③ 在哈斯图中,如果集合 B 的某个元素不存在 B 的其他元素从上(下)方与其相通,则该元素就是 B 的极大元(极小元)。

定义 3.9.5 ▶ 设$<X,\preccurlyeq>$是偏序集,$B\subseteq X,b\in B$,如果对任意 $x\in B$,都有 $x\preccurlyeq b(b\preccurlyeq x)$,则称 b 是 B 的最大元(最小元)。

当 $B=X$ 时,B 的最大元(最小元)称为偏序集$<X,\preccurlyeq>$的最大元(最小元)。在例 3.9.4 中,集合 A 没有最大元和最小元。

在例 3.9.3 中,令 $B=\{\varnothing,\{a\}\}$,B 的最大元是 $\{a\}$,B 的最小元是 $\varnothing$。令 $B=\{\{a\},\{b\}\}$,B 没有最大元和最小元。

定理 3.9.2 ▶ 设$<X,\preccurlyeq>$是偏序集,$B\subseteq X$,如 B 有最大元(最小元),则必唯一。

证明 设 a,b 都是 B 的最大元,由 a 是 B 的最大元得 $b\preccurlyeq a$,由 b 是 B 的最大元得 $a\preccurlyeq b$,因为$\preccurlyeq$是反对称的,所以 $a=b$。即 B 的最大元唯一。

类似地可证明最小元唯一。

从以上例题和定理 3.9.2 可以得出以下结论:

① 最大元和最小元不一定存在。如果存在,一定唯一。

② 在哈斯图中,如果集合 B 的某个元素向下(上)通向 B 的所有元素,则该元素就是 B 的最大元(最小元)。

定义 3.9.6 ▶ 设$<X,\preccurlyeq>$是偏序集,$B\subseteq X,b\in X$,对任意 $x\in B$,都有 $x\preccurlyeq b(b\preccurlyeq x)$,则称 b 是 B 的上界(下界)。

例 3.9.5 设 $A=\{2,3,6,12,24,36\}$,其上的整除关系 $R=\{<a,b>|a\in A \wedge b\in A \wedge a$ 能整除 $b\}$ 是 A 上的偏序关系,试求盖住关系 COV A,画出哈斯图,确定下列集合的上界和下界。

(1) $B_1=\{2,3,6\}$;(2) $B_2=\{12,24,36\}$。

解 用列举法表示 R:

$R=\{<2,6>,<2,12>,<2,24>,<2,36>,<3,6>,<3,12>,<3,24>,<3,36>,$
$\qquad <6,12>,<6,24>,<6,36>,<12,24>,<12,36>\}\bigcup I_A$

盖住关系 COV $A=\{<2,6>,<3,6>,<6,12>,<12,24>,<12,36>\}$

其哈斯图如图 3.9.4 所示。

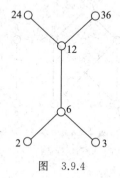

图 3.9.4

B_1 的上界是 $6,12,24,36$,没有下界。

B_2 的下界是 $2,3,6,12$,没有上界。

从本例可以看出:

① 上界和下界并不唯一。

② 在哈斯图中,如果集合 X 的某个元素向下(上)通向 B 的所有元素,则该元素就是 B 的上界(下界)。

定义 3.9.7 ▶ 设 $<X,\leqslant>$ 是偏序集,$B\subseteq X$,如果存在 B 的一个上界(下界)b,对 B 的任意一个上界(下界)y,都有 $b\leqslant y(y\leqslant b)$,则称 b 是 B 的最小上界(最大下界),也叫上确界(下确界),记为 LUB B(GLB B)。

在例 3.9.5 中,B_1 的最小上界是 6;B_1 无下界,当然无最大下界。B_2 的最大下界是 12;B_2 无上界,当然无最小上界。

定义 3.9.8 ▶ 设 $<X,\leqslant>$ 为偏序集,如果 $\forall x\in X$,$\forall y\in X$,都有 $x\leqslant y$ 或 $y\leqslant x$,则称偏序集 $<X,\leqslant>$ 为全序集,也称为线序集。偏序关系 $\leqslant$ 称为 X 上的全序关系或线序关系。

例 3.9.6 设 $P=\{\varnothing,\{a\},\{a,b\},\{a,b,c\}\}$,$P$ 上的包含关系 $R_\subseteq=\{<x,y>|x\in P \wedge y\in P \wedge x\subseteq y\}$,验证 $R_\subseteq$ 是全序关系。

证明 用列举法表示 P 上的包含关系 $R_\subseteq$:

$R_\subseteq=\{<\varnothing,\{a\}>,<\varnothing,\{a,b\}>,<\varnothing,\{a,b,c\}>,<\{a\},\{a,b\}>,<\{a\},\{a,b,c\}>,$
$\qquad <\{a,b\},\{a,b,c\}>\}\bigcup I_P$

容易验证 $R_\subseteq$ 是 P 上的偏序关系。

从 $R_\subseteq$ 的序偶集合可以看出,对于 P 中的任意两个元素 x 和 y,必有 $x\subseteq y$ 或 $y\subseteq x$,所以 $R_\subseteq$ 是 P 上的全序关系。其盖住关系为

COV $P=\{<\varnothing,\{a\}>,<\{a\},\{a,b\}>,<\{a,b\},\{a,b,c\}>\}$

其哈斯图如图 3.9.5 所示。

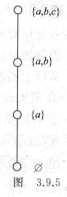

图 3.9.5

定义 3.9.9 ▶ 设 $<X,\leqslant>$ 为偏序集,如果 X 的每一个非空子集存在最小元,则偏序集 $<X,\leqslant>$ 叫作良序集。偏序关系 $\leqslant$ 称为 X 上的良序关系。

类似例 3.9.1 可以证明自然数集合 N 上的小于或等于关系是偏序关系,而自然数集合的每一个非空子集都存在最小数即最小元。所以自然数集合上的小于或等于关系 $<N,R_\leqslant>$ 是良序集。

定理 3.9.3 ▶ 每一个良序集一定是全序集。

证明　设$<X,\leqslant>$是良序集，x 和 y 是 X 中的任意两个元素，显然$\{x,y\}\subseteq X$。由良序集的定义可知，集合$\{x,y\}$必有最小元，即 x 是最小元或 y 是最小元。如果 x 是最小元，则有 $x\leqslant y$。如果 y 是最小元，则有 $y\leqslant x$。所以$<X,\leqslant>$是全序集。

定理 3.9.4 ▶有限全序集一定是良序集。

3.10　集合与关系的应用

集合论被应用在计算机科学研究的各个方面。集合是构造离散结构的基础，离散结构是计算机的基本结构。集合论在人工智能领域、逻辑学及程序设计语言等方面都有重要的应用。

集合论在数学理论中起着"地基"作用，在解析几何、解集、矩阵运算等方面都有重要意义。例如，几何元素间的各种结合关系，平行与垂直是集合间的某种关系；从自然数集到整数集→有理数集→实数集的扩充过程中都可以通过对前一个集按集合的某种等价关系分类得出。

平面几何图形的平移、旋转、反射、相似等几何变换都是 $\mathbf{R}^2$ 的子集中集合间满足一定条件的对应关系等。

1. 在几何学中的应用

例如，平面图形中以原点为圆心，1 为半径的圆周、开（闭）圆面可以分别表示为

$$A=\{(x,y)\mid x^2+y^2=1,(x,y)\in \mathbf{R}^2\}$$
$$A'=\{(x,y)\mid x^2+y^2<1,(x,y)\in \mathbf{R}^2\}$$
$$A''=\{(x,y)\mid x^2+y^2\leqslant 1,(x,y)\in \mathbf{R}^2\}$$

又如，解析几何中关于曲线方程"纯粹性"和"完备性"的叙述，"曲线上的点的坐标都是这个方程的解""以这个方程的解为坐标的点都是曲线的点"。用集合论的语言可以简单地表述为 $C=\{(x,y)\mid f(x,y)=0,(x,y)\in \mathbf{R}^2\}$，即曲线上的点满足方程 $C\subset\{(x,y)\mid f(x,y)=0,(x,y)\in \mathbf{R}^2\}$，且满足方程的点在曲线上 $C\supset\{(x,y)\mid f(x,y)=0,(x,y)\in \mathbf{R}^2\}$。

2. 在运算中的应用

（1）几何图形性质运算。

$$\{x\mid x\text{ 为矩形}\}\bigcap\{x\mid x\text{ 为菱形}\}=\{x\mid x\text{ 为正方形}\}$$

（2）数轴上数的运算。

$$\{x\mid x-1>0\}\bigcap\{x\mid 0<x<10\}=\{x\mid 1<x<10\}$$

（3）解方程组。

$$\begin{cases} x-y+1=0 \\ 3x+y-9=0 \end{cases}$$

即两条直线的交点坐标：$\{(x,y)\mid x-y+1=0 \text{ 且 } 3x+y-9=0\}$。

（4）解不等式组。

$$\begin{cases} x^2+y^2\geqslant 1 \\ y\leqslant x \end{cases}$$

（5）差集和补集的运算。

$$A-B=\{x\mid x\in A \text{ 且 } x\notin B\}$$

由定义显然：$A-B\neq B-A$。

设 $A=\{x\,|\,x$ 是江苏大学大一学生$\}$，$B=\{x\,|\,x$ 是江苏大学大一的女学生$\}$，$C=\{x\,|\,x$ 是江苏大学的女学生$\}$，$D=\{x\,|\,x$ 是江苏大学的学生$\}$，则有下列运算：

$$A-B=\{x\,|\,x\text{ 是江苏大学大一的男学生}\}$$
$$C-B=\{x\,|\,x\text{ 是江苏大学除了大一的女学生}\}$$
$$D-B=\{x\,|\,x\text{ 是江苏大学除了大一女学生的学生}\}$$

(6) 包容互斥问题。

如例 3.2.1，利用容斥原理推算满足特定条件的人员数量，结合文氏图可以更加清晰、快速地得出结果。

关系是集合的元素之间存在的某种关系，也就是说关系实际上是某个集合的子集(关系是笛卡儿积的子集)。例如，员工与其工资之间的关系。

3. 数据库技术中的应用

二元关系最广泛的应用领域是在数据库技术中。

数据库技术被广泛应用于社会各个领域，关系数据库(如 MySQL)已经成为数据库的主流，集合论中的笛卡儿积是一个纯数学理论，是研究关系数据库的一种重要方法，显示出不可替代的作用。不仅为其提供理论和方法上的支持，更重要的是推动了数据库技术的研究和发展。关系数据模型建立在严格的集合代数的基础上，其数据的逻辑结构用行和列组成的二维表来描述关系数据模型。在研究实体集中的域和域之间的可能关系、表结构的确定与设计、关系操作的数据查询和维护功能的实现、关系分解的无损连接性分析、连接依赖等问题中都用到二元关系理论。

关系中的每个元素是关系中的元组，通常用 t 表示，关系中元组的个数是关系的基数。

由于关系是笛卡儿积的子集，因此，也可以把关系看成一个二维表，具有相同关系框架的关系称为同类关系。

关系的性质：一种规范化的二维表中行的集合。

(1) 每一列中的分量必须来自同一个域，必须是同一类型的数据。

(2) 不同的列可来自同一个域，每一列为一个属性，不同的属性必须有不同的名字。

(3) 列的顺序可任意交换，名字同时换。

(4) 关系中元组的顺序(行序)可任意。

(5) 关系中每一分量必须是不可分的数据项。

习 题

1. 设 $A=\{a,\{a\}\}$，下列命题中错误的是(　　　　)。

(1) $\{a\}\in P(A)$

(2) $\{a\}\subseteq P(A)$

(3) $\{\{a\}\}\in P(A)$

(4) $\{\{a\}\}\subseteq P(A)$

2. 在 0(　　　)$\varnothing$ 之间填上正确的符号。

(1) $=$　　　　　　(2)$\subseteq$　　　　　(3) $\in$　　　　　　(4) $\notin$

3. 若集合 S 的基数 $|S|=5$，则 S 的幂集的基数 $|P(s)|=($　　　　)。

4. 设 $P=\{x\,|\,(x+1)^2\leqslant4,$ 且 $x\in R\},Q=\{x\,|\,5\leqslant x^2+16\text{ 且 }x\in R\},$ 则下列命题中正确的是(　　)。

(1) $Q\subset P$　　　　(2) $Q\subseteq P$　　　　(3) $P\subset Q$　　　　(4) $P=Q$

5. 下列集合中,哪几个分别相等?

(1) $A_1=\{a,b\}$　　(2) $A_2=\{b,a\}$　　(3) $A_3=\{a,b,a\}$　　(4) $A_4=\{a,b,c\}$

6. 若 $A-B=\varnothing,$ 则下列选项中不可能正确的是(　　)。

(1) $A=\varnothing$　　　　(2) $B=\varnothing$　　　　(3) $A\subset B$　　　　(4) $B\subset A$

7. 设 $A\bigcap B=A\bigcap C,\overline{A}\bigcap B=\overline{A}\bigcap C,$ 则 B(　　)C。

8. 下列命题中为真的是(　　)。

(1) $A-B=B-A\Rightarrow A=B$

(2) 空集是任何集合的真子集

(3) 空集是任何非空集合的子集

(4) 若 A 的一个元素属于 $B,$ 则 $A=B$

9. 下列命题中正确的是(　　)。

(1) 所有空集都不相等

(2) $\{\varnothing\}\neq\varnothing$

(3) 若 A 为非空集,则 $A\subset A$ 成立

10. 设 $A=\{1,2,3,4,5,6\},B=\{1,2,3\},$ 从 A 到 B 的关系 $R=\{<x,y>\,|\,x=y^2\},$ 求:(1)R;(2)R^{-1}。

11. 举出集合 A 上既是等价关系又是偏序关系的一个例子。

12. 集合 A 上的等价关系的三个性质是什么?

13. 设 $S=\{1,2,3,4\},A$ 上的关系 $R=\{<1,2>,<2,1>,<2,3>,<3,4>\},$ 求:(1)$R\circ R$;(2)R^{-1}。

14. 设 $A=\{1,2,3,4,5,6\},B=\{1,2,3\},$ 从 A 到 B 的关系 $R=\{<x,y>\,|\,x=2y\},$ 求:(1)R;(2)R^{-1}。

15. 设 $A=\{1,2,3,4,5,6\},B=\{1,2,3\},$ 从 A 到 B 的关系 $R=\{<x,y>\,|\,x=y^2\},$ 求 R、R^{-1} 的关系矩阵。

16. 集合 $A=\{1,2,3,\cdots,10\}$ 上的关系 $R=\{<x,y>\,|\,x+y=10,x,y\in A\},$ 则 R 的性质为(　　)。

(1) 自反的　　　　(2) 对称的　　　　(3) 传递的,对称的　　　　(4) 传递的

17. 如果 r_1 和 r_2 是 a 上的自反关系,则 $r_1\bigcup r_2,r_1\bigcap r_2,r_1-r_2$ 中自反关系有(　　)个。

18. 设 F,G,R 为 A 上的二元关系,证明:

(1) $R\circ(F\bigcup G)=R\circ F\bigcup R\circ G$

(2) $R\circ(F\bigcap G)=R\circ F\bigcap R\circ G$

(3) $R\circ(F\circ G)=R\circ(F\circ G)$

Chapter 4　Function

Function is a basic mathematical concept. In the general definition of function, $y = f(x)$ is discussed on the real number set. We generalize the concept of function here and regard function as a special relationship. For example, the relationship between input and output is regarded as a function in the computer; similarly, in the fields of switch theory, automaton theory and computability theory, functions have been widely used.

4.1　The Concept and Representation of Function

From the perspective of set theory, function is a special case of binary relation. Generally speaking, a function is a many to one relationship, that is to say, where each element in the domain of the relation corresponds to exactly one element in the range. Therefore, this section will further explain the function from the perspective of set theory.

Definition 4.1.1　Let X and Y be any two sets and f be a binary relation from X to Y. If for every element x in X, there is a unique element y in Y to get $<x, y> \in f$, then f is called a function or mapping from X to Y, denoted as:

$$f: X \rightarrow Y$$

$<x, y> \in f$, often recorded as $y = f(x)$. Where x is called the independent variable or image source, and y is the function value or image of x under the action of f.

From the definition of function, we can see that function is a special binary relation. Here, if f is a function of X to Y. Then, it differs from the general binary relation as follows.

① The domain of a function is X, not a proper subset of X.

② The function also emphasizes that the image y is unique, which is called the uniqueness of the image, that is, an $x \in X$ can only correspond to a unique y. It is simply understood as: suppose $f(x_1) = y_1$ and $f(x_2) = y_2$. If $x_1 = x_2$, then $y_1 = y_2$. If $y_1 \neq y_2$, then $x_1 \neq x_2$.

According to the definition of binary relation, all binary relations from X to Y are $X \times Y$ (the Cartesian product of X and Y), but $X \times$ Subsets of Y cannot all be functions of X to Y.

Such is the question. Let $X = \{a, b, c\}$, $Y = \{1, 2\}$, so $X \times Y = \{<a, 1>, <a, 2>, <b, 1>, <b, 2>, <c, 1>, <c, 2>\}$. There are 26 possible subsets of $X \times Y$, but only 2^3 of these subsets can be defined as functions from X to Y:

$$f_0 = \{<a, 1>, <b, 1>, <c, 1>\}$$
$$f_1 = \{<a, 1>, <b, 1>, <c, 2>\}$$
$$f_2 = \{<a, 1>, <b, 2>, <c, 1>\}$$
$$f_3 = \{<a, 1>, <b, 2>, <c, 2>\}$$
$$f_4 = \{<a, 2>, <b, 1>, <c, 1>\}$$
$$f_5 = \{<a, 2>, <b, 1>, <c, 2>\}$$
$$f_6 = \{<a, 2>, <b, 2>, <c, 1>\}$$
$$f_7 = \{<a, 2>, <b, 2>, <c, 2>\}$$

Let X and Y be finite sets, in which there are a and b different elements respectively. Because the definition domain of any function from X to Y is X, then each of these functions has exactly a order pair. At the same time, for any element $x \in X$ in the function, any one of the b elements of Y can be used as its image, so there are b^a different functions. In the above example, $a = 3$, $b = 2$, so 2^3 different functions can be formed. In the future, we use the symbol Y^X to represent the set of all functions from X to Y, and it also applies when X and Y are infinite.

Definition 4.1.2　If f is a function from X to Y, then X is the domain of f, denoted $\mathrm{dom}\, f = X$, and Y is the companion domain of f. If $X_1 \subseteq X$, the set $\{f(x) \mid x \in X_1\}$ is called the image of set X_1 under f, which is called $f(X_1)$. The image $f(X) = \{f(x) \mid x \in X\}$ of set X under f is called the image of function f. Obviously, the image $f(X)$ of function f is the range of binary relation f, that is, $f(X) = \mathrm{ran}\, f$.

Example 4.1.1　Let $X = \{p, 2, 1, \text{XiaoYun}\}$, $Y = \{1, 2, t, q, s\}$, $f = \{<p, 1>,$ $<2, 2>, <1, t>, <\text{XiaoYun}, s>\}$. That is, $f(p) = 1$, $f(2) = 2$, $f(1) = t$, $f(\text{XiaoYun}) = s$. Therefore, $\mathrm{dom}\, f = X$, $\mathrm{ran}\, f = \{1, 2, t, s\}$.

Special note: here $\mathrm{ran}\, f \subseteq Y$, not Y itself.

Example 4.1.2　Determine whether the following relationships form a function.

(1) $f = \{<x_1, x_2> \mid x_1, x_2 \in \mathbf{Z}, \text{and } x_1 + x_2 = 20\}$

This relationship can not form a function. Obviously, x_1 corresponds to many x_2; x_1 cannot take all the values in the definition field.

(2) $f = \{<x_1, x_2> \mid x_1, x_2 \in \mathbf{N}, \text{ and } x_2 \text{ is the number of prime numbers less than } x_1\}$

This relationship can form a function.

Definition 4.1.3　Let $f: A \to B$, $g: C \to D$, if $A = B$, $C = D$ and $\forall x \in A$, there is $f(x) = g(x)$, then the function f and g are equal, which is denoted as $f = g$.

For example, the function $f: \mathbf{N} \to \mathbf{N}$, $f(x) = x^2$, the function $g: \{1, 2\} \to \mathbf{N}$, $g(x) = x^2$.

According to the above definition, although the functions f and g have the same expression x^2, they are two different functions.

From the perspective of binary relation, f and g, $f \in \mathbf{N} \times \mathbf{N}$, the set of f can be obtained by the enumeration method:

$$\{<0, 0>, <1, 1>, <2, 4>, <3, 9>, \ldots\}$$

$g \in \{1, 2\} \times \mathbf{N}$, it is also expressed by enumeration:

$$\{<1,1>, <2, 4>\}$$

According to the condition that binary relations are equal, they are also unequal. The equality of functions is consistent with the equality of binary relations.

In the following, we will discuss some special cases of functions.

Definition 4.1.4 Let $f: X \to Y$, if the value range of f is ran $f = Y$, that is, every element of Y is an image point of one or more elements in X, then f is called surjection (or upper mapping).

Let f be a function from X to Y. it is easy to see from the definition that if $\forall y \in Y$, there exists $x \in A$ such that $f(x) = y$, then f is a surjective function.

For example, $X = \{a, b, c, d, e\}$, $Y = \{1, 2, 3\}$, f is a function from X to Y, $f = \{<a, 1>, <b, 1>, <c, 3>, <d, 2>, <e, 3>\}$, then f is surjective.

Definition 4.1.5 Let $f: X \to Y$, if $\forall y \in$ ran f, there exists a unique $x \in X$ such that $f(x) = y$, that is, no two elements in X have the same image, then f is called an injunction.

Let f be a function from X to Y. It is easy to see from the definition that if for $x_1 \in X$, $x_2 \in X$, $f(x_1) = y_1$, $f(x_2) = y_2$.

① If $y_1 = y_2$, there must be $x_1 = x_2$, then f is an injunction.

② If $x_1 \neq x_2$, there must be $y_1 \neq y_2$, then f is an injunction.

For example, if the function $f: \{m, n\} \to \{1, 2, 3\}$, there exists $f = \{<m, 2>, <n, 3>\}$, then the function is an injunction but not a surjection.

Definition 4.1.6 Let $X \to Y$. if f is both incident and surjective, then f is called bijection.

For example: let $[a, b]$ denote the closed interval of real number, that is, $[a, b] = \{x \mid a \leqslant x \leqslant b\}$, let $f: [0, 1] \to [a, b]$, where $f(x) = (b-a)x + a$, this function is bijective.

Theorem 4.1.1 Let X and Y be finite sets. If the number of elements of X and Y are the same, that is, $|X| = |Y|$, then $f: X \to Y$ is an injunction if and only if it is a surjection.

Proof

(1) If f is an injunction, then $|X| = |f(X)|$. Because of $|X| = |Y|$, from the definition of f, we can know that $f(X) \subseteq Y$, and $|f(X)| = |Y|$ and because $|Y|$ is finite, so $f(X) = Y$. So f is an injunction which is to know that f is a surjection.

(2) If f is a surjection, according to the definition of surjection, $f(X) = Y$, then $|X| = |Y| = |f(X)|$. Because $|X| = |f(X)|$ and $|X|$ are finite, so f is an injunction. So f is a surjection to know that f is an injunction.

This theorem can only be true in the case of finite sets. It is not necessarily valid on infinite sets. Such as $f: I \to I$, where $f(x) = 2x$. In this case, integers are mapped to even numbers. Obviously, this is an incidence, but not a surjection.

4.2　Inverse Function and Compound Function

From the relationship in the previous chapter, we can see that the relationship from X to Y is R, and the corresponding inverse relationship R^C is the relationship from Y to X. That is $<y, x> \in R^C \Leftrightarrow <x, y> \in R$. But for functions, it is impossible to deduce inverse functions by simply exchanging elements of ordered pairs. For example, if the mapping of $X \to Y$ is a many to one correspondence, $<x_1, y> \in f$, $<x_2, y> \in f$, the inverse relationship is $<y, x_1> \in f^C$, $<y, x_2> \in f^C$. Obviously, this violates the requirement of the uniqueness of function values. Therefore, the definition of inverse function needs some conditions.

Theorem 4.2.1　Let $f: X \to Y$ be a bijection, then the inverse relation f^C of f is a bijection from Y to X.

Proof

The following first prove that the inverse relation f^C is a function of Y to X.

Since f is a function, we can know that f^C is a binary relation from Y to X. Let's prove that the binary relation f^C from Y to X is a function of Y to X:

(1) $\forall y \in Y$. Since $f: X \to Y$ is a surjection, then $\exists x \in X$ makes $<x, y> \in f$. From the definition of inverse relation, we can know $<y, x> \in f^C$.

(2) Let $<y_1, x_1> \in f^C$, $<y_2, x_2> \in f^C$, $y_1 = y_2$, and according to the definition of inverse relation, we can know $<x_1, y_1> \in f$, $<x_2, y_2> \in f$. Because f is an injunction, $x_1 = x_2$.

Therefore, according to the definition of function, we can know that f^C is a function from Y to X.

It is proved that f^C is a surjection.

According to the theorem in the previous chapter, we can know ran $f^C = $ dom f. Since f is a function from X to Y, we can know dom $f = X$. So ran $f^C = X$. Therefore, we can know that f^C is a surjection from Y to X.

Finally, it is proved that f^C is an injunction:

Let $<y_1, x_1> \in f^C$, $<y_2, x_2> \in f^C$, $x_1 = x_2$, from the definition of inverse relation, we can know $<x_1, y_1> \in f$, $<x_2, y_2> \in f$. And because f is a function, there must be $y_1 = y_2$. So f^C is an injunction.

To sum up, it can be proved that f^C is a bijection.

Definition 4.2.1　Let $f: X \to Y$ be a bijection and the inverse relation f^C of f be a bijection from Y to X. The bijection f^C is called the inverse function of f, denoted as f^{-1}.

For example, let $X = \{m, n, t\}$, $Y = \{0, 1, 2\}$, $f = \{<m, 0>, <n, 1>, <t, 2>\}$. Obviously, if f is a bijection from X to Y, then the inverse relation $f^{-1} = \{<0, m>, <1, n>, <2, t>\}$ is a $Y \to X$ function.

If $f = \{<m, 0>, <n, 1>, <t, 1>\}$, then the inverse relation $f^{-1} = \{<0, m>,$

$<1, n>, <1, t>\}$ is not a function.

Theorem 4.2.2 Let $f:X \rightarrow Y$ be a bijection and $f^{-1}:Y \rightarrow X$ be the inverse function of f, then $(f^{-1})^{-1}=f$.

Proof

From Theorem 4.2.1 and Definition 4.2.1, we can know that $(f^{-1})^{-1}:X \rightarrow Y$, and it is a bijection.

$\forall x \in X$, Let $f(x)=y$, then $f^{-1}(y)=x$, $(f^{-1})^{-1}(x)=y=f(x)$, so $(f^{-1})^{-1}=f$.

As we have said before, if a function is a binary relation satisfying certain conditions, it can also be used for compound operations. Then the composition relation of the function can be simply described as follows.

Let $f:X \rightarrow Y$, $g:W \rightarrow Z$, if $f(X) \subseteq W$, the set $g \circ f=\{<x, z>|x \in X \wedge z \in Z \wedge (\exists y)$ $(y \in Y \wedge <x, y> \in f \wedge <y, z> \in g)\}=\{<x, z>|x \in X \wedge z \in Z \wedge (\exists y)(y \in Y \wedge y= f(x) \wedge z=g(y))\}$.

Definition 4.2.2 Let $f:X \rightarrow Y$, $g:W \rightarrow Z$, if $f(X) \subseteq W$, then the set $\{<x, z>|x \in X \wedge z \in Z \wedge (\exists y)(y \in Y \wedge y=f(x) \wedge z=g(y))\}$ is called the composition relation of function g on the left of function f, which is denoted as $g \circ f$.

Theorem 4.2.3 Let $f:X \rightarrow Y$, $g:W \rightarrow Z$, $f(X) \subseteq W$, then the composition relation $g \circ f$ of function g on the left side of function f is a function from X to Z, and the simple understanding is that the composition of two functions is a function.

Proof

$\forall x \in X$, because f is a function from X to Y, there must be a unique $y \in Y$ such that $<x, y> \in f$, that is, $y=f(x)$. And $y=f(x) \in f(X) \subseteq W$, so $y \in W$.

And because g is a function from W to Z, there must be a unique $z \in Z$ such that $<y, z> \in g$, that is, $z=g(y)$. Therefore, according to Definition 4.2.2, we can know $<x, z> \in g \circ f$, that is, $g \circ f(x)=z$.

Therefore, we can know that $g \circ f$ is a function from X to Z.

Then according to the definition of composite function, there is obviously $g \circ f(x)= g(f(x))$.

Example 4.2.1 Let $A=\{1, 2, 3\}$, $B=\{a, b\}$, $C=\{m, n\}$, $f=\{<1, a>, <2, b>, <3, a>\}$, $g=\{<a, m>, <b, n>\}$, and find $g \circ f$.

Solution

$$g \circ f=\{<1, m>, <2, n>, <3, m>\}$$

Theorem 4.2.4 Let $f:X \rightarrow Y$, $g:Y \rightarrow Z$, and composite function $g \circ f:X \rightarrow Z$.

(1) If f and g are surjections, then $g \circ f$ is also a surjection.

(2) If f and g are both injunctions, then $g \circ f$ is also an injunction.

(3) If f and g are bijections, then $g \circ f$ is also a bijection.

Proof

(1) $\forall z \in Z$, because g is a surjection from Y to Z, $\exists y \in Y$, so that $z = g(y)$. And because f is a surjection from X to Y, $\exists x \in X$, $y = f(x)$. Then, $g \circ f(x) = g(f(x)) = g(y) = z$, so $g \circ f$ is a surjection.

(2) Let $x_1 \in X$, $x_2 \in X$ and $x_1 \neq x_2$, because f is a simple projection from X to Y, $f(x_1) \neq f(x_2)$, $f(x_1) \in Y$, $f(x_2) \in Y$. Because g is a Y to Z injectivity, $g(f(x_1)) \neq g(f(x_2))$, that is, $g \circ f(x_1) \neq g \circ f(x_2)$, so $g \circ f$ is an injunction.

(3) If both f and g are bijections, then f and g are surjections and injunctions. From (1) and (2), we can know that $g \circ f$ is a surjection and an injunction, so $g \circ f$ is a bijection.

Example 4.2.2　Let $x \in \mathbf{N}$, $f(x) = x + 3$, $g(x) = x - 3$, $t(x) = x^2$, find $g \circ f$ and $t \circ (g \circ f)$.

Solution　$g \circ f = \{<x, x> | x \in \mathbf{N}\}$

$$t \circ (g \circ f) = \{<x, x^2> x \in \mathbf{N}\}$$

Theorem 4.2.5　Let $f: X \rightarrow Y$, $g: Y \rightarrow Z$, $h: Z \rightarrow W$, then $h \circ (g \circ f) = (h \circ g) \circ f$.

Definition 4.2.3　The function $f: X \rightarrow Y$ is called a constant function. If there is a $y_0 \in Y$, there is $f(x) = y_0$ for every $x \in X$, that is, $f(X) = \{y_0\}$.

Definition 4.2.4　If function $I_X = \{<x, x> | x \in X\}$, the function $I_X: X \rightarrow X$ is called an identity function.

Theorem 4.2.6　Let $f: X \rightarrow Y$, I_X be an identity function on X and I_Y be an identity function on Y, then $I_Y \circ f = f \circ I_X = f$.

The proof of this theorem can be obtained directly from the definition.

Theorem 4.2.7　Let $f: X \rightarrow Y$ be a bijection function and $f^{-1}: Y \rightarrow X$ be the inverse function of F, then

$$f^{-1} \circ f = I_X \quad \text{and} \quad f \circ f^{-1} = I_Y$$

Proof

It is proved that $f^{-1} \circ f = I_X$.

According to the definition of a composite function, we can know $f^{-1} \circ f: X \rightarrow X$, $I_X: X \rightarrow X$. $\forall x \in X$, Let $f(x) = y$, then $f^{-1}(y) = x$.

$$f^{-1} \circ f(x) = f^{-1}(f(x)) = f^{-1}(y) = x = I_X(x)$$

So $f^{-1} \circ f = I_X$.

Similarly, it can be proved that $f \circ f^{-1} = I_Y$.

Theorem 4.2.8　If the function $f: X \rightarrow Y$ is a one-to-one corresponding function, then $(f^{-1})^{-1} = f$.

Proof

(1) Since the function $f: X \rightarrow Y$ is a one-to-one corresponding function, it can be seen that $f^{-1}: Y \rightarrow X$ is also a one-to-one corresponding function, so $(f^{-1})^{-1}: X \rightarrow Y$ is also a

one-to-one corresponding function:
$$\text{dom } f = \text{dom } (f^{-1})^{-1} = X$$

(2) $x \Rightarrow X \Rightarrow f : x \rightarrow f(x) \Rightarrow f^{-1} : f(x) \rightarrow x \Rightarrow (f^{-1})^{-1} : x \rightarrow f(x)$.

From (1) (2), we know that $(f^{-1})^{-1} = f$.

Theorem 4.2.9　Let $f : X \rightarrow Y$, $g : Y \rightarrow Z$ and both f and g are bijections, then $(g \circ f)^{-1} = f^{-1} \circ g^{-1}$.

Proof

(1) Because f and g are bijections, f^{-1} and g^{-1} exist and
$$f^{-1} : Y \rightarrow X, \quad g^{-1} : Z \rightarrow Y$$

So $f^{-1} \circ g^{-1} : Z \rightarrow X$.

On the other hand, from Definition 4.2.3, $g \circ f : X \rightarrow Z$.

So $(g \circ f)^{-1} : Z \rightarrow X$.

(2) $\forall z \in Z$, because g is a bijection, $\exists y \in Y$ makes $g(y) = z$, and because f is a bijection, $\exists x \in X$, so that $f(x) = y$.

So, $g \circ f(x) = g(f(x)) = g(y) = z$, and we can know $(g \circ f)^{-1}(z) = x$.

On the other hand, $f^{-1}(y) = x$, $g^{-1}(z) = y$.

So $f^{-1} \circ g^{-1}(z) = f^{-1} \circ (g^{-1}(z)) = f^{-1}(y) = x$.

In the end, we know $(g \circ f)^{-1} = f^{-1} \circ g^{-1}$.

4.3　The Concept of Characteristic Function and Fuzzy Subset

Through the analysis, we can know that some functions and sets can establish some special relations, then with the help of these functions, we can operate the set, and at the same time, we can promote the expression of the concept of fuzzy sets.

Definition 4.3.1　Let E be a complete set and A be a subset of E. Thus $\psi_A : E \rightarrow \{0, 1\}$ is defined as:

$$\psi_A(x) = \begin{cases} 1, & x \in A \\ 0, & x \notin A \end{cases}$$

And we call $\psi_A(x)$ the characteristic function of set A.

For example, if E is the set of all students in a school, and A is the set of all male students, then ψ_A is the characteristic function of male students.

Theorem 4.3.1　Given the complete sets E, $A \subseteq E$ and $B \subseteq E$, for all $x \in E$, the characteristic functions have the following properties.

(1) $\psi_A(x) = 0 \Leftrightarrow A = \phi$

(2) $\psi_A(x) = 1 \Leftrightarrow A = E$

(3) $\psi_A(x) \leqslant \psi_B(x) \Leftrightarrow A \subseteq B$

(4) $\psi_A(x) = \psi_B(x) \Leftrightarrow A = B$

(5) $\psi_{A \cap B}(x) = \psi_A(x) * \psi_B(x)$

(6) $\psi_{A\cup B}(x)=\psi_A(x)+\psi_B(x)-\psi_{A\cap B}(x)$

(7) $\psi_{\sim A}(x)=1-\psi_A(x)$

(8) $\psi_{A-B}(x)=\psi_{A\cap\sim B}(x)=\psi_A(x)-\psi_{A\cap B}(x)$

The operations $+$, $-$, $*$ in the characteristic function are the usual arithmetic operations $+$, $-$, $\times$.

Proof Here we only prove partial relation, others can prove by themselves.

For example (3): For any $x\in A$, because $A\subseteq B$, then $x\in B$, according to the definition of the characteristic function, we can know that $\psi_A(x)=1$, $\psi_B(x)=1$, so we can know that $\psi_A(x)\leqslant\psi_B(x)$ holds. If $x\notin A$, obviously there is also $\psi_A(x)\leqslant\psi_B(x)$.

On the contrary, if $\psi_A(x)\leqslant\psi_B(x)$, and $x\notin B$, then $\psi_B(x)=0$, then $\psi_A(x)=0$, thus $x\notin A$, so $A\subseteq B$.

Example 4.3.1 Try to prove $\psi_{(A-C)\cup(B-C)}(x)=\psi_{(A\cup B)-C}(x)$.

Proof

$$\begin{aligned}
\psi_{(A-C)\cup(B-C)}(x)&=\psi_{A-C}(x)+\psi_{B-C}(x)-\psi_{A-C}(x)*\psi_{B-C}(x)\\
&=\psi_A(x)-\psi_A(x)*\psi_C(x)+\psi_B(x)-\psi_B(x)*\psi_C(x)-\psi_A(x)*\psi_B(x)+\\
&\quad \psi_A(x)*\psi_B(x)*\psi_C(x)\\
&=\psi_{(A\cup B)}(x)-(\psi_A(x)+\psi_B(x)-\psi_A(x)*\psi_B(x))*\psi_C(x)\\
&=\psi_{(A\cup B)}(x)-\psi_{(A\cup B)}(x)*\psi_C(x)\\
&=\psi_{(A\cup B)-C}(x)
\end{aligned}$$

The concept of the fuzzy subset can be derived by generalizing the characteristic function.

Let $E=\{x_1, x_2, ..., x_n\}$, We can express any subset A of E as follows:

$$A: \{<x_1, \psi_A(x_1)>, <x_2, \psi_A(x_2)>, ..., <x_n, \psi_A(x_n)>\}$$

When $\psi_A(x_i)=1$, $x_i\in A$.

When $\psi_A(x_i)=0$, $x_i\in A$.

If we do not limit the value range of $\psi_A(x_i)$ to 0 and 1, but take any number between 0 and 1, for example:

$$A^*: \{<x_1, 0.1>, <x_2, 0>, <x_3, 0.4>, <x_4, 1>, <x_5, 0.9>\}$$

Then, A^* can be understood as follows: it means that x_1 belongs to a small amount of A^*, x_2 does not belong to A^*, x_3 also belongs to A^* (but more than x_1), x_4 must belong to A^*, and x_5 belongs to A^* basically. Such a set A^* is a fuzzy subset, where 0.1, 0.4, 0.9, ... are called the membership degree of corresponding elements in the set.

Definition 4.3.2 Let E be a complete set (or domain), and a fuzzy subset $\underline{A}$ on E means that every $x\in E$ has a membership degree $\mu=\psi_{\underline{A}}(x)(0\leqslant\mu\leqslant1)$ corresponding to it, and $\psi_{\underline{A}}(x)$ is called the membership function of $\underline{A}$.

From the definition of fuzzy subset, we can see that when $\psi_{\underline{A}}(x)$ only takes two values of 0 and 1, $\underline{A}$ becomes an ordinary subset.

Example 4.3.2 Let $E=\{0, 1, 2, ..., 100\}$, The fuzzy subset $\underline{A}$ of E is the set of "large numbers" in E. Obviously, 100 is the "big number" in E, and 0 is not the "big number" in E. Is 80 the "big number" in E? Therefore, we define the membership function of A as follows:

$$\psi_A(x)=x/100$$

Therefore, we can know that the degree of 80 belonging to the "big number" is 0.8, that is, the degree of 80's membership is 0.8. Similarly, the membership degree of 99 was 0.99.

4.4 Common Functions

Next, we will introduce two important functions in discrete mathematics, i.e. Ceiling and Floor function and hash function. Among them, the integer function plays an important role in analyzing the number of steps needed in the process of solving a certain scale problem.

Definition 4.4.1   For a rational number x, if $f(x)$ is the smallest integer greater than or equal to x, then $f(x)$ is called the Ceiling function (forced integral function), which is denoted as $f(x)=\lceil x\rceil$. For the rational number x, if $f(x)$ is the largest integer less than or equal to x, it is called $f(x)$ as the floor function (weak rounding function), which is denoted as $f(x)=\lfloor x\rfloor$.

Note: The floor function is also known as the maximum integer function, which is recorded as $f(x)=[x]$.

Attribute:

(1) $\lfloor x\rfloor\leqslant x<\lfloor x\rfloor+1$, $x\leqslant\lceil x\rceil<x+1$, in this case, the equal sign is taken if and only if x is an integer.

(2) The floor function is equal power function: $\lfloor\lfloor x\rfloor\rfloor=\lfloor x\rfloor$.

(3) For any integer k and any real number x, there is $\lfloor k+x\rfloor=k+\lfloor x\rfloor$.

(4) For integer x, there are $\lfloor x/2\rfloor+\lceil x/2\rceil=x$.

(5) $\lceil x\rceil=-\lfloor -x\rfloor$.

Example 4.4.1   Data that exists on a computer disk or is transmitted over a data network is usually represented as a string of bytes. Each byte consists of 8 bits. So how many bytes are needed to represent 300 bits of data?

Solution Since the number of bytes required is in integer form, all 300 bits must be included. So the data representing 300 bits needs $\lceil 300/8\rceil=38$ bytes.

Example 4.4.2   It is proved that for any $x\in\mathbf{R}$, $x-1<\lfloor x\rfloor\leqslant x<\lfloor x\rfloor+1$.

Proof

Let $x=\lfloor x\rfloor+r(0\leqslant r<1)$, so $0\leqslant r=x-\lfloor x\rfloor<1$.

So $x-1<\lfloor x\rfloor\leqslant x<\lfloor x\rfloor+1$.

Example 4.4.3   Prove that for any x, $y\in\mathbf{R}$, if $x\leqslant y$, then $\lfloor x\rfloor\leqslant\lfloor y\rfloor$.

The proof to the contradiction is as follows:

If $\lfloor x\rfloor>\lfloor y\rfloor$, then $\lfloor x\rfloor-\lfloor y\rfloor>0$.

So the integer $\lfloor x\rfloor-\lfloor y\rfloor\geqslant1$, that is $\lfloor x\rfloor\geqslant\lfloor y\rfloor+1$.

Then we can know that $x\geqslant\lfloor x\rfloor\geqslant\lfloor y\rfloor+1>y$.

This is in contradiction with the known $x \leqslant y$, so $\lfloor x \rfloor \leqslant \lfloor y \rfloor$.

Definition 4.4.2   Hash function, also known as a hash function, is mainly used for data storage and retrieval.

For example, in a relatively large company, there are many employees (the number of employees is basically determined). If the information of these employees is stored in the computer in order (i.e., the order is stored in a single linked list), it is very inconvenient to find the information of a person. It can only be compared one by one from the beginning to the end. It takes a lot of time to put the information of the person to be searched at the end. Obviously, this kind of storage is very time-consuming The way is not appropriate.

If multiple linked lists are used, when a new employee comes in, which linked list should his information be stored in?

Suppose that a person's code (called a key) is represented by seven digits, and it is the unique identification of an employee. And the company has 10000 people, $C = \{c_1, c_2, c_3, \ldots, c_{10000}, \ldots\}$, and each C_i has a key. At the same time, the company's computer can check a list containing 100 items in an acceptable time, that is, each list can store 100 personal information. Thus, 101 linked lists can be used to store the information of all employees. So, in which list does an employee's information go? We can use a hash function $h : C \rightarrow \{0, 1, 2, \ldots, 100\}$ to allocate?

Let i (is a seven digit number of words) be an employee's code (key), then

$$h(i) = i \pmod{101}$$

For example, if $i = 2473871$, then

$$h(2473871) = 2473871 \pmod{101} = 78$$

Indicates that the record of the employee is in list No. 78.

Of course, we hope that the function h is incident, that is, the function values $h(i_1)$ and $h(i_2)$ corresponding to the codes i_1 and i_2 of different employees are different, to facilitate the search of employee information. But it is not easy to design such a function, especially when the amount of data is very large. For example, when i is 2474073 and $h(2474073) = 78$, it is obvious that both positions are 78, then there is a conflict. When designing a good hash function, conflict should be as little as possible.

There are many ways to resolve conflicts, such as:

(1) Open addressing method: the so-called open addressing method is to find the next empty hash address in case of conflict. As long as the hash table is large enough, the empty hash address can always be found and the records will be saved. The formula is:

$$h(\text{key}) = (h(\text{key}) + d) \bmod m (d = 1, 2, 3, \ldots, m - 1)$$

(2) Chain address method: hash function values are placed in the same linear table.

(3) Create a common overflow area: dedicated to storing conflict data.

(4) Again Hash method: use another positive integer m and hash it again.

In addition, there are many ways to construct hash functions, not necessarily for key (mod m), but as follows.

(1) Direct address method: $h(\text{key}) = \text{key}$ or $h(\text{key}) = a.\text{key} + b$.

(2) Random number method: $h(\text{key}) = \text{random}(\text{key})$, where $\text{random}()$ is a random function. This method is often used when the length of the key is not equal.

Example **4.4.4**　Suppose that there are memory units numbered from 0 to 10 in the computer memory, as shown in Figure 4.4.1. The figure shows the case of storing the cells in order of 15, 558, 32,132,102 and 5 in which the initial time is empty. Now we hope to store any nonnegative integers in these storage units and retrieve them. We try to use the hash function to store 257 and retrieve 558, and use the open address method to solve the conflict.

132			102	15	5	259		558		32
0	1	2	3	4	5	6	7	8	9	10	0

Figure　4.4.1

Solution

Because $h(257) = 257 (\bmod\ 11) = 4$, but position 4 is occupied by No. 15, a conflict occurs. According to the open address method, position 7 just has a space to store 257.

Because $h(558) = 8$, the check position 8,558 is exactly at position 8.

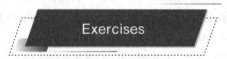

Exercises

1. Which of the following functions are injunction, surjection, or bijection?

(1) $f: I \to I, f(x) = x \bmod 4$

(2) $f: I \to \mathbf{N}, f(x) = |2x| + 1$

(3) $f: \mathbf{R} \to \mathbf{R}, f(x) = 3x + 10$

(4) $f: \mathbf{N} \to \{0,1\}, f(x) = \begin{cases} 1, & x \text{ is odd} \\ 0, & x \text{ is even} \end{cases}$

2. Let $X = \{a, b, c\}$, $Y = \{1, 2, 3\}$, as $f: X \to Y$, how many relations f and how many surjections are there?

3. Which of the following are functions, which are injunctions, which are surjections, and for any bijections, write its inverse function.

(1) $f: \mathbf{Z} \to \mathbf{N}, f(x) = x^2 - 2$

(2) $h: \mathbf{N} \to Q, h(x) \dfrac{1}{x}$

(3) $g: \mathbf{Z} \times \mathbf{N} \to Q, h(z,n) = \dfrac{z}{n+1}$

(4) $g: \mathbf{N} \to \mathbf{N}, g$ 定义为 $g(x) = 2^x$.

4. Try to give examples of functions satisfying the following conditions.

(1) It's an injunction, not a surjection.

(2) It's a surjection, but not an injunction.

(3) It's not an injunction or a surjection.

(4) It is an injunction and a surjection.

5. If f and g are functions, please prove that $f \cap g$ is also a function.

6. Suppose that A and B are finite sets, what is the necessary condition for A and B to have incidence?

7. Let X and Y be finite sets. How many different injunctions and different bijections are there?

8. Let $g : N \rightarrow N$, then define the following functions:
$$\begin{cases} f(x) = x - 10, & x > 10 \\ f(x) = f(f(x) + 11), & x \leqslant 100 \end{cases}$$

Prove: (1) $f(99) = 91$;

(2) $f(x) = 91, (0 \leqslant x \leqslant 100)$。

9. Let Z be the set of integers, $f : Z \times Z \rightarrow Z$, and $f(x, y) = x + y$. is f an injunction or a surjection? Why? And find $f(x, x)$, $f(x, -x)$.

10. Let $f : X \rightarrow Y$ and $g : Y \rightarrow Z$ such that $g \circ f$ is an injunction and f is a surjection. It is proved that g is an injunction. At the same time, an example is given to show that if f is not a surjection, then g is not necessarily an injunction.

11. Let $f : X \rightarrow Y$, $g : Y \rightarrow X$, and prove that if $f^{-1} = g$, $f = g^{-1}$, then $g \circ f = I_X$, $f \circ g = I_Y$.

12. It is proved by the characteristic function that:
$$A \cup (B \cap C) = (A \cup B) \cap (A \cup C)$$

第4章 函 数

函数是一个基本的数学概念,在通常的函数定义中,$y=f(x)$是在实数集合上讨论,我们这里把函数的概念予以推广,把函数看作一种特殊的关系。例如,计算机中把输入、输出间的关系看成一种函数;类似地,在开关理论、自动机理论和可计算性理论等领域,函数有着极其广泛的应用。

4.1 函数的概念与表示

从集合论的角度来看,函数是二元关系中的一个特例。通俗地说,一个函数是一个多对一的关系,即对该关系定义域中的任何元素恰恰对应值域中的一个元素。所以本节将从集合论角度进一步对函数进行解释。

定义 4.1.1 ▶ 设X和Y是任何两个集合,f是X到Y的二元关系,如果对于X中的每一个元素x,在Y中都有唯一元素y,使$<x,y>\in f$,则称f是X到Y的函数或映射,记作

$$f:X \rightarrow Y$$

$<x,y>\in f$,常记为$y=f(x)$。其中x称为自变元或像源,y称为在f作用下x的函数值或像。

从函数的定义中可以看出,函数是一种特殊的二元关系。这里,若f是X到Y的函数,那么,它与一般的二元关系的区别如下。

① 函数的定义域是X,而不能是X的某个真子集。

② 函数还强调像y是唯一的,称为像的唯一性,也就是说一个$x \in X$只能对应唯一的一个y。可简单理解:假设$f(x_1)=y_1$且$f(x_2)=y_2$,如果$x_1=x_2$,那么必有$y_1=y_2$;如果$y_1 \neq y_2$,则$x_1 \neq x_2$。

根据二元关系定义可知,X到Y的所有二元关系是$X \times Y$(X与Y的笛卡尔积)的子集,但$X \times Y$的子集并不能都成为X到Y的函数。

例如,设$X=\{a,b,c\}$,$Y=\{1,2\}$,则$X \times Y=\{<a,1>,<a,2>,<b,1>,<b,2>,<c,1>,<c,2>\}$,$X \times Y$有$2^6$个可能的子集,但是其中只有$2^3$个子集可以定义为$X$到$Y$的函数。

$$f_0=\{<a,1>,<b,1>,<c,1>\}$$
$$f_1=\{<a,1>,<b,1>,<c,2>\}$$
$$f_2=\{<a,1>,<b,2>,<c,1>\}$$
$$f_3=\{<a,1>,<b,2>,<c,2>\}$$
$$f_4=\{<a,2>,<b,1>,<c,1>\}$$
$$f_5=\{<a,2>,<b,1>,<c,2>\}$$
$$f_6=\{<a,2>,<b,2>,<c,1>\}$$

$$f_7 = \{<a,2>,<b,2>,<c,2>\}$$

设 X 和 Y 都是有限集合,它们集合中分别有 a 个和 b 个不同的元素。因为从 X 到 Y 的任何一个函数的定义域都是 X,那么,在这些所能组成的函数中每一个都恰好有 a 个序偶。同时,对于函数中的任何一个元素 $x \in X$,可以有 Y 的 b 个元素中的任何一个作为它的像,故共有 b^a 个不同的函数。在上面的例子中 $a=3,b=2$,故可以组成 2^3 个不同的函数。今后我们用符号 Y^X 表示从 X 到 Y 的所有函数的集合,并且当 X 和 Y 都是无限集合时也适用。

定义 4.1.2 ▶ 如果 f 是 X 到 Y 的函数,我们说 X 是 f 的定义域,记作 dom $f=X$,而 Y 是 f 的陪域。如果 $X_1 \subseteq X$,集合 $\{f(x) \mid x \in X_1\}$ 称为集合 X_1 在 f 下的像,记为 $f(X_1)$。集合 X 在 f 下的像 $f(X)=\{f(x) \mid x \in X\}$ 称为函数 f 的像。显然,函数 f 的像 $f(X)$ 就是二元关系 f 的值域,即 $f(X)=$ ran f。

例 4.1.1 设 $X=\{p,2,1,小云\}$,$Y=\{1,2,t,q,s\}$,$f=\{<p,1>,<2,2>,<1,t>,<小云,s>\}$,即 $f(p)=1,f(2)=2,f(1)=t,f(小云)=s$,由此可知,dom $f=X$,ran $f=\{1,2,t,s\}$。

特别注意:这里 ran $f \subseteq Y$,并不是 Y 本身。

例 4.1.2 判断下列关系能否构成函数。

(1) $f=\{<x_1,x_2> \mid x_1,x_2 \in \mathbf{Z},且 x_1+x_2=20\}$

这个关系不能构成函数,很明显的一点就是 x_1 对应很多个 x_2;x_1 并不能取定义域中的所有值。

(2) $f=\{<x_1,x_2> \mid x_1,x_2 \in \mathbf{N},且 x_2 为小于 x_1 的质数个数\}$

这个关系能够构成函数。

定义 4.1.3 ▶ 设函数 $f:A \to B$,$g:C \to D$,若 $A=B$,$C=D$ 且 $\forall x \in A$,有 $f(x)=g(x)$,则称函数 f 和 g 相等,记为 $f=g$。

例如,函数 $f:\mathbf{N} \to \mathbf{N}$,$f(x)=x^2$,函数 $g:\{1,2\} \to \mathbf{N}$,$g(x)=x^2$。

根据上述的定义可以知道,虽然函数 f 和 g 有相同的表达式 x^2,但是它们是两个不同的函数。

以二元关系的角度来看 f 和 g,$f \in \mathbf{N} \times \mathbf{N}$,用列举法可得 f 的集合为

$$\{<0,0>,<1,1>,<2,4>,<3,9>,\cdots\}$$

$g \in \{1,2\} \times \mathbf{N}$,同样用列举法表示为

$$\{<1,1>,<2,4>\}$$

按二元关系相等的条件衡量,它们也是不等的。函数相等和二元关系相等是一致的。

下面,我们来讨论函数的几类特殊情况。

定义 4.1.4 ▶ 设函数 $f:X \to Y$,如果 f 的值域 ran $f=Y$,也就是 Y 的每一个元素是 X 中一个或多个元素的像点,则称 f 是满射(或者上映射)的。

设 f 是 X 到 Y 的函数,由定义不难看出,若 $\forall y \in Y$,都存在 $x \in A$,使 $f(x)=y$,则 f 是满射的。

例如,$X=\{a,b,c,d,e\}$,$Y=\{1,2,3\}$,f 是由 X 到 Y 的函数,$f=\{<a,1>,<b,1>,<c,3>,<d,2>,<e,3>\}$,则 f 是满射的。

定义 4.1.5 ▶ 设函数 $f:X \to Y$,若 $\forall y \in$ ran f,存在唯一的 $x \in X$,使 $f(x)=y$,也就是说 X 中没有两个元素有相同的像,则称 f 是单射的。

设 f 是 X 到 Y 的函数,由定义不难看出,如果对于 $x_1 \in X, x_2 \in X, f(x_1) = y_1, f(x_2) = y_2$。

① 当 $y_1 = y_2$ 时,一定有 $x_1 = x_2$,则 f 是单射的。

② 当 $x_1 \neq x_2$ 时,一定有 $y_1 \neq y_2$,则 f 是单射的。

例如,函数 $f:\{m, n\} \to \{1,2,3\}$,存在 $f = \{<m,2>,<n,3>\}$,则这个函数是单射的,但不是满射的。

定义 4.1.6 ▶ 设 $f: X \to Y$,若 f 既是单射的,又是满射的,则称 f 为双射的。

例如,令 $[a,b]$ 表示实数的闭区间,即 $[a,b] = \{x \mid a \leqslant x \leqslant b\}$;令 $f:[0,1] \to [a,b]$,这里 $f(x) = (b-a)x + a$,这个函数是双射的。

定理 4.1.1 ▶ 令 X 和 Y 为有限集,若 X 和 Y 的元素个数相同,即 $|X| = |Y|$,则 $f: X \to Y$ 是单射的,当且仅当它是满射的。

证明 (1) 如果 f 是单射的,则 $|X| = |f(X)|$。因为 $|X| = |Y|$,从 f 的定义中我们可以知道 $f(X) \subseteq Y$,而 $|f(X)| = |Y|$,又因为 $|Y|$ 是有限的,故 $f(X) = Y$。因此,f 是单射的推出 f 是满射的。

(2) 若 f 是满射的,根据满射定义有 $f(X) = Y$,于是 $|X| = |Y| = |f(X)|$。因为 $|X| = |f(X)|$ 和 $|X|$ 是有限的,故 f 是单射的。因此,f 是满射的推出 f 是单射的。

这个定理必须在有限集情况下才能成立,在无限集上不一定有效。如 $f: I \to I$,这里 $f(x) = 2x$,在这种情况下整数映射到偶数,显然这是单射的,但不是满射的。

4.2 逆函数与复合函数

从上一节的关系中可以了解到,从 X 到 Y 的关系 R,其对应的逆关系 R^c 是从 Y 到 X 的关系,即 $<y,x> \in R^c \Leftrightarrow <x,y> \in R$。但是对于函数而言,不能用简单的交换序偶的元素推导出逆函数。例如,如果 $X \to Y$ 的映射是一个多对一对应时,即有 $<x_1,y> \in f, <x_2, y> \in f$,其逆关系就有 $<y, x_1> \in f^c, <y, x_2> \in f^c$。很显然,这违反了函数值唯一性的要求。所以,逆函数的定义需要一些条件。

定理 4.2.1 ▶ 设 $f: X \to Y$ 是双射的,则 f 的逆关系 f^c 是 Y 到 X 的双射的。

证明 下面首先证逆关系 f^c 是 Y 到 X 的函数。

由于 f 是函数,那么可以得知 f^c 是 Y 到 X 的二元关系。下面再来证明 Y 到 X 的二元关系 f^c 是 Y 到 X 的函数。

(1) $\forall y \in Y$,由于 $f: X \to Y$ 是满射的,那么,$\exists x \in X$,使 $<x,y> \in f$。再由逆关系的定义可知,$<y, x> \in f^c$。

(2) 设 $<y_1,x_1> \in f^c, <y_2,x_2> \in f^c, y_1 = y_2$,由逆关系的定义可知,$<x_1,y_1> \in f, <x_2,y_2> \in f$,因为 f 是单射的,所以 $x_1 = x_2$。

故根据函数的定义可以得知 f^c 是 Y 到 X 的函数。

下面再证 f^c 是满射的。

由上一节的定理可知,$\mathrm{ran}\, f^c = \mathrm{dom}\, f$。又由于 f 是 X 到 Y 的函数,则 $\mathrm{dom}\, f = X$。故 $\mathrm{ran}\, f^c = X$。故可以得知 f^c 是 Y 到 X 的满射。

之后再证 f^c 是单射的。

设 $<y_1,x_1> \in f^c, <y_2,x_2> \in f^c, x_1 = x_2$,由逆关系的定义可知 $<x_1,y_1> \in f, <x_2,$

$y_2>\in f$。又因为 f 是函数,必有 $y_1=y_2$。故 f^C 是单射的。

综上所述,f^C 是双射的。

定义 4.2.1 ▶ 设 $f:X\to Y$ 是双射的,f 的逆关系 f^C 是 Y 到 X 的双射函数,称双射函数 f^C 为 f 的反函数,记为 f^{-1}。

例如,设 $X=\{m,n,t\},Y=\{0,1,2\},f=\{<m,0>,<n,1>,<t,2>\}$。

显然,f 是 X 到 Y 的双射函数,那么其逆关系 $f^{-1}=\{<0,m>,<1,n>,<2,t>\}$ 就是 $Y\to X$ 函数。

如果 $f=\{<m,0>,<n,1>,<t,1>\}$,则 f 的逆关系 $f^{-1}=\{<0,m>,<1,n>,<1,t>\}$ 就不是函数。

定理 4.2.2 ▶ 设 $f:X\to Y$ 为双射的,$f^{-1}:Y\to X$ 是 f 的反函数,则 $(f^{-1})^{-1}=f$。

证明 由定理 4.2.1 和定义 4.2.1 可以知道 $(f^{-1})^{-1}:X\to Y$,且为双射的。

$\forall x\in X$,设 $f(x)=y$,则 $f^{-1}(y)=x,(f^{-1})^{-1}(x)=y=f(x)$,所以 $(f^{-1})^{-1}=f$。

前面我们已经讲到,函数是满足一定条件的二元关系,那么它同样可以进行复合运算。函数的复合关系可以简单地描述如下。

设 $f:X\to Y,g:W\to Z$,若 $f(X)\subseteq W$,集合 $g\circ f=\{<x,z>\mid x\in X\wedge z\in Z\wedge(\exists y)(y\in Y\wedge<x,y>\in f\wedge<y,z>\in g)\}=\{<x,z>\mid x\in X\wedge z\in Z\wedge(\exists y)(y\in Y\wedge y=f(x)\wedge z=g(y))\}$。

定义 4.2.2 ▶ 设函数 $f:X\to Y,g:W\to Z$,若 $f(X)\subseteq W$,则集合 $\{<x,z>\mid x\in X\wedge z\in Z\wedge(\exists y)(y\in Y\wedge y=f(x)\wedge z=g(y))\}$,称为函数 g 在函数 f 左边的复合关系,记为 $g\circ f$。

定理 4.2.3 ▶ 设函数 $f:X\to Y,g:W\to Z,f(X)\subseteq W$,那么函数 g 在函数 f 左边的复合关系 $g\circ f$ 是 X 到 Z 的函数,简单理解就是两个函数的复合是一个函数。

证明 $\forall x\in X$,因为 f 是 X 到 Y 的函数,必存在唯一的 $y\in Y$,使 $<x,y>\in f$,即 $y=f(x)$。而 $y=f(x)\in f(X)\subseteq W$,故 $y\in W$。

又因为 g 是 W 到 Z 的函数,必存在唯一的 $z\in Z$,使 $<y,z>\in g$,即 $z=g(y)$。故由定义 4.2.2 可知,$<x,z>\in g\circ f$,即 $g\circ f(x)=z$。

故可以得知 $g\circ f$ 是 X 到 Z 的函数。

那么根据复合函数的定义,显然有 $g\circ f(x)=g(f(x))$。

例 4.2.1 设 $A=\{1,2,3\},B=\{a,b\},C=\{m,n\},f=\{<1,a>,<2,b>,<3,a>\},g=\{<a,m>,<b,n>\}$,求 $g\circ f$。

解 $g\circ f=\{<1,m>,<2,n>,<3,m>\}$

定理 4.2.4 ▶ 设函数 $f:X\to Y,g:Y\to Z$,复合函数 $g\circ f:X\to Z$。

(1) 如果 f 和 g 都是满射的,则 $g\circ f$ 也是满射的。

(2) 如果 f 和 g 都是单射的,则 $g\circ f$ 也是单射的。

(3) 如果 f 和 g 都是双射的,则 $g\circ f$ 也是双射的。

证明 (1) $\forall z\in Z$,因为 g 是满射的,$\exists y\in Y$,使 $z=g(y)$。又因为 f 是满射的,$\exists x\in X$,使 $y=f(x)$。那么,$g\circ f(x)=g(f(x))=g(y)=z$,所以 $g\circ f$ 是满射的。

(2) 设 $x_1 \in X, x_2 \in X$ 且 $x_1 \neq x_2$,因为 f 是单射的,故 $f(x_1) \neq f(x_2)$,$f(x_1) \in Y$,$f(x_2) \in Y$。又因为 g 是单射的,所以 $g(f(x_1)) \neq g(f(x_2))$,即 $g \circ f(x_1) \neq g \circ f(x_2)$,故 $g \circ f$ 是单射的。

(3) f 和 g 都是双射的,则 f 和 g 都是满射的和单射的。从(1)和(2)可知,$g \circ f$ 是满射的和单射的,故 $g \circ f$ 是双射的。

例 4.2.2 设 $x \in \mathbf{N}$,有函数 $f(x) = x+3$,$g(x) = x-3$,$t(x) = x^2$,求 $g \circ f$ 与 $t \circ (g \circ f)$。

解 $g \circ f = \{<x,x> \mid x \in \mathbf{N}\}$

$t \circ (g \circ f) = \{<x,x^2> \mid x \in \mathbf{N}\}$

定理 4.2.5 设函数 $f: X \to Y$,$g: Y \to Z$,$h: Z \to W$,则 $h \circ (g \circ f) = (h \circ g) \circ f$。

定义 4.2.3 函数 $f: X \to Y$ 叫作常函数,如果存在某个 $y_0 \in Y$,对于每个 $x \in X$ 都有 $f(x) = y_0$,即 $f(X) = \{y_0\}$。

定义 4.2.4 如果函数 $I_X = \{<x,x> \mid x \in X\}$,则称函数 $I_X: X \to X$ 为恒等函数。

定理 4.2.6 设函数 $f: X \to Y$,I_X 是 X 上的恒等函数,I_Y 是 Y 上的恒等函数,则 $I_Y \circ f = f$,$I_X = f$。

这个定理的证明可以由定义直接得到。

定理 4.2.7 设函数 $f: X \to Y$ 是双射的,$f^{-1}: Y \to X$ 是 f 的反函数,则

$$f^{-1} \circ f = I_X \quad 且 \quad f \circ f^{-1} = I_Y$$

证明 先来证明 $f^{-1} \circ f = I_X$。

由复合函数定义可知,$f^{-1} \circ f: X \to X$,$I_X: X \to X$。

$\forall x \in X$,设 $f(x) = y$,则 $f^{-1}(y) = x$。

$$f^{-1} \circ f(x) = f^{-1}(f(x)) = f^{-1}(y) = x = I_X(x)$$

所以 $f^{-1} \circ f = I_X$。

类似地,可以证明 $f \circ f^{-1} = I_Y$。

定理 4.2.8 如果函数 $f: X \to Y$ 是一一对应的函数,则 $(f^{-1})^{-1} = f$。

证明 (1) 由于函数 $f: X \to Y$ 是一一对应的函数,可知 $f^{-1}: Y \to X$ 同样也是一一对应的函数,因此 $(f^{-1})^{-1}: X \to Y$ 也为一一对应的函数,那么显然:

$$\mathrm{dom}\, f = \mathrm{dom}\,(f^{-1})^{-1} = X$$

(2) $x \in X \Rightarrow f: x \to f(x) \Rightarrow f^{-1}: f(x) \to x \Rightarrow (f^{-1})^{-1}: x \to f(x)$。

由(1)(2)可知 $(f^{-1})^{-1} = f$。

定理 4.2.9 设函数 $f: X \to Y$,$g: Y \to Z$ 且 f 和 g 均是双射的,则 $(g \circ f)^{-1} = f^{-1} \circ g^{-1}$。

证明 (1) 因为 f 和 g 均是双射的,f^{-1} 和 g^{-1} 存在且

$$f^{-1}: Y \to X, \quad g^{-1}: Z \to Y$$

所以 $f^{-1} \circ g^{-1}: Z \to X$。

由定义 4.2.3 可知,$g \circ f: X \to Z$。

所以 $(g \circ f)^{-1}: Z \to X$。

(2) $\forall z \in Z$,因为 g 是双射的,$\exists y \in Y$ 使 $g(y) = z$,又因为 f 是双射的,$\exists x \in X$,使 $f(x) = y$。

于是,$g \circ f(x) = g(f(x)) = g(y) = z$,故 $(g \circ f)^{-1}(z) = x$。

另一方面,$f^{-1}(y)=x$,$g^{-1}(z)=y$。

于是,$f^{-1}\circ g^{-1}(z)=f^{-1}\circ(g^{-1}(z))=f^{-1}(y)=x$。

所以$(g\circ f)^{-1}=f^{-1}\circ g^{-1}$。

4.3　特征函数与模糊子集的概念

通过分析可知,有些函数与集合之间可以建立一些特殊的联系,可以借助于这些函数对集合进行运算,同时推广表达模糊集合的概念。

定义 4.3.1 ▶ 设 E 是全集,A 是 E 的子集,即 $A\subseteq E$,于是把 $\psi_A:E\rightarrow\{0,1\}$ 定义为

$$\psi_A(x)=\begin{cases}1, & x\in A \\ 0, & x\notin A\end{cases}$$

并称 $\psi_A(x)$ 为集合 A 的特征函数。

例如,E 是某学校全体学生的集合,A 是全体男学生的集合,则 ψ_A 为男学生的特征函数。

定理 4.3.1 ▶ 给定全集 E,$A\subseteq E$ 和 $B\subseteq E$,对于所有的 $x\in E$,特征函数有如下性质。

(1) $\psi_A(x)=0\Leftrightarrow A=\phi$

(2) $\psi_A(x)=1\Leftrightarrow A=E$

(3) $\psi_A(x)\leqslant\psi_B(x)\Leftrightarrow A\subseteq B$

(4) $\psi_A(x)=\psi_B(x)\Leftrightarrow A=B$

(5) $\psi_{A\cap B}(x)=\psi_A(x)*\psi_B(x)$

(6) $\psi_{A\cup B}(x)=\psi_A(x)+\psi_B(x)-\psi_{A\cap B}(x)$

(7) $\psi_{\sim A}(x)=1-\psi_A(x)$

(8) $\psi_{A-B}(x)=\psi_{A\cap\sim B}(x)=\psi_A(x)-\psi_{A\cap B}(x)$

其中特征函数中的运算＋、－、＊就是通常的算术运算＋、－、×。

证明 这里仅证明部分关系式,其余可自行证明。

如(3):对任意的 $x\in A$,因为 $A\subseteq B$,故 $x\in B$,由特征函数的定义可知 $\psi_A(x)=1$,$\psi_B(x)=1$,所以可知 $\psi_A(x)\leqslant\psi_B(x)$ 成立。若 $x\notin A$,显然也有 $\psi_A(x)\leqslant\psi_B(x)$。

反之,若 $\psi_A(x)\leqslant\psi_B(x)$,$x\notin B$,则 $\psi_B(x)=0$,于是 $\psi_A(x)=0$,从而 $x\notin A$,故 $A\subseteq B$。

例 4.3.1 试证明 $\psi_{(A-C)\cup(B-C)}(x)=\psi_{(A\cup B)-C}(x)$。

证明

$$\begin{aligned}
\psi_{(A-C)\cup(B-C)}(x)&=\psi_{A-C}(x)+\psi_{B-C}(x)-\psi_{A-C}(x)*\psi_{B-C}(x)\\
&=\psi_A(x)-\psi_A(x)*\psi_C(x)+\psi_B(x)-\psi_B(x)*\psi_C(x)-\psi_A(x)*\psi_B(x)+\\
&\quad\psi_A(x)*\psi_B(x)*\psi_C(x)\\
&=\psi_{(A\cup B)}(x)-(\psi_A(x)+\psi_B(x)-\psi_A(x)*\psi_B(x))*\psi_C(x)\\
&=\psi_{(A\cup B)}(x)-\psi_{(A\cup B)}(x)*\psi_C(x)\\
&=\psi_{(A\cup B)-C}(x)
\end{aligned}$$

对特征函数进行推广可以导出模糊子集的概念。

设 $E=\{x_1,x_2,\cdots,x_n\}$,我们可以将 E 的任一子集 A 表示为

$$A:\{<x_1,\psi_A(x_1)>,<x_2,\psi_A(x_2)>,\cdots,<x_n,\psi_A(x_n)>\}$$

当 $\psi_A(x_i)=1$ 时，$x_i\in A$。

当 $\psi_A(x_i)=0$ 时，$x_i\in A$。

如果我们将 $\psi_A(x_i)$ 的取值范围不局限于 0 和 1，而是取 0 和 1 之间的任何数，比如

$$A^*:\{<x_1,0.1>,<x_2,0>,<x_3,0.4>,<x_4,1>,<x_5,0.9>\}$$

那么，对 A^* 可以做如下理解：它表示 x_1 是少量属于 A^*，x_2 是不属于 A^*，x_3 也是少量属于 A^*（但比 x_1 要多），x_4 是必定属于 A^*，x_5 是基本上属于 A^*。这样的一个集合 A^* 就是一个模糊子集，其中 $0.1,0.4,0.9,\cdots$ 分别称为该集合中对应元素的隶属程度。

定义 4.3.2 ▶ 设 E 是全集（或称论域），指定 E 上的一个模糊子集 $\underline{A}$ 是指对任一 $x\in E$ 都有一个隶属程度 $\mu=\psi_{\underline{A}}(x)(0\leqslant\mu\leqslant1)$ 与它对应，称 $\psi_{\underline{A}}(x)$ 为 $\underline{A}$ 的隶属函数。

从模糊子集的定义可以看出，当 $\psi_{\underline{A}}(x)$ 只取 0、1 两值时，$\underline{A}$ 便成为普通子集。

例 4.3.2 设 $E=\{0,1,2,\cdots,100\}$，E 的模糊子集 $\underline{A}$ 是 E 中的"大数"的集合。显然，100 是 E 中的"大数"，0 不是 E 中的"大数"。那么 80 是否是 E 中的"大数"呢？为此我们定义 $\underline{A}$ 的隶属函数：

$$\psi_{\underline{A}}(x)=x/100$$

于是，就可以知道 80 属于"大数"的程度为 0.8，也就是 80 的隶属程度为 0.8。同样，99 的隶属程度为 0.99。

4.4 常用函数

下面介绍离散数学中的两个重要的函数，即上下取整函数与哈希函数。其中取整函数在分析求解一定规模问题的计算步骤数中起着重要作用。

定义 4.4.1 ▶ 对于有理数 x，$f(x)$ 为大于或等于 x 的最小的整数，则称 $f(x)$ 为上取整函数（强取整函数），记为 $f(x)=\lceil x\rceil$。而对于有理数 x，若 $f(x)$ 为小于或等于 x 的最大整数，则称为 $f(x)$ 为下取整函数（弱取整函数），记为 $f(x)=\lfloor x\rfloor$。

注意：下取整函数也常称为最大整数函数，此时则记为 $f(x)=[x]$。

性质：

(1) $\lfloor x\rfloor\leqslant x<\lfloor x\rfloor+1$，$x\leqslant\lceil x\rceil<x+1$，此时取等号当且仅当 x 为整数。

(2) 下取整函数为等幂函数：$\lfloor\lfloor x\rfloor\rfloor=\lfloor x\rfloor$。

(3) 对任意的整数 k 和任意实数 x，有 $\lfloor k+x\rfloor=k+\lfloor x\rfloor$。

(4) 对于整数 x 有 $\lfloor x/2\rfloor+\lceil x/2\rceil=x$。

(5) $\lceil x\rceil=-\lfloor -x\rfloor$

例 4.4.1 存在计算机磁盘上的数据或者数据网络上传输的数据通常表示为字节串。每个字节由 8 位组成。那么要表示 300 字位的数据需要多少字节？

解 由于需要的字节数是整数形式，同时 300 字位必须全部包括，所以表示 300 字位的数据需要 $\lceil 300/8\rceil=38$ 字节。

例 4.4.2 证明对任意 $x\in\mathbf{R}$，$x-1<\lfloor x\rfloor\leqslant x<\lfloor x\rfloor+1$。

证明 设 $x=\lfloor x\rfloor+r(0\leqslant r<1)$，于是 $0\leqslant r=x-\lfloor x\rfloor<1$。所以 $x-1<\lfloor x\rfloor\leqslant x<\lfloor x\rfloor+1$。

例 4.4.3 证明对任意 $x,y\in\mathbf{R}$，若 $x\leqslant y$，则 $\lfloor x\rfloor\leqslant\lfloor y\rfloor$。

利用反证法证明如下：

若$\lfloor x \rfloor > \lfloor y \rfloor$，则$\lfloor x \rfloor - \lfloor y \rfloor > 0$。

所以整数$\lfloor x \rfloor - \lfloor y \rfloor \geqslant 1$，即$\lfloor x \rfloor \geqslant \lfloor y \rfloor + 1$。

那么可以得知，$x \geqslant \lfloor x \rfloor \geqslant \lfloor y \rfloor + 1 > y$。

这与已知的$x \leqslant y$矛盾，故$\lfloor x \rfloor \leqslant \lfloor y \rfloor$。

定义 4.4.2 ▶哈希函数也称为散列函数，主要用于数据存储和检索。

例如，一个比较大的公司有很多职员（人数基本确定），如果将这些职员的信息顺序存放到计算机中（即顺序存放到一个单链表中），那么查找某个人的信息会很不方便，只能从头到尾逐个进行比较，特别是当要查找的那个人的信息放在最后时，要耗时很多，显然这样的存储方式是不合适的。

如果采用多链表，当新来一个职员时，他的信息要存放到哪个链表中呢？

假定用七位数字表示一个人的代码（称为关键字（key）），它是一个职员的唯一标识，并且该公司有1万人，$C = \{c_1, c_2, c_3, \cdots, c_{10000}, \cdots\}$，每个$C_i$有一个key。同时该公司的计算机在可接受的时间内可以查含有100项的一个链表，即每个链表中可以存放100个人的信息。于是可以用101个链表存放所有职员的信息。那么一个职员的信息应放在哪个链表里呢？可以用哈希函数$h: C \to \{0, 1, 2, \cdots, 100\}$来分配？

令i（是一个七位数字数）是一个职员的代码（key），则

$$h(i) = i \pmod{101}$$

例如，$i = 2473871$，则

$$h(2473871) = 2473871 \pmod{101} = 78$$

表明这个职员的记录应存放在第78号链表中。

当然我们希望函数h是入射的，即不同职员的代码i_1, i_2对应的函数值$h(i_1), h(i_2)$不同，这样便于查找职员信息。但是设计这样的函数并不容易，特别是当数据量特别大时。比如，当i取2474073时，$h(2474073) = 78$，很明显两个位置都是78，那么就会产生冲突。设计一个好的哈希函数，应该是冲突越少越好。

解决冲突的方法很多。

（1）开放定址法：一旦发生冲突，就去寻找下一个空的散列地址，只要散列表足够大，空的散列地址总能找到，并将其记录存入。其公式为

$$h(\text{key}) = (h(\text{key}) + d) \bmod m \, (d = 1, 2, 3, \cdots, m-1)$$

（2）链地址法：哈希函数值相同的放在同一个线性表中。

（3）建立一个公共溢出区：专门存放冲突数据。

（4）再哈希法：用另一个正整数m再散列一次。

另外，构造哈希函数的方法也很多，不一定是对key（mod m），还有如下几种。

（1）直接地址法：$h(\text{key}) = \text{key}$或者$h(\text{key}) = a.\text{key} + b$。

（2）随机数法：$h(\text{key}) = \text{random(key)}$，其中random()是一个随机函数。当key的长度不等时常用这个方法。

例 4.4.4　假设在计算机内存中有编号从0到10的存储单元，如图4.4.1所示。它表示在初始时刻全为空的单元中，按次序15、558、32、132、102和5存入后的情形。现希望能在这些存储单元中存储任意的非负整数并能进行检索，试用哈希函数方法完成257的存储和558

的检索,并利用开放地址法解决冲突。

132			102	15	5	259		558		32	
0	1	2	3	4	5	6	7	8	9	10	0

图 4.4.1

解 因为 $h(257)=257(\bmod 11)=4$,但是第 4 号位被 15 号占据,此时产生了冲突。根据开放地址法可知位置 7 刚好可以有空位用于存储 257。

又因为 $h(558)=8$,所以检查位置 8,558 恰好在位置 8。

习 题

1. 下列函数中哪些是单射的、满射的、双射的?

(1) $f:I \to I, f(x)=x \bmod 4$

(2) $f:I \to \mathbf{N}, f(x)=|2x|+1$

(3) $f:\mathbf{R} \to \mathbf{R}, f(x)=3x+10$

(4) $f:\mathbf{N} \to \{0,1\}, f(x)=\begin{cases} 1, & x \text{ 是奇数} \\ 0, & x \text{ 是偶数} \end{cases}$

2. 设 $X=\{a,b,c\}, Y=\{1,2,3\}$,作 $f:X \to Y$,问有多少关系 f,其中有多少是满射的?

3. 以下哪些是函数?有哪些是单射的?有哪些是满射的?写出双射函数的逆函数。

(1) $f:\mathbf{Z} \to \mathbf{N}, f(x)=x^2-2$

(2) $h:\mathbf{N} \to Q, h(x) \dfrac{1}{x}$

(3) $g:\mathbf{Z} \times \mathbf{N} \to Q, h(z,n)=\dfrac{z}{n+1}$

(4) $g:\mathbf{N} \to \mathbf{N}, g$ 定义为 $g(x)=2^x$

4. 试分别给出满足下列条件的函数例子。

(1) 函数是单射的但不是满射的。

(2) 函数是满射的但不是单射的。

(3) 函数不是单射的也不是满射的。

(4) 函数既是单射的又是满射的。

5. 假设 f 和 g 是函数,试证明 $f \cap g$ 也是函数。

6. 假定 A 和 B 都是有穷集合,找出 A 和 B 存在单射的必要条件。

7. 设 X 和 Y 是有穷集合,有多少不同函数是单射的?有多少不同函数是双射的?

8. 设 $g:\mathbf{N} \to \mathbf{N}$,定义如下函数:

$$\begin{cases} f(x)=x-10, & \text{若 } x>10 \\ f(x)=f(f(x)+11), & \text{若 } x \leqslant 100 \end{cases}$$

证明:(1) $f(99)=91$;

(2) $f(x)=91,(0 \leqslant x \leqslant 100)$。

9. 设 $\mathbf{Z}$ 为整数集,函数 $f:\mathbf{Z} \times \mathbf{Z} \to \mathbf{Z}$,且 $f(x,y)=x+y$,问 f 是单射的还是满射的?为什么?并求 $f(x,x), f(x,-x)$。

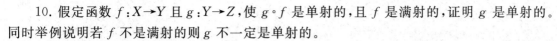

10. 假定函数 $f:X{\to}Y$ 且 $g:Y{\to}Z$，使 $g{\circ}f$ 是单射的，且 f 是满射的，证明 g 是单射的。同时举例说明若 f 不是满射的则 g 不一定是单射的。

11. 设函数 $f:X{\to}Y,g:Y{\to}X$，证明：若 $f^{-1}=g,f=g^{-1}$，则 $g{\circ}f=I_X,f{\circ}g=I_Y$。

12. 用特征函数证明：

$$A\bigcup(B\bigcap C)=(A\bigcup B)\bigcap(A\bigcup C)$$

代数结构

The Algebraic Structure

Chapter 5　Algebra System

An algebraic system is a system of operations based on sets. It is a method of constructing a mathematical system with operations, so it is called an algebraic system.

5.1　The Introduction of Algebraic Systems

Before introducing the algebraic system, we introduce an operation concept on set A. For example, if every number a ($a \neq 0$) on the set of real numbers $\mathbf{R}$ is mapped to its reciprocal $1/a$, or every number y on $\mathbf{R}$ is mapped to $[y]$, these mappings can be called unary operations on the set $\mathbf{R}$; On set $\mathbf{R}$, ordinary addition and multiplication on any two numbers are binary operations on set $\mathbf{R}$, which can also be regarded as mapping every two numbers on $\mathbf{R}$ to a number in $\mathbf{R}$; The common feature of the above examples is that the results are all in the original set $\mathbf{R}$, and we call the operations that have this characteristic closed, closed operations for short. Conversely, operations without such characteristics are not closed.

It is easy to give the example of an unclosed computation: a vending machine will accept fifty-cent coins and one-yuan coins for products such as Ice Dew (bottles), NongFu Spring (bottles) and Coca-cola(tin). When people put in any two of the above, the vending machine will supply the corresponding goods as shown in Table 5.1.1.

Table　5.1.1

*	Fifty-cent coin	One-yuan coin
Fifty-cent coin	Ice Dew	NongFu Spring
One-yuan coin	NongFu Spring	Coca-cola

The symbol * in the upper left corner of the table can be understood as an operator of a binary operation. The binary operation * in this example is an unclosed operation on the set {fifty-cent coin, one-yuan coin}.

Definition 5.1.1　For a set A, a mapping from set A^n to B is called an n-element operation on set A. If $B \subseteq A$, then the n-element operation is said to be closed.

Definition 5.1.2　A nonempty set A consists of several operations f_1, f_2, ..., f_n defined on the set. The system is called an algebraic system, which we usually call the system $<A, f_1, f_2, ..., f_n>$.

For example, the set of positive integers I_+ and the addition operation "$+$" on that set make up an algebraic system $<I_+, +>$. Another example is a finite set S, and an algebraic system $<P(S), \bigcup, \bigcap, \sim>$ is constituted by the power set of S $P(S)$ and the power set of S operating "$\bigcup$", "$\sim$", "$\bigcap$" on the set. Although some algebraic systems take different forms, there may be some common laws among them.

For example, consider an algebraic system$<I, +>$, where I is a set of integers, and $+$ is a common addition operation. Obviously, in this algebraic system, the addition operation has the following three operation rules, that is, any $x, y, z \in I$, has

(1) $x+y \in I$　　　　　　　　　　(Closed law)

(2) $x+y=y+x$　　　　　　　　　(Commutative law)

(3) $(x+y)+z=x+(y+z)$　　　　(Associative law)

5.2　Operations and Properties of Algebraic Systems

There are four operations of grade $+, -, \times, \div$ in elementary algebra, which are expanded to include vector operation, matrix operation and determinant operation. Here we make the operation not only include all the preceding operations but also have a more general meaning.

Definition 5.2.1　Let $*$ be a binary operation defined on set A. If for any $x, y \in A$, there is $x * y \in A$, then the binary operation $*$ is closed on A.

Example 5.2.1　If $A=\{x \mid x=2^n, n \in \mathbf{N}\}$. Is the multiplication operation closed? What about the addition operation?

Solution　For any $2^r, 2^s \in A, r, s \in \mathbf{N}$, because $2^r * 2^s = 2^{r+s} \in A$, so the multiplication is closed. But it's not closed for additional operations, because there are at least $2+2^2=6 \notin A$.

Definition 5.2.2　Let $*$ be a binary operation defined on set A. If for any $x, y \in A$, there is $x * y = y * x$, then the binary operation $*$ is said to be interchangeable.

Example 5.2.2　Let Q be a set of rational numbers and let $\triangle$ be a binary operation on Q. For any $a, b \in \mathbf{R}$, $a \triangle b = a+b-a * b$, ask whether the operation is interchangeable.

Solution　Because

$$a \triangle b = a+b-a * b = b+a-b * a = b \triangle a$$

So operation $\triangle$ is interchangeable.

Definition 5.2.3　Let $*$ be a binary operation defined on set A, if for any $x, y, z \in A$, we have $(x * y) * z = x * (y * z)$, then the operation $*$ is said to be associative.

Example 5.2.3　Let A be a non-empty set, $\bigstar$ be a binary operation on A, for any $a, b \in A$, there is $a \bigstar b = b$. Prove $\bigstar$ be a associative operation.

Proof　Because for any $x, y, z \in A$, both

$$(a \bigstar b) \bigstar c = b \bigstar c = c$$

but　　　　　　　　$a \bigstar (b \bigstar c) = a \bigstar c = c$

so　　　　　　　　$(a \bigstar b) \bigstar c = a \bigstar (b \bigstar c)$

Definition 5.2.4 Let $*$, Δ be two binary operations defined on set A , if for any x , y , $z \in A$, there are

$$x * (y \Delta z) = (x * y) \Delta (x * z)$$
$$(y \Delta z) * x = (y * x) \Delta (z * x)$$

Then the operation $*$ pair operation Δ is called assignable.

Example 5.2.4 Set $A = \{\alpha, \beta\}$, define two binary operations $*$ and Δ on A , as shown in Table 5.2.1. Is the operation $*$ assignable for the operation Δ?

Table 5.2.1

$*$	α	β		Δ	α	β
α	α	β		α	α	β
β	β	α		β	β	α
(a)				(b)		

Solution The $*$ operation is not assignable because

$$\beta * (\alpha \Delta \beta) = \beta * \alpha = \beta$$

But

$$(\beta * \alpha)(\beta * \beta) = \beta \Delta \alpha = \alpha$$

Definition 5.2.5 Let $*$ and Δ be two commutative binary operations defined on set A , if for any x , $y \in A$, we have

$$x * (x \Delta y) = x$$
$$x \Delta (x * y) = x$$

Then the operation $*$ and Δ satisfies the absorptivity.

Example 5.2.5 Let set $\mathbf{N}$ to be all-natural Numbers, define two binary operations $*$ and $\bigstar$ on $\mathbf{N}$, for any x , $y \in \mathbf{N}$, we have

$$x * y = \max(x, y)$$
$$x \bigstar y = \min(x, y)$$

Verify the absorption law of $*$ and $\bigstar$.

Solution For any a , $b \in \mathbf{N}$

$$a * (a \bigstar b) = \max(a, \min(a, b)) = a$$
$$a \bigstar (a * b) = \min(a, \max(a, b)) = a$$

Therefore, $*$ and $\bigstar$ satisfy the absorption law.

Definition 5.2.6 Let $*$ be a binary operation defined on set A. If for any $x \in A$, there is $x * x = x$, then the operation $*$ is idempotent.

Example 5.2.6 Let $P(S)$ be a powerful set of set S , and the two binary operations defined on $P(S)$, the union operation $\bigcup$ of set, the intersection operation $\bigcap$ and verify operation $\bigcup$ and $\bigcap$ are equipotent.

Solution For any $A \in P(S)$, there is $A \bigcup A = A$ and $A \bigcap A = A$, therefore, $\bigcap$ and $\bigcup$ both satisfy the equal power law.

Definition 5.2.7 Let $*$ be a binary operation defined on set A. If there is an element

$e_1 \in A$ and for any element $x \in A$, there is $e_1 * x = x$, then e_1 is called the left identity element of operation $*$ in A. If there is an element $e_r \in A$ and for any element $x \in A$, there is $x * e_r = x$, it is called the right identity element in A about operation $*$; If e is an element in A, is both the left and right identity element, then e is called the identity element in A with respect to operation $*$. Obviously, for any $x \in A$, there's $e * x = x * e = x$.

Example　5.2.7　Let set $S = \{\alpha, \beta, \gamma, \delta\}$. Two binary operations $*$ and $\bigstar$ defined on S are shown in Table 5.2.2. Try to point out the left unitary element and the right unitary element.

Table　5.2.2

$*$	α	β	γ	δ
α	δ	α	β	γ
β	α	β	γ	δ
γ	α	β	γ	γ
δ	α	β	γ	δ

(a)

$\bigstar$	α	β	γ	δ
α	α	β	δ	γ
β	β	α	γ	δ
γ	γ	δ	α	β
δ	δ	δ	β	γ

(b)

Solution　It can be seen from Table 5.2.2 that they are all the left identify element in S about $*$, while α is the right identity element in S about operation $\bigstar$.

Theorem 5.2.1　Let $*$ be a binary operation defined on set A, and in A there are the left identity element e_l and right identity element e_r of the operation $*$, then $e_l = e_r = e$, and the identity element in A is unique.

Proof　Since e_l and e_r are the left and right identity elements of operation $*$ in A respectively, so

$$e_l = e_l * e_r = e_r = e$$

Set another identity element $e_1 \in A$, then

$$e_1 = e_1 * e = e$$

Definition 5.2.8　Let $*$ be a binary operation defined on set A. If there is an element $\theta_l \in S$, for any element $x \in A$, we always have $\theta_l * x = \theta_l$, it is called the left zero element of operation θ_l in A. If there is an element $\theta_r \in A$, for any element $x \in A$, we always have $\theta_r * x = \theta_r$, it is called the right zero element of operation $*$ in A; If an element θ in A is both a left and a right zero element, θ is said to be a zero of the operation $*$. Obviously, for either one $x \in A$, there is

$$\theta * x = x * \theta = \theta$$

Example　5.2.8　Set $S = \{\text{light color, dark color}\}$, a binary operation $*$ defined on S is shown in Table 5.2.3.

Try pointing out the zero element and the identify element.

Solution　The dark color is the zero element in S with respect to operation $*$, and the light color is the identity element in S with respect to operation $*$.

Table 5.2.3

*	Light color	Dark color
Light color	Light color	Dark color
Dark color	Dark color	Dark color

Theorem 5.2.2　Let * be a binary operation defined on set A, and there are left and right zeros on operation * in A, then $\theta_r = \theta_1 = \theta$, and the zeros in A are unique.

The proof of this theorem is similar to Theorem 5.2.1.

Theorem 5.2.3　Let that $<A, *>$ is an algebraic system, and that the number of elements in set A is greater than 1. If there are identity element e and zero element θ in the algebraic system, then $\theta \neq e$.

Proof　By contradiction. Let $\theta = e$, then for any $x \in A$, there must be

$$x = e * x = \theta * x = \theta = e$$

So, all the elements in A are the same, which contradicts the fact that there are multiple elements in A.

Definition 5.2.9　Let the algebraic system $<A, *>$, where * is a binary operation defined on A, and e is the identity element of operation * in A. If A is a member of A and b is a member of A, such that $b * a = e$, then b is the left inverse of a; If $a * b = e$, then b is the right inverse of a; If a member b is both the left inverse of A and the right inverse of A, then b is an inverse of A.

And obviously, if b is an inverse of a, then a is also an inverse of b, or simply a and b are inverses of each other. So in the future, the inverse of x is going to be x^{-1}.

In general, the left inverse of an element is not necessarily equal to the right inverse of the element. Moreover, an element can have a left inverse element and no right inverse element, and even the left (right) inverse element of an element can not be unique.

Theorem 5.2.4　Let the algebraic system $<A, *>$, where * is a binary operation defined on A, where there is a identity element e, and each element has a left inverse element. If * is a associative operation, then the left inverse of any element in the algebraic system must also be the right inverse of that element, and the inverse of each element is unique.

Proof　Let a, b, $c \in A$, where b is the left inverse element of a and c is the right inverse element of b.

Because　　　　　　　$(b * a) * b = e * b = b$

So　$e = c * b = c * ((b * a) * b) = (c * (b * a)) * b = ((c * b) * a) * b = (e * a) * b = a * b$

So b is also the inverse of a.

Let's say element a has two inverse elements b and c, so

$$b = b * e = b * (a * c) = (b * a) * c = e * c = c$$

So the inverse element of a is unique.

Example 5.2.9　Try to construct an algebraic system in which only one element has an inverse element.

Solution   Let m, $n \in I$, $T = \{x \mid x \in I, m \leqslant x \leqslant n\}$, then, in an algebraic system $<T, \max>$ has a identity element m, and only m has an inverse element, because $m = \max(m, m)$.

Example 5.2.10   For an algebraic system $<\mathbf{R}, \cdot>$, where $\mathbf{R}$ is the whole of the real numbers, $\cdot$ is the ordinary multiplication, whether every element has an inverse element.

Solution   The identity element in this algebraic system is 1, and all elements except for the zero element 0 have an inverse element.

It can be pointed out that $<A, *>$ is an algebraic system, $*$ is a binary operation on A, then some properties of the operation can be seen directly from the operation table. That is:

(1) The operation $*$ is closed if and only if every element in the operation table belongs to A.

(2) The operation $*$ is commutative if and only if the operation table is symmetric with respect to the main diagonal.

(3) The operation $*$ is idempotent, If and only if each element on the main diagonal of the operation table is identical to the corresponding header element for that element.

(4) A has zero elements for $*$, if and only if the elements in the row and column corresponding to that element are identical to that element.

(5) A has an identity element about $*$, if and only if the row and column corresponding to that element corresponds in turn to the row and column of the operand table.

(6) Suppose that A has an identity element, and a and b are inverting each other, If and only if the element in the row of a, the element in the column of b and the element in the row of b, the element in the column of a are all identity elements.

5.3   Homomorphism and Isomorphism of Algebraic Systems

Using homomorphism and isomorphism, we can compare unknown algebraic systems to known algebraic systems. This can speed up our understanding and cognitive process.

Definition 5.3.1   Let $<A, \bigstar>$ and $<B, *>$ be two algebraic systems, $\bigstar$ and $*$ be binary (n) operations on A and B respectively, let f be a mapping from A to B, so that for any $a_1, a_2 \in A$, there is

$$f(a_1 \bigstar a_2) = f(a_1) * f(a_2)$$

Then f is called a homomorphic mapping from $<A, \bigstar>$ to $<B, *>$, called $<A, \bigstar>$ homomorphism in $<B, *>$, denoted as $A \sim B$. We call $<f(A), *>$ a homomorphism image of $<A, \bigstar>$. Among them

$$f(A) = \{x \mid x = f(a), a \in A\} \subseteq B$$

Example 5.3.1   Look at the algebraic system $<I, \cdot>$, where I is the set of integers, $\cdot$ is the ordinary multiplication. If we are only interested in the feature difference between positive, negative and zero, then the feature of the operation result in the algebraic system $<I, \cdot>$ can be described by the operation result of another algebraic system

$<B,\odot>$, where $B=\{$positive, negative, zero$\}$, and $\odot$ are binary operations defined on B, as shown in Table 5.3.1.

Table 5.3.1

$\odot$	positive	negative	zero
positive	positive	negative	zero
negative	negative	positive	zero
zero	zero	zero	zero

As a mapping $f: I \to B$

$$f(n)=\begin{cases}\text{positive,} & \text{if } n>0 \\ \text{negative,} & \text{if } n<0 \\ \text{zero,} & \text{if } n=0\end{cases}$$

Obviously, for any $a,b \in I$, there is

$$f(a \cdot b)=f(a)\odot f(b)$$

Therefore, the mapping f is a homomorphism from $<I,\cdot>$ to $<B,\odot>$.

Definition 5.3.2 Let f be a homomorphism from $<A,\bigstar>$ to $<B,*>$. If f is a surjective from A to B, then f is a surjective homomorphism. If f is an incidence from A to B, then f is called a single homomorphism; If f is a bijection from A to B, then f is called an isomorphic mapping, and $<A,\bigstar>$ and $<B,*>$ are isomorphic, denoted as $A\cong B$.

Example 5.3.2 Let $f: \mathbf{R}\to \mathbf{R}$ be defined as for any $x\in \mathbf{R}$, there is

$$f(x)=5^x$$

So, f is a single homomorphism from $<\mathbf{R},+>$ to $<\mathbf{R},\cdot>$.

Example 5.3.3 Set $f: \mathbf{N}\to N_k$ defined as for any $x\in \mathbf{N}$, there is

$$f(x)=x \bmod k$$

So, f is a full homomorphism of $<\mathbf{N},+>$ to $<N_k,+_k>$.

Example 5.3.4 Set $H=\{x\mid x=dn, d$ is some positive integer, $n\in I\}$, define the mapping $f: I\to H$ as for any $n\in I$, $f(n)=dn$. So, f is an isomorphism from $<I,+>$ to $<H,+>$. So, $I\cong H$.

Example 5.3.5 Set $A=\{a,b,c,d\}$, and define a binary operation on A, as shown in Table 5.3.2. Set $B=\{\alpha,\beta,\gamma,\delta\}$ and define a binary operation on B as shown in Table 5.3.3. Prove that $<A,\bigstar>$ and $<B,*>$ are isomorphic.

Proof Let's look at the mapping f, so that

$$f(a)=\alpha, \quad f(b)=\beta, \quad f(c)=\gamma, \quad f(d)=\delta$$

Table 5.3.2

$\bigstar$	a	b	c	d
a	a	b	c	d
b	b	a	a	c
c	b	d	d	c
d	a	b	c	d

Table 5.3.3

$*$	α	β	γ	δ
α	α	β	γ	δ
β	β	α	α	γ
γ	β	δ	δ	γ
δ	α	β	γ	δ

Obviously, f is a bijection from A to B, and it is easy to verify from Tables 5.3.2 and 5.3.3 that f is a homomorphism from $<A, \bigstar>$ to $<B, *>$. Therefore, $<A, \bigstar>$ and $<B, *>$ are isomorphic.

If I look at the mapping g, so that
$$g(a) = \delta, \quad g(b) = \gamma, \quad g(c) = \beta, \quad g(d) = \alpha$$
So, g is also an isomorphism of $<A, \bigstar>$ to $<B, *>$.

We know from Example 5.3.5 that when two algebraic systems are isomorphic, the isomorphic mapping between them can be non-unique.

Example 5.3.6 The algebraic system in Table 5.3.4 $<B, \bigoplus>$(Table 5.3.4(b)) and $<C, *>$(Table 5.3.4(c)) are painted $<A, \bigstar>$(Table 5.3.4(a)) isomorphism with algebraic systems.

The idea of isomorphism is important. As we can see from the example above, algebraic systems of different forms, if they are isomorphisms, can be abstractly treated as essentially the same algebraic systems.

Table 5.3.4

$\bigstar$	a	b
a	a	b
b	b	a

(a)

$\bigoplus$	even	odd
even	even	odd
odd	odd	even

(b)

$*$	even	odd
even	even	odd
odd	odd	even

(c)

Definition 5.3.3 Let that $<A, \bigstar>$ is an algebraic system. If f is a homomorphism from $<A, \bigstar>$ to $<A, \bigstar>$, then f is an endomorphism. If g is an isomorphism from $<A, \bigstar>$ to $<A, \bigstar>$, then g is an automorphism.

Theorem 5.3.1 If G is a set of algebraic systems, the isomorphic relation among algebraic systems in G is equivalent.

Proof Because any algebraic system $<A, \bigstar>$ can be isomorphic to itself by identity mapping, that is, reflexivity holds. Set about symmetry, $<A, \bigstar> \cong <B, *>$ and a corresponding isomorphic mapping f, because the inverse f is from $<B, *>$ to $<A, \bigstar>$ isomorphic mapping, namely $<B, *> \cong <A, \bigstar>$. Finally, if f is an isomorphic mapping from $<A, \bigstar>$ to $<B, *>$, and g is an isomorphic mapping from $<B, *>$ to $<C, \triangle>$, $g \circ f$ is an isomorphic mapping from $<A, \bigstar>$ to $<C, \triangle>$. That's $<A, \bigstar> \cong <C, \triangle>$, so the isomorphism is self-inverse. Therefore, the isomorphism relation is an equivalence relation.

Example 5.3.7 Set $V = <Z_n, \bigoplus>$, which $Z_n = \{0, 1, \ldots, n-1\}$, $\bigoplus$ is magnitude n plus. Prove that the endomorphism contains exactly n V.

Proof So let's show that there are n V endomorphisms, set $f_p: Z_n \to Z_n$, $f_p(x) = (px) \bmod n$, $p = 0, 1, \ldots, n-1$.

f_p is the endomorphism of V, because for arbitrary $x, y \in Z_n$, have
$$f_p(x \bigoplus y) = (p(x \bigoplus y)) \bmod n = (px) \bmod n \bigoplus (py) \bmod n = f_p(x) \bigoplus f_p(y)$$

Since there are n values for p, different p defines different mapping f_p, so there are n endomorphisms for V.

It is shown that any endomorphism of V is one of the n endomorphisms described above. Let f be an endomorphism of V and $f(1)=i$, $i \in \{0, 1, \ldots, n-1\}$. Let's prove that for arbitrary $x \in Z_n$, $f(x)=(ix) \bmod n$.

Obviously, $f(1)=i=(i \cdot 1) \bmod n$. Suppose that $x \in \{1, 2, \ldots, n-2\}$ have $f(x)=(ix) \bmod n$ is established, then

$$f(x+1)=f(x \oplus 1)=f(x) \oplus f(1)=(ix) \bmod n \oplus i=(ix+i) \bmod n=(i(x+1)) \bmod n$$

Finally

$$f(0)=f((n-1) \oplus 1)=f(n-1) \oplus f(1)=(i(n-1)) \bmod n \oplus i=(in) \bmod n=0$$
$$=(i \cdot 0) \bmod n$$

Example 5.3.8 (1) Let's say the algebraic system $V_1=<\mathbf{Z}, +>$, $V_2=<Z_n, \oplus>$. Where $\mathbf{Z}$ is the set of integers and $+$ is ordinary addition, $Z_n=\{0, 1, \ldots, n-1\}$, $\oplus$ is magnitude n plus. Let

$$f: \mathbf{Z} \to Z_n, \quad f(x)=(x) \bmod n$$

So f is a full homomorphism from V_1 to V_2. Obviously, f is surjective and it's arbitrary, and for arbitrary x, $y \in \mathbf{Z}$, we can know

$$f(x+y)=(x+y) \bmod n=(x) \bmod n \oplus (y) \bmod n=f(x) \oplus f(y)$$

(2) Set $V_1=<\mathbf{R}, +>$, $V_2=<\mathbf{R}^*, \cdot>$, were $\mathbf{R}$ and $\mathbf{R}^*$ are the set of real numbers and the set of non-zero real numbers respectively, and $+$ and $\cdot$ represent ordinary addition and multiplication respectively. Let

$$f: \mathbf{R} \to \mathbf{R}^*, \quad f(x)=e^x$$

So f is a single homomorphism from V_1 to V_2, it is easy to see that f is injective, and for arbitrary x, $y \in \mathbf{R}$ have

$$f(x+y)=e^{x+y}=e^x \cdot e^y=f(x) \cdot f(y)$$

5.4　Congruence and Quotient Algebra

Congruence relation is the equivalence relation on the set of algebraic systems, and under the action of operation, can maintain the equivalence relation. In addition, congruence is closely related to the operation. If there are multiple operations in an algebraic structure, it is necessary to examine whether the equivalence relation has substitution properties for all of these operations. An equivalence relation is not a congruence relation on an algebraic system if it does not satisfy the substitution property on an operation. Next, we further discuss the correspondence between homomorphism and congruence and the concept of quotient algebra.

Definition 5.4.1 Let $<A, \bigstar>$ be an algebraic system, and let R be an equivalence relation over A. If an algebraic system $<a_1, a_2>, <b_1, b_2> \in \mathbf{R}$, implies $<a_1 \bigstar b_1, a_2 \bigstar b_2> \in \mathbf{R}$, $\mathbf{R}$ is said to be the congruent relation on A about $\bigstar$. The equivalence classes that A is divided into by this congruence relation are called congruences.

Example 5.4.1 Set $A = \{a, b, c, d\}$, for the algebraic system $<A, \bigstar>$ determined by Table 5.4.1, the equivalence relation R defined by Table 5.4.2 over A. Known that R is congruence over A. This congruence divides A into congruence classes $\{a, b\}$ and $\{c, d\}$.

Table 5.4.1

$\bigstar$	a	b	c	d
a	a	a	d	c
b	b	a	c	d
c	c	d	a	b
d	d	d	b	a

Table 5.4.2

R	a	b	c	d
a	$\checkmark$	$\checkmark$		
b	$\checkmark$	$\checkmark$		
c			$\checkmark$	$\checkmark$
d			$\checkmark$	$\checkmark$

Example 5.4.2 Let $A = \{a, b, c, d\}$, for the algebraic system $<A, \bigstar>$ as determined by Table 5.4.3 and the equivalence relation R on A as defined by Table 5.4.2.

Table 5.4.3

$\bigstar$	a	b	c	d
a	a	a	d	c
b	b	a	d	a
c	c	b	a	b
d	c	d	b	a

Due to $<a, b>$, $<c, d> \in R$ we have
$$<a \bigstar c, b \bigstar d> = <d, a> \notin R$$

So, the equivalence relation on A defined by Table 5.4.2 is not an A congruence relation.

From the above two examples, we can see that the equivalence relation R defined on A is not necessarily the congruence relation on A, because the congruence relation must be closely related to the binary operation defined on A.

Theorem 5.4.1 Let that $<A, \bigstar>$ be an algebraic system, R be a congruent relation on A, and $B = \{A_1, A_2, ..., A_r\}$ is a division of A induced by R, then there must be a new algebraic system $<B, *>$, which is the homomorphism of $<A, \bigstar>$.

Proof Define binary operation $*$ on B as: for any $A_i, A_j \in B$, any $a_1 \in A_i, a_2 \in A_j$, if $a_1 \bigstar a_2 \in A_k$, then $A_i * A_j = A_k$. Since R is A congruence over A, the $A_i * A_j = A_k$ of the above definition is unique.

Do the mapping $f(a) = A_i, a \in A_i$.

Obviously, f is a full mapping from A to B.

For any $x, y \in A$, x, y must belong to one of two congruent classes in B, let's say $x \in A_i$, $y \in A_j$, $1 \leq i, j \leq r$; At the same time, $x \bigstar y$ must belong to some congruent class in B, so let's say $x \bigstar y \in A_k$, then there is
$$f(x \bigstar y) = A_k = A_i * A_j = f(x) * f(y)$$

Therefore, f is the full homomorphism from $<A, \bigstar>$ to $<R, *>$, that is, $<B, *>$ is the homomorphism image of $<A, \bigstar>$.

Theorem 5.4.2 Let f be a homomorphic mapping from $<A$, $\bigstar>$ to $<B$, $*>$. If the binary relation R is defined on A is: $<a$, $b>\in R$ if and only if

$$f(a)=f(b)$$

So R is a congruence on A.

Proof Because $f(a)=f(a)$, so $<a$, $a>\in R$. If $<a$, $b>\in R$, then $f(a)=f(b)$ that is $f(b)=f(a)$, so $<b$, $a>\in R$. If $<a$, $b>\in R$, $<b$, $c>\in R$, then $f(a)=f(b)=f(c)$, so $<a$, $c>\in R$. Finally, because if $<a$, $b>\in R$, $<c$, $d>\in R$, there is

$$f(a\bigstar c)=f(a)*f(c)=f(b)*f(d)=f(b\bigstar d)$$

So, $<a\bigstar c$, $b\bigstar d>\in R$.

So, R is the congruence of A.

Definition 5.4.2 Let's say $U=\{X$, $*$, $+$, V, ..., $W\}$ is an algebraic system, and E is the congruence in U. If the quotient set $X/E=\{[x]_E\mid x\in E\}$ introduces the corresponding operation $*'$,$+'$,V',...,W':

$$[x_1]_E*'[x_2]_E=[x_1*x_2]_E$$
$$[x_1]_E+'[x_2]_E=[x_1+x_2]_E$$
$$V'([x_1]_E)=[V(x_1)]_E$$

Then $<X/E$, $*'$,$+'$,V',...,$W'>$ is the algebraic system, called the U system of the quotient of E or simply the quotient of E, called U/E.

It should be noted that in the above definition, precisely because E is congruent, those equations are independent of the choice of "representative elements" in the congruent class, that is $*'$,$+'$,V',...,W', the corresponding operations.

Theorem 5.4.3 Let's say $U=\{X$, $*$, $+$, V, ..., $W\}$ is an algebraic system, where E is the congruence in U, the quotient $U/E=<X/E$, $*'$,$+'$,V',...,$W'>$ of U for E. If the mapping from X to X/E is defined:

$$g: x\to[x]_E$$

Then g is the full homomorphism of the algebraic system U to the quotient algebra U/E, called the natural homomorphism. Thus $U\sim U/E$.

Proof g is obviously surjective from X to X/E, and for any member x_1,x_2 of X

$$g(x_1*x_2)=[x_1*x_2]_E=[x_1]_E*'[x_2]_E=g(x_1)*g(x_2)$$
$$g(x_1+x_2)=[x_1+x_2]_E=[x_1]_E+'[x_2]_E=g(x_1)+'g(x_2)$$
$$g(V(x_1))=[V(x_1)]_E=V'([x_1])_E=V'(g(x_1))$$

$$...$$

So g is a full homomorphism of $U\sim U/E$.

Theorem 5.4.4 Let you have an algebraic system $U=\{X$, $*$, $+$, V, ..., $W\}\sim V=<Y$, e, $\oplus$, $\otimes$, ..., $\Theta>$ in which f is a full homomorphism and E is a congruent relatationship of U corresponding to f. Given is the natural homomorphism $U\sim U/E=<X/E$, $*'$,$+'$,V',...,$W'>$ of its algebraic system, then $U/E\cong V$,and $U/E\cong V$.

Proof Define the mapping X/E to Y:

$$h: [x]_E \rightarrow f(x)$$

On the one hand, since any element of Y can be represented as a form, h is surjective; on the other hand, if there are two elements in Y, it can be obtained that, thus h is injective, so h is bijective.

And for any elements $[x_1]_E$ and $[x_2]_E$ in X/E, have

$$h([x_1]_E *' [x_2]_E) = h([x_1 * x_2]_E) = f(x_1 * x_2) = f(x_1)ef(x_2) = h([x_1]_E)eh([x_2]_E)$$

$$h([x_1]_E +' [x_2]_E) = h([x_1 + x_2]_E) = f(x_1 + x_2) = f(x_1) \oplus f(x_2) = h([x_1]_E) \oplus h([x_2]_E)$$

$$h(V'([x_1]_E)) = h([V(x_1)_E]) = f(V(x_1)) = \otimes(h([x_1]_E))$$

So h is the isomorphism of U/E to V, and $U/E \cong V$.

5.5　Product Algebra

This section attempts to extend the cartesian product concept to the homogeneous algebraic system, thus producing a new algebraic system, called the product algebraic system of the algebraic system, or the product algebra for short.

Definition 5.5.1　Let $U = \{X, *, +, V, ..., W\}$ and $V = \langle Y, \odot, \oplus, \circledcirc, ..., \odot \rangle$ be an algebraic system of isomorphism (The number of operations, the number of operations and the number of algebraic constants in the algebraic system are the same). Introduce the corresponding operation $\bar{*}, \bar{+}, \bar{V}, ..., \bar{W}$ in $X \times Y$:

$$\langle x_1, y_1 \rangle \bar{*} \langle x_2, y_2 \rangle = \langle x_1 * x_1, y_1 \odot y_2 \rangle$$

$$\langle x_1, y_1 \rangle \bar{+} \langle x_2, y_2 \rangle = \langle x_1 + x_2, y_1 \oplus y_2 \rangle$$

$$V(\langle x_1, y_1 \rangle) = \langle V(x_1), \circledcirc(y_1) \rangle$$

$$...$$

Then $\langle X \times Y, \bar{*}, \bar{+}, \bar{V}, ..., \bar{W} \rangle$ is the algebraic system, called the product algebraic system of U and V or abbreviated product algebra, denoted by $U \times V$, and U and V are called the factor algebraic system of $U \times V$ or abbreviated factor algebra.

Example 5.5.1　Set $A = \{a_1, a_2\}$, $B = \{b_1, b_2, b_3\}$. If binary operations $*$ and $\odot$ are provided in A and B respectively (as shown in Table 5.5.1).

Table　5.5.1

$*$	a_1	a_2
a_1	a_1	a_2
a_2	a_2	a_1

(a)

$\odot$	b_1	b_2	b_3
b_1	b_1	b_1	b_3
b_2	b_2	b_2	b_3
b_3	b_1	b_1	b_3

(b)

Are $U = \langle A, * \rangle$ and $V = \langle B, \odot \rangle$ the same type of algebraic system, its product algebra for

$$U \times V = \langle A \times B, \bar{*} \rangle,$$ including $A \times B = \{\langle a_1, b_1 \rangle, \langle a_1, b_2 \rangle, \langle a_1, b_3 \rangle, \langle a_2, b_1 \rangle, \langle a_2, b_2 \rangle, \langle a_2, b_3 \rangle\}$ and the operation $\bar{*}$ table as shown in Table 5.5.2.

Table 5.5.2

*	$<a_1,b_1>$	$<a_1,b_2>$	$<a_1,b_3>$	$<a_2,b_1>$	$<a_2,b_2>$	$<a_2,b_3>$
$<a_1,b_1>$	$<a_1,b_1>$	$<a_1,b_1>$	$<a_1,b_3>$	$<a_2,b_1>$	$<a_2,b_1>$	$<a_2,b_3>$
$<a_1,b_2>$	$<a_1,b_2>$	$<a_1,b_2>$	$<a_1,b_3>$	$<a_2,b_2>$	$<a_2,b_2>$	$<a_2,b_3>$
$<a_1,b_3>$	$<a_1,b_1>$	$<a_1,b_3>$	$<a_1,b_3>$	$<a_2,b_1>$	$<a_2,b_3>$	$<a_2,b_3>$
$<a_2,b_1>$	$<a_2,b_1>$	$<a_2,b_1>$	$<a_2,b_3>$	$<a_1,b_1>$	$<a_1,b_1>$	$<a_1,b_3>$
$<a_2,b_2>$	$<a_2,b_2>$	$<a_2,b_2>$	$<a_2,b_3>$	$<a_1,b_2>$	$<a_1,b_2>$	$<a_1,b_3>$
$<a_2,b_3>$	$<a_2,b_1>$	$<a_2,b_3>$	$<a_2,b_3>$	$<a_1,b_1>$	$<a_1,b_3>$	$<a_1,b_3>$

It is easy to generalize Definition 5.5.1 to the product algebra of several homogeneous algebraic systems.

Exercises

1. Set $A=\{1, 2, 3, \ldots, 10\}$, ask whether the binary operation defined below is closed about set A?

(1) $x * y = \max(x, y)$

(2) $x * y = \min(x, y)$

(3) $x * y = GCD(x, y)$

(4) $x * y = LCM(x, y)$

(5) $x * y$ is equal to the number of prime numbers p, such that $x \leqslant p \leqslant y$

2. Try to illustrate some of the algebraic systems that you are familiar with.

3. For $\mathbf{R}$, the set of real Numbers, whether the binary operations listed in the following table have the properties of the left column, fill in "Yes" or "no" in the corresponding position in the Exercise 3 Table.

Exercise 3   Table

Quality	$+$	$-$	$\cdot$	max	min	$|x-y|$
associative						
commutative						
exist identity element						
exist null element						

4. Let the algebraic system $<A, *>$, where $A=\{a, b, c\}$, $*$ be a binary operation on A. For the operations determined by the Exercise 4 Table, discuss their commutativity, idempotency, and whether there is A iden + tity in A, respectively. If there is a single element, then does each element in A have an inverse element?

Exercise 4 Table

*	a	b	c
a	a	b	c
b	b	c	a
c	c	a	b

(a)

*	a	b	c
a	a	b	c
b	a	a	c
c	a	c	c

(b)

*	a	b	c
a	a	b	c
b	b	b	c
c	c	c	b

(c)

*	a	b	c
a	a	b	c
b	b	b	c
c	c	c	b

(d)

5. Take everyday examples to illustrate identity element, zero element and inverse element respectively.

6. The two binary operations defined are:
$$a * b = a^b, \quad a \triangle b = a \cdot b \quad a, b \in I_+$$
Try to prove that the pair $*$ is not assignable to $\triangle$.

7. prove: If f is a homomorphism mapping from $<A, \bigstar>$ to $<B, *>$, and g is a homomorphism mapping from $<B, *>$ to $<C, \triangle>$, then it is a homomorphism mapping from $<A, \bigstar>$ to $<C, \triangle>$.

8. Prove: Any intersection of any two congruences on a set is also a congruence.

9. Considering algebraic systems $<I, +>$, is the following binary relation R defined on I congruent?

(1) $<x, y> \in R$, if and only if $(x < 0 \wedge y < 0) \vee (x \geqslant 0 \wedge y \geqslant 0)$.

(2) $<x, y> \in R$, if and only if $|x - y| < 10$.

(3) $<x, y> \in R$, if and only if $(x = y = 0) \vee (x \neq 0 \wedge y \neq 0)$.

(4) $<x, y> \in R$, if and only if $x \geqslant y$.

10. Given an algebraic system $U = <N_2, +_2, \times_2>$ and $V = <N_3, +_3, \times_3>$. A trial quadrature algebra operation table of $U \times V$, $V \times U$.

11. Given an algebraic system $U = <X, \circ>$ and $V = <Y, *>$, where $\circ$ and $*$ are binary operations, the product of U and V is $U \times V = <X \times Y, \otimes>$. Try to prove:

(1) If $\circ$ and $*$ are commutative, then $\otimes$ is also commutative.

(2) If $\circ$ and $*$ are interchangeable, then $\otimes$ is also interchangeable.

12. Given the algebraic system $U_m = <N_m, +_m>$.

(1) Test the isomorphism of product algebras $U_2 \times U_3$ and U_6.

(2) Try to prove all congruences of product algebra $U_2 \times U_3$.

第 5 章　代数系统

代数系统是建立在集合上的一种运算系统。它是用运算构造数学系统的一种方法,因此称为代数系统。

5.1　代数系统的引入

在介绍代数系统之前,先引进一个集合 A 上的运算概念。例如,将实数集合 $\mathbf{R}$ 上每一个数 $a \neq 0$ 映射成它的倒数 $1/a$,或者将 $\mathbf{R}$ 上的每一个数 y 映射成 $[y]$,就可以将这些映射称为在集合 $\mathbf{R}$ 上的一元运算;而在集合 $\mathbf{R}$ 上,对任意两个数所进行的普通加法和乘法,都是集合 $\mathbf{R}$ 上的二元运算,也可以看作是将 $\mathbf{R}$ 上的每两个数映射成 $\mathbf{R}$ 中的一个数。上述的例子有一个共同的特征,那就是其运算结果都是在原来的集合 $\mathbf{R}$ 中,我们称具有这种特征的运算是封闭的,简称闭运算。相反地,没有这种特征的运算就是不封闭的。

很容易举出不封闭运算的例子:一台自动售货机能接受五角硬币和一元硬币,而所对应的商品是冰露(瓶)、农夫山泉(瓶)和可口可乐(罐)。当人们投入对应金额的硬币时,自动售货机将按表 5.1.1 所示供应相应的商品。

表　5.1.1

*	五角硬币	一元硬币
五角硬币	冰露	农夫山泉
一元硬币	农夫山泉	可口可乐

表格左上角的记号 * 可以理解为一个二元运算的运算符。这个例子中的二元运算 * 就是集合{五角硬币,一元硬币}上的不封闭运算。

定义 5.1.1 ▶ 对于一个集合 A,一个从 A^n 到 B 的映射称为集合 A 上的一个 n 元运算。如果 $B \subseteq A$,则称该 n 元运算是封闭的。

定义 5.1.2 ▶ 一个非空集合 A 连同若干个定义在该集合上的运算 $f_1, f_2, \cdots, f_n$ 所组成的系统称为一个代数系统,我们通常把这个系统记作 $<A, f_1, f_2, \cdots, f_n>$。

例如,正整数集合 I_+ 以及在该集合上的加法运算"+"组成一个代数系统 $<I_+, +>$。再比如,一个有限集合 S,由 S 的幂集 $P(S)$ 以及在该集合上的幂集运算"∪""∩""~"组成一个代数系统 $<P(S), \cup, \cap, \sim>$。虽然一些代数系统的形式不同,但是它们之间可能有一些共同的规律。

例如,考察代数系统 $<I, +>$,这里 I 是整数集合,+ 是普通的加法运算。很明显,在这个代数系统中,加法运算有以下三个运算规律,即对任意 $x, y, z \in I$,有

(1) $x+y\in I$　　　　　　　　（封闭性）

(2) $x+y=y+x$　　　　　　（交换律）

(3) $(x+y)+z=x+(y+z)$　　（结合律）

5.2　代数系统的运算及其性质

在初等代数中有 $+,-,\times,\div$ 四则运算,将其扩充后有向量运算、矩阵运算及行列式运算等。在这里,运算不仅包含前面的所有运算,还具有更普遍的含义。

定义 5.2.1 ▶设 $*$ 是定义在集合 A 上的二元运算,如果对于任意的 $x,y\in A$,都有 $x*y\in A$,则称二元运算 $*$ 在 A 上是封闭的。

例 5.2.1 设 $A=\{x\mid x=2^n,n\in\mathbf{N}\}$,问乘法运算是否封闭?加法运算呢?

解 对于任意的 $2^r,2^s\in A,r,s\in\mathbf{N}$,因为 $2^r*2^s=2^{r+s}\in A$,所以乘法运算是封闭的。但是对于加法运算来说是不封闭的,因为至少有 $2+2^2=6\notin A$。

定义 5.2.2 ▶设 $*$ 是定义在集合 A 上的二元运算,如果对于任意的 $x,y\in A$,都有 $x*y=y*x$,则称该二元运算 $*$ 是可交换的。

例 5.2.2 设 Q 是有理数集合,$\triangle$ 是 Q 上的二元运算,对任意的 $a,b\in\mathbf{R}$,都有 $a\triangle b=a+b-a*b$,问运算 $\triangle$ 是否可交换?

解 因为
$$a\triangle b=a+b-a*b=b+a-b*a=b\triangle a$$
所以运算 $\triangle$ 是可以交换的。

定义 5.2.3 ▶设 $*$ 是定义在集合 A 上的二元运算,如果对于任意的 $x,y,z\in A$ 都有 $(x*y)*z=x*(y*z)$,则称该二元运算 $*$ 是可结合的。

例 5.2.3 设 A 是一个非空集合,$\bigstar$ 是 A 上的二元运算,对于任意的 $a,b\in A$,有 $a\bigstar b=b$。证明 $\bigstar$ 是可结合的运算。

证明 因为对于任意的 $a,b,c\in A$,都有
$$(a\bigstar b)\bigstar c=b\bigstar c=c$$
而
$$a\bigstar(b\bigstar c)=a\bigstar c=c$$
所以
$$(a\bigstar b)\bigstar c=a\bigstar(b\bigstar c)$$

定义 5.2.4 ▶设 $*,\triangle$ 都是定义在集合 A 上的二元运算,如果对于任意的 $x,y,z\in A$,都有
$$x*(y\triangle z)=(x*y)\triangle(x*z)$$
$$(y\triangle z)*x=(y*x)\triangle(z*x)$$
则称运算 $*$ 对运算 $\triangle$ 是可分配的。

例 5.2.4 设 $A=\{\alpha,\beta\}$,在 A 上定义二元运算 $*$ 和 $\triangle$,如表 5.2.1 所示。运算 $*$ 对于运算 $\triangle$ 可分配吗?

表　5.2.1

$*$	α	β
α	α	β
β	β	α

(a)

$\triangle$	α	β
α	α	β
β	β	α

(b)

解 运算 $*$ 对运算 $\triangle$ 是不可分配的，因为

$$\beta * (\alpha \triangle \beta) = \beta * \alpha = \beta$$

而

$$(\beta * \alpha)(\beta * \beta) = \beta \triangle \alpha = \alpha$$

定义 5.2.5 设 $*$ 和 $\triangle$ 是定义在集合 A 上的两种可交换的二元运算，如果对于任意的 $x, y \in A$，都有

$$x * (x \triangle y) = x$$
$$x \triangle (x * y) = x$$

则称运算 $*$ 和 $\triangle$ 满足吸收率。

例 5.2.5 设集合 $\mathbf{N}$ 为自然数全体，在 $\mathbf{N}$ 上定义两种二元运算 $*$ 和 $\star$，对于任意 $x, y \in \mathbf{N}$，有

$$x * y = \max(x, y)$$
$$x \star y = \min(x, y)$$

验证 $*$ 和 $\star$ 的吸收律。

解 对于任意 $a, b \in \mathbf{N}$

$$a * (a \star b) = \max(a, \min(a, b)) = a$$
$$a \star (a * b) = \min(a, \max(a, b)) = a$$

因此，$*$ 和 $\star$ 满足吸收律。

定义 5.2.6 设 $*$ 是定义在集合 A 上的二元运算，如果对于任意的 $x \in A$，都有 $x * x = x$，则称运算 $*$ 是等幂的。

例 5.2.6 设 $P(S)$ 是集合 S 的幂集，在 $P(S)$ 上定义两种二元运算：集合的并运算 $\cup$ 和集合的交运算 $\cap$，验证 $\cup$，$\cap$ 是等幂的。

解 对于任意的 $A \in P(S)$，有 $A \cup A = A$ 和 $A \cap A = A$，因此 $\cap$ 和 $\cup$ 都满足等幂律。

定义 5.2.7 设 $*$ 是定义在集合 A 上的二元运算，如果有一个元素 $e_l \in A$，对于任意的元素 $x \in A$ 都有 $e_l * x = x$，则称 e_l 为 A 中关于运算 $*$ 的左幺元；如果有一个元素 $e_r \in A$，对于任意的元素 $x \in A$ 都有 $x * e_r = x$，则称 e_r 为 A 中关于运算 $*$ 的右幺元；如果 A 中的一个元素 e，它既是左幺元又是右幺元，则称 e 为 A 中关于运算 $*$ 的幺元。很明显，对于任一 $x \in A$，都会有 $e * x = x * e = x$。

例 5.2.7 设集合 $S = \{\alpha, \beta, \gamma, \delta\}$，在 S 上定义两种二元运算 $*$ 和 $\star$，如表 5.2.2 所示。试指出其左幺元和右幺元。

解 由表 5.2.2 可知，β, δ 都是 S 中关于 $*$ 的左幺元，而 α 是 S 中关于运算 $\star$ 的右幺元。

表 5.2.2

$*$	α	β	γ	δ
α	δ	α	β	γ
β	α	β	γ	δ
γ	α	β	γ	γ
δ	α	β	γ	δ

(a)

$\star$	α	β	γ	δ
α	α	β	δ	γ
β	β	α	γ	δ
γ	γ	δ	α	β
δ	δ	δ	β	γ

(b)

定理 5.2.1 设 $*$ 是定义在集合 A 上的二元运算，且在 A 中有关于运算 $*$ 的左幺元 e_l 和右幺元 e_r，则 $e_l = e_r = e$，且 A 中的幺元是唯一的。

证明　因为 e_1 和 e_r 分别是 A 中关于运算 $*$ 的左幺元和右幺元,所以

$$e_1 = e_1 * e_r = e_r = e$$

设另一个幺元 $e_1 \in A$,则

$$e_1 = e_1 * e = e$$

定义 5.2.8　设 $*$ 是定义在集合 A 上的二元运算,如果有一个元素 $\theta_1 \in S$,对于任意的元素 $x \in A$ 都有 $\theta_1 * x = \theta_1$,则称 θ_1 为 A 中关于运算 $*$ 的左零元;如果有一个元素 $\theta_r \in A$,对于任意的元素 $x \in A$ 都有 $\theta_r * x = \theta_r$,则称 θ_r 为 A 中关于运算 $*$ 的右零元;如果 A 中的一个元素 θ,它既是左零元又是右零元,则称 θ 为中关于运算 $*$ 的零元。显然,对于任一 $x \in A$,有

$$\theta * x = x * \theta = \theta$$

例 5.2.8　设集合 $S = \{浅色, 深色\}$,定义在 S 上的二元运算 $*$ 如表 5.2.3 所示。

表　5.2.3

$*$	浅色	深色
浅色	浅色	深色
深色	深色	深色

试指出其零元和幺元。

解　深色是 S 中关于运算 $*$ 的零元,浅色是 S 中关于运算 $*$ 的幺元。

定理 5.2.2　设 $*$ 是定义在集合 A 上的二元运算,且在 A 中有关于运算 $*$ 的左零元 θ_1 和右零元 θ_r,那么,$\theta_r = \theta_1 = \theta$,且 A 中的零元是唯一的。

这个定理的证明与定理 5.2.1 相似。

定理 5.2.3　设 $<A, *>$ 是一个代数系统,且集合 A 中元素的个数大于 1。如果该代数系统中存在幺元 e 和零元 θ,则 $\theta \neq e$。

证明　用反证法。设 $\theta = e$,那么对于任意的 $x \in A$,必有

$$x = e * x = \theta * x = \theta = e$$

于是,A 中的所有元素都是相同的,这与 A 中含有多个元素相矛盾。

定义 5.2.9　设代数系统 $<A, *>$,这里 $*$ 是定义在 A 上的二元运算,且 e 是 A 中关于运算 $*$ 的幺元。如果对于 A 中的一个元素 a,存在 A 中的某个元素 b,使 $b * a = e$,那么称 b 为 a 的左逆元;如果 $a * b = e$ 成立,那么称 b 为 a 的右逆元;如果一个元素 b,它既是 a 的左逆元又是 a 的右逆元,那么就称 b 是 a 的一个逆元。

很明显,如果 b 是 a 的逆元,那么 a 也是 b 的逆元,简称 a 与 b 互为逆元。一个元素 x 的逆元记为 x^{-1}。

一般地说,一个元素的左逆元不一定等于该元素的右逆元。而且,一个元素可以有左逆元而没有右逆元,甚至一个元素的左(右)逆元还可以不是唯一的。

定理 5.2.4　设代数系统 $<A, *>$,这里 $*$ 是定义在 A 上的二元运算,A 中存在幺元 e,且每一个元素都有左逆元。如果 $*$ 是可结合的运算,那么,这个代数系统中任何一个元素的左逆元必定也是该元素的右逆元,且每个元素的逆元是唯一的。

证明　设 $a, b, c \in A$,且 b 是 a 的左逆元,c 是 b 的右逆元。

因为　　　　　　　　　　$(b * a) * b = e * b = b$

所以 $e = c * b = c * ((b * a) * b) = (c * (b * a)) * b = ((c * b) * a) * b = (e * a) * b$
$= a * b$

因此，b 也是 a 的逆元。

设元素 a 有两个逆元 b 和 c，那么

$$b = b * e = b * (a * c) = (b * a) * c = e * c = c$$

因此，a 的逆元是唯一的。

例 5.2.9 试构造一个代数系统，使其中只有一个元素具有逆元。

解 设 $m, n \in I$，$T = \{x \mid x \in I, m \leqslant x \leqslant n\}$ 那么代数系统 $<T, \max>$ 中有一个幺元是 m，且只有 m 有逆元，因为 $m = \max(m, m)$。

例 5.2.10 对于代数系统 $<\mathbf{R}, \cdot>$，这里 $\mathbf{R}$ 是实数全体，$\cdot$ 是普通的乘法运算，判断是否每个元素都有逆元。

解 该代数系统中的幺元是 1，除零元素 0 外，所有的元素都有逆元。

可以指出：$<A, *>$ 是一个代数系统，$*$ 是 A 上的二元运算，那么该运算的有些性质可以从运算表中直接看出。

（1）运算 $*$ 具有封闭性，当且仅当运算表中的每个元素都属于 A。

（2）运算 $*$ 具有可交换性，当且仅当运算表关于主对角线是对称的。

（3）运算 $*$ 具有等幂性，当且仅当运算表的主对角线上的每个元素与该元素对应的表头元素一致。

（4）A 中关于 $*$ 有零元，当且仅当该元素所对应的行和列中的元素都与该元素相同。

（5）A 中关于 $*$ 有幺元，当且仅当该元素所对应的行和列依次与运算表的行和列一致。

（6）设 A 中有幺元，a 和 b 互逆，当且仅当位于 a 所在行、b 所在列的元素以及 b 所在行、a 所在列的元素都是幺元。

5.3 代数系统的同态与同构

借助同态和同构，我们可以把未知的代数系统通过同态、同构的方式类比为已知的代数系统。这样便可以加快我们的理解和认知过程。

定义 5.3.1 设 $<A, \bigstar>$ 和 $<B, *>$ 是两个代数系统，$\bigstar$ 和 $*$ 分别是 A 和 B 上的二元（n 元）运算，设 f 是从 A 到 B 的一个映射，使对于任意的 $a_1, a_2 \in A$，有

$$f(a_1 \bigstar a_2) = f(a_1) * f(a_2)$$

则称 f 为由 $<A, \bigstar>$ 到 $<B, *>$ 的一个同态映射（简称同态），称 $<A, \bigstar>$ 同态于 $<B, *>$，记作 $A \sim B$。我们把 $<f(A), *>$ 称为 $<A, \bigstar>$ 的一个同态像。其中

$$f(A) = \{x \mid x = f(a), a \in A\} \subseteq B$$

例 5.3.1 考察代数系统 $<I, \cdot>$，这里 I 是整数集，$\cdot$ 是普通乘法运算。如果我们对运算结果的兴趣仅在于正、负、零之间的特征区别，那么，代数系统 $<I, \cdot>$ 中运算结果的特征可以用另一个代数系统 $<B, \odot>$ 的运算结果来描述，其中 $B = \{$正，负，零$\}$，$\odot$ 是定义在 B 上的二元运算，如表 5.3.1 所示。

作映射 $f: I \rightarrow B$ 如下：

$$f(n)=\begin{cases}正，&若\ n>0\\负，&若\ n<0\\零，&若\ n=0\end{cases}$$

显然，对于任意的 $a,b\in I$，有

$$f(a\cdot b)=f(a)\odot f(b)$$

因此，f 是由 $<I,\cdot>$ 到 $<B,\odot>$ 的一个同态映射。

表　5.3.1

$\odot$	正	负	零
正	正	负	零
负	负	正	零
零	零	零	零

定义 5.3.2 ▶ 设 f 是由 $<A,\bigstar>$ 到 $<B,*>$ 的一个同态映射，如果 f 是从 A 到 B 的一个满射，则称 f 为满同态；如果 f 是从 A 到 B 的一个入射，则 f 称为单同态；如果 f 是从 A 到 B 的一个双射，则 f 称为同构映射，并称 $<A,\bigstar>$ 和 $<B,*>$ 是同构的，记作 $A\cong B$。

例 5.3.2　将 $f:\mathbf{R}\to\mathbf{R}$ 定义为对任意 $x\in\mathbf{R}$，有

$$f(x)=5^x$$

那么，f 是从 $<\mathbf{R},+>$ 到 $<\mathbf{R},\cdot>$ 的一个单同态。

例 5.3.3　将 $f:\mathbf{N}\to N_k$ 定义为对任意的 $x\in\mathbf{N}$，有

$$f(x)=x\bmod k$$

那么，f 是 $<\mathbf{N},+>$ 到 $<N_k,+_k>$ 的一个满同态。

例 5.3.4　设 $H=\{x\,|\,x=dn,d\ 是某一个正整数,n\in I\}$，定义映射 $f:I\to H$ 为对任意 $n\in I,f(n)=dn$，那么，f 是 $<I,+>$ 到 $<H,+>$ 的一个同构。所以 $I\cong H$。

例 5.3.5　设 $A=\{a,b,c,d\}$，在 A 上定义一个二元运算，如表 5.3.2 所示。再设 $B=\{\alpha,\beta,\gamma,\delta\}$，在 B 上定义一个二元运算，如表 5.3.3 所示。证明 $<A,\bigstar>$ 和 $<B,*>$ 是同构的。

表　5.3.2

$\bigstar$	a	b	c	d
a	a	b	c	d
b	b	a	a	c
c	b	d	d	c
d	a	b	c	d

表　5.3.3

$*$	α	β	γ	δ
α	α	β	γ	δ
β	β	α	α	γ
γ	β	δ	δ	γ
δ	α	β	γ	δ

证明　考察映射 f，使

$$f(a)=\alpha,\quad f(b)=\beta,\quad f(c)=\gamma,\quad f(d)=\delta$$

显然，f 是一个从 A 到 B 的双射，由表 5.3.2 和表 5.3.3 容易验证 f 是由 $<A,\bigstar>$ 到 $<B,*>$ 的一个同态。因此，$<A,\bigstar>$ 和 $<B,*>$ 是同构的。

如果考察映射 g，使

$$g(a)=\delta,\quad g(b)=\gamma,\quad g(c)=\beta,\quad g(d)=\alpha$$

那么，g 也是 $<A,\bigstar>$ 到 $<B,*>$ 的一个同构。

由例 5.3.5 可知，若两个代数系统是同构的，它们之间的同构映射可以是不唯一的。

例 5.3.6　表 5.3.4 中的代数系统$<A,\bigstar>$(表 5.3.4(a))与代数系统$<B,\oplus>$(表 5.3.4 (b))和$<C,*>$(表 5.3.4(c))都是同构的。

表 5.3.4

$\bigstar$	a	b
a	a	b
b	b	a

$\oplus$	偶	奇
偶	偶	奇
奇	奇	偶

$*$	偶	奇
偶	偶	奇
奇	奇	偶

(a)　　　　　　　　　　(b)　　　　　　　　　　(c)

同构这个概念很重要。从上例中可以看到,形式上不同的代数系统,如果它们是同构的,那么,就可抽象地把它们看作本质上相同的代数系统。

定义 5.3.3▶设$<A,\bigstar>$是一个代数系统,如果 f 是由$<A,\bigstar>$到$<A,\bigstar>$的同态,则称 f 为自同态。如果 g 是由$<A,\bigstar>$到$<A,\bigstar>$的同构,则称 g 为自同构。

定理 5.3.1▶设 G 是代数系统的集合,则 G 中代数系统之间的同构关系是等价关系。

证明　因为任何一个代数系统$<A,\bigstar>$可以通过恒等映射与它自身同构,即自反性成立。关于对称性,设$<A,\bigstar>\cong<B,*>$且有对应的同构映射 f,因为 f 的逆是由$<B,*>$到$<A,\bigstar>$的同构映射,即$<B,*>\cong<A,\bigstar>$。最后,如果 f 是由$<A,\bigstar>$到$<B,*>$的同构映射,g 是由$<B,*>$到$<C,\triangle>$的同构映射,则 $g\circ f$ 是$<A,\bigstar>$到$<C,\triangle>$的同构映射,即$<A,\bigstar>\cong<C,\triangle>$,所以同构关系满足自逆。因此,同构关系是等价关系。

例 5.3.7　设 $V=<Z_n,\oplus>$,其中 $Z_n=\{0,1,\cdots,n-1\}$,$\oplus$ 为模 n 加。证明自同态中恰含 n 个 V。

证明　先证明存在 n 个 V 的自同态,令 $f_p:Z_n\to Z_n$,$f_p(x)=(px)\bmod n$,$p=0,1,\cdots,n-1$。

则 f_p 是 V 的自同态,因为任意的 $x,y\in Z_n$,有
$$f_p(x\oplus y)=(p(x\oplus y))\bmod n=(px)\bmod n\oplus(py)\bmod n=f_p(x)\oplus f_p(y)$$
由于 p 有 n 种取值,不同的 p 确定了不同的映射 f_p,所以存在 n 个 V 的自同态。

下面证明任何 V 的自同态都是上述 n 个自同态中的一个。设 f 是 V 的自同态,且 $f(1)=i,i\in\{0,1,\cdots,n-1\}$。下面证明任意 $x\in Z_n$ 有 $f(x)=(ix)\bmod n$。

显然 $f(1)=i=(i\cdot 1)\bmod n$。假设对一切 $x\in\{1,2,\cdots,n-2\}$ 有 $f(x)=(ix)\bmod n$ 成立,那么
$$f(x+1)=f(x\oplus 1)=f(x)\oplus f(1)=(ix)\bmod n\oplus i=(ix+i)\bmod n=(i(x+1))\bmod n$$
最后有
$$f(0)=f((n-1)\oplus 1)=f(n-1)\oplus f(1)=(i(n-1))\bmod n\oplus i=(in)\bmod n=0$$
$$=(i\cdot 0)\bmod n$$

例 5.3.8　(1) 设代数系统 $V_1=<\mathbf{Z},+>$,$V_2=<Z_n,\oplus>$。其中 $\mathbf{Z}$ 为整数集合,$+$ 为普通加法,$Z_n=\{0,1,\cdots,n-1\}$,$\oplus$ 为模 n 加。令
$$f:\mathbf{Z}\to Z_n,\quad f(x)=(x)\bmod n$$
那么 f 是 V_1 到 V_2 的满同态。显然,f 是满射的,且任意 $x,y\in\mathbf{Z}$ 有
$$f(x+y)=(x+y)\bmod n=(x)\bmod n\oplus(y)\bmod n=f(x)\oplus f(y)$$
(2) 设 $V_1=<\mathbf{R},+>$,$V_2=<\mathbf{R}^*,\cdot>$,其中 $\mathbf{R}$ 和 $\mathbf{R}^*$ 分别为实数集和非零实数集,$+$ 和 $\cdot$

分别表示普通加法和乘法。令

$$f:\mathbf{R} \rightarrow \mathbf{R}^*, \quad f(x) = e^x$$

则 f 是 V_1 到 V_2 的单同态,易见 f 是单射的,且任意的 $x, y \in \mathbf{R}$ 有

$$f(x+y) = e^{x+y} = e^x \cdot e^y = f(x) \cdot f(y)$$

5.4　同余关系与商代数

同余关系是代数系统的集合上的等价关系,并且在运算的作用下,能够保持关系的等价类。此外,同余关系与运算密切相关。如果一个代数结构中有多个运算,则需要考察等价关系对于所有这些运算是否都有代换性质。如果等价关系在一个运算上不满足代换性质,该等价关系就不是代数系统上的同余关系。下面,我们进一步讨论同态与同余关系的对应,以及商代数的概念。

定义 5.4.1 ▶ 设 $<A, \bigstar>$ 是一个代数系统,并设 R 是 A 上的一个等价关系。如果代数系统 $<a_1, a_2>, <b_1, b_2> \in R$ 时,蕴涵着 $<a_1 \bigstar b_1, a_2 \bigstar b_2> \in R$,则称 R 为 A 上关于 $\bigstar$ 的同余关系。由这个同余关系将 A 划分成的等价类就称为同余类。

例 5.4.1　设 $A = \{a, b, c, d\}$,对于由表 5.4.1 所确定的代数系统 $<A, \bigstar>$,以及由表 5.4.2 所定义的在 A 上的等价关系 R 可知,R 是 A 上的同余关系。这个同余关系将 A 划分成同余类 $\{a, b\}$ 和 $\{c, d\}$。

表　5.4.1

$\bigstar$	a	b	c	d
a	a	a	d	c
b	b	a	c	d
c	c	d	a	b
d	d	d	b	a

表　5.4.2

R	a	b	c	d
a	$\checkmark$	$\checkmark$		
b	$\checkmark$	$\checkmark$		
c			$\checkmark$	$\checkmark$
d			$\checkmark$	$\checkmark$

例 5.4.2　设 $A = \{a, b, c, d\}$,对于由表 5.4.3 所确定的代数系统 $<A, \bigstar>$,以及由表 5.4.2 所定义的在 A 上的等价关系 R。

表　5.4.3

$\bigstar$	a	b	c	d
a	a	a	d	c
b	b	b	d	a
c	c	b	a	b
d	c	d	b	a

由于对 $<a, b>, <c, d> \in R$ 有

$$<a \bigstar c, b \bigstar d> = <d, a> \notin R$$

因此,由表 5.4.2 所定义的在 A 上的等价关系 R 不是 A 上的一个同余关系。

由上述两例可知:在 A 上定义的等价关系 R,不一定是在 A 上的同余关系,这是因为同余关系必须与定义在 A 上的二元运算密切相关。

定理 5.4.1 ▶ 设 $<A, \bigstar>$ 是一个代数系统,R 是 A 上的一个同余关系,$B = \{A_1, A_2, \cdots, A_r\}$ 是由 R 诱导的 A 的一个划分,那么,必定存在新的代数系统 $<B, *>$ 是 $<A, \bigstar>$ 的同态像。

证明 在 B 上定义二元运算 $*$：对于任意的 $A_i,A_j\in B$，任取 $a_1\in A_i,a_2\in A_j$，如果 $a_1\bigstar a_2\in A_k$，则 $A_i*A_j=A_k$。由于 R 是 A 上的同余关系，所以，以上定义的 $A_i*A_j=A_k$ 是唯一的。

作映射 $f(a)=A_i,a\in A_i$。

显然，f 是从 A 到 B 的满映射。

对于任意的 $x,y\in A$，x,y 必属于 B 中的某两个同余类，不妨设 $x\in A_i,y\in A_j,1\leqslant i,j\leqslant r$；同时，$x\bigstar y$ 必属于 B 中某个同余类，不妨设 $x\bigstar y\in A_k$，于是有

$$f(x\bigstar y)=A_k=A_i*A_j=f(x)*f(y)$$

因此，f 是由 $<A,\bigstar>$ 到 $<R,*>$ 的满同态，即 $<B,*>$ 是 $<A,\bigstar>$ 的同态像。

定理 5.4.2 设 f 是由 $<A,\bigstar>$ 到 $<B,*>$ 的一个同态映射，如果在 A 上定义二元关系 R：$<a,b>\in R$ 当且仅当

$$f(a)=f(b)$$

那么，R 是 A 上的一个同余关系。

证明 因为 $f(a)=f(a)$，所以 $<a,a>\in R$。若 $<a,b>\in R$，则 $f(a)=f(b)$ 即 $f(b)=f(a)$，所以 $<b,a>\in R$。若 $<a,b>\in R,<b,c>\in R$，则 $f(a)=f(b)=f(c)$，所以 $<a,c>\in R$。最后，又因为若 $<a,b>\in R,<c,d>\in R$，则有

$$f(a\bigstar c)=f(a)*f(c)=f(b)*f(d)=f(b\bigstar d)$$

所以，$<a\bigstar c,b\bigstar d>\in R$。

因此，R 是 A 上的同余关系。

定义 5.4.2 设 $U=\{X,*,+,V,\cdots,W\}$ 是代数系统，E 是 U 中的同余关系。若商集 $X/E=\{[x]_E\mid x\in E\}$ 中引入相应的运算 $*',+',V',\cdots,W'$：

$$[x_1]_E*'[x_2]_E=[x_1*x_2]_E$$
$$[x_1]_E+'[x_2]_E=[x_1+x_2]_E$$
$$V'([x_1]_E)=[V(x_1)]_E$$

则 $<X/E,*',+',V',\cdots,W'>$ 是代数系统，称为 U 对 E 的商代数系统或简称商代数，记作 U/E。

应特别注意，在上述定义中，正因为 E 是同余关系，所以那些等式与同余类中“代表元素”的选择无关，即 $*',+',V',\cdots,W'$ 确是相应的运算。

定理 5.4.3 设 $U=\{X,*,+,V,\cdots,W\}$ 是代数系统，E 是 U 中的同余关系，$U/E=<X/E,*',+',V',\cdots,W'>$ 是 U 对 E 的商代数。若定义 X 到 X/E 的映射：

$$g:x\to[x]_E$$

则 g 是代数系统 U 到商代数 U/E 的满同态，称为自然同态。因此，$U\sim U/E$。

证明 g 显然是 X 到 X/E 的满射，且对 X 中任意元素 x_1,x_2 有

$$g(x_1*x_2)=[x_1*x_2]_E=[x_1]_E*'[x_2]_E=g(x_1)*g(x_2)$$
$$g(x_1+x_2)=[x_1+x_2]_E=[x_1]_E+'[x_2]_E=g(x_1)+'g(x_2)$$
$$g(V(x_1))=[V(x_1)]_E=V'([x_1]_E)=V'(g(x_1))$$

$$\cdots$$

故 g 是 $U\sim U/E$ 的满同态。

定理 5.4.4 ▶ 设代数系统 $U=\{X,*,+,V,\cdots,W\}\sim V=<Y,e,\oplus,\otimes,\cdots,\Theta>$，$f$ 是满同态，E 是 U 中对应于 f 的同余关系。已知 $U\sim U/E=<X/E,*',+',V',\cdots,W'>$，是其代数系统的自然同态，则 $U/E\cong V$。

证明　定义 X/E 到 Y 的映射

$$h:[x]_E\to f(x)$$

一方面，因 Y 的任意元素都可表示为 $f(x)$ 的形式，所以 h 是满射的。另一方面，若 Y 中两个元素 $f(x_1)=f(x_2)$，则 x_1Ex_2，可得 $[x_1]_E=[x_2]_E$，从而 h 又是单射的，故 h 是双射的。

又对 X/E 中任意元素 $[x_1]_E$ 和 $[x_2]_E$，有

$$h([x_1]_E*'[x_2]_E)=h([x_1*x_2]_E)=f(x_1*x_2)=f(x_1)ef(x_2)=h([x_1]_E)eh([x_2]_E)$$
$$h([x_1]_E+'[x_2]_E)=h([x_1+x_2]_E)=f(x_1+x_2)=f(x_1)\oplus f(x_2)=h([x_1]_E)\oplus h([x_2]_E)$$
$$h(V'([x_1]_E))=h([V(x_1)_E])=f(V(x_1))=\otimes(h([x_1]_E))$$

故 h 是 U/E 到 V 的同构，因此 $U/E\cong V$。

5.5　积代数

本节试图将笛卡尔积的概念推广到同型的代数系统中，从而产生一个新的代数系统，称为积代数系统，简称积代数。

定义 5.5.1 ▶ 设 $U=\{X,*,+,V,\cdots,W\}$ 和 $V=<Y,\odot,\oplus,\copyright,\cdots,\odot>$ 是同型（两个代数系统中运算的个数，运算的元数且代数常数的个数都相同）的代数系统。在 $X\times Y$ 中引入相应的运算 $\bar{*},\bar{+},\bar{V},\cdots,\bar{W}$：

$$<x_1,y_1>\bar{*}<x_2,y_2>=<x_1*x_1,y_1\odot y_2>$$
$$<x_1,y_1>\bar{+}<x_2,y_2>=<x_1+x_2,y_1\oplus y_2>$$
$$\bar{V}(<x_1,y_1>)=<V(x_1),\copyright(y_1)>$$
$$\cdots$$

则 $<X\times Y,\bar{*},\bar{+},\bar{V},\cdots,\bar{W}>$ 是代数系统，称为 U 和 V 的积代数系统，简称积代数，记作 $U\times V$。而 U 和 V 称为 $U\times V$ 的因子代数系统，简称因子代数。

例 5.5.1　设集合 $A=\{a_1,a_2\}$，$B=\{b_1,b_2,b_3\}$。若在 A 和 B 中分别规定二元运算 $*$ 和 $\odot$，如表 5.5.1 所示。

表　5.5.1

$*$	a_1	a_2
a_1	a_1	a_2
a_2	a_2	a_1

$\odot$	b_1	b_2	b_3
b_1	b_1	b_1	b_3
b_2	b_2	b_2	b_3
b_3	b_1	b_1	b_3

(a)　　　　　　　　　　　(b)

则 $U=<A,*>$ 和 $V=<B,\odot>$ 是同型的代数系统，其积代数为 $U\times V=<A\times B,\bar{*}>$，其中 $A\times B=\{<a_1,b_1>,<a_1,b_2>,<a_1,b_3>,<a_2,b_1>,<a_2,b_2>,<a_2,b_3>\}$。$\bar{*}$ 的运算表如表 5.5.2 所示。

表 5.5.2

$\bar{*}$	$<a_1,b_1>$	$<a_1,b_2>$	$<a_1,b_3>$	$<a_2,b_1>$	$<a_2,b_2>$	$<a_2,b_3>$
$<a_1,b_1>$	$<a_1,b_1>$	$<a_1,b_1>$	$<a_1,b_3>$	$<a_2,b_1>$	$<a_2,b_1>$	$<a_2,b_3>$
$<a_1,b_2>$	$<a_1,b_2>$	$<a_1,b_2>$	$<a_1,b_3>$	$<a_2,b_2>$	$<a_2,b_2>$	$<a_2,b_3>$
$<a_1,b_3>$	$<a_1,b_1>$	$<a_1,b_3>$	$<a_1,b_3>$	$<a_2,b_1>$	$<a_2,b_3>$	$<a_2,b_3>$
$<a_2,b_1>$	$<a_2,b_1>$	$<a_2,b_1>$	$<a_2,b_3>$	$<a_1,b_1>$	$<a_1,b_1>$	$<a_1,b_3>$
$<a_2,b_2>$	$<a_2,b_2>$	$<a_2,b_2>$	$<a_2,b_3>$	$<a_1,b_2>$	$<a_1,b_2>$	$<a_1,b_3>$
$<a_2,b_3>$	$<a_2,b_1>$	$<a_2,b_3>$	$<a_2,b_3>$	$<a_1,b_1>$	$<a_1,b_3>$	$<a_1,b_3>$

可以容易地将定义 5.5.1 推广成几个同型代数系统的积代数。

习 题

1. 设集合 $A=\{1,2,3,\cdots,10\}$，下面定义的二元运算关于集合 A 是否封闭？

(1) $x * y=\max(x,y)$

(2) $x * y=\min(x,y)$

(3) $x * y=GCD(x,y)$

(4) $x * y=LCM(x,y)$

(5) $x * y=$ 质数 p 的个数，使 $x\leqslant p\leqslant y$

2. 试列举一些你熟悉的代数系统。

3. 对于实数集合 $\mathbf{R}$，习题 3 表所列的二元运算是否具有左边一列中所表述的性质，请在相应的位置上填写"是"或"否"。

习题 3 表

性 质	$+$	$-$	$\cdot$	max	min	$|x-y|$
可结合性						
可交换性						
存在幺元						
存在零元						

4. 设代数系统 $<A,*>$，其中 $A=\{a,b,c\}$，$*$ 是 A 上的一个二元运算。对于由习题 4 表所确定的运算，试分别讨论它们的交换性、等幂性，以及在 A 中是否有幺元；如果有幺元，那么 A 中的每个元素是否有逆元。

习题 4 表

$*$	a	b	c
a	a	b	c
b	b	c	a
c	c	a	b

(a)

$*$	a	b	c
a	a	b	c
b	a	a	c
c	a	c	c

(b)

$*$	a	b	c
a	a	b	c
b	b	b	c
c	c	c	b

(c)

$*$	a	b	c
a	a	b	c
b	b	b	b
c	c	c	b

(d)

5. 列举日常生活中的例子,分别说明幺元、零元和逆元。

6. 定义 I_+ 上的两个二元运算:

$$a * b = a^b, \quad a \triangle b = a \cdot b, \quad a, b \in I_+$$

试证明 * 对 $\triangle$ 是不可分配的。

7. 证明:如果 f 是由 $<A, \bigstar>$ 到 $<B, *>$ 的同态映射,g 是由 $<B, *>$ 到 $<C, \triangle>$ 的同态映射,那么,$g \circ f$ 是由 $<A, \bigstar>$ 到 $<C, \triangle>$ 的同态映射。

8. 证明:一个集合上任意两个同余关系的交也是一个同余关系。

9. 考察代数系统 $<I, +>$,以下定义在 I 上的二元关系 R 是同余关系吗?

(1) $<x, y> \in \mathbf{R}$,当且仅当 $(x < 0 \wedge y < 0) \vee (x \geqslant 0 \wedge y \geqslant 0)$。

(2) $<x, y> \in R$,当且仅当 $|x - y| < 10$。

(3) $<x, y> \in R$,当且仅当 $(x = y = 0) \vee (x \neq 0 \wedge y \neq 0)$。

(4) $<x, y> \in R$,当且仅当 $x \geqslant y$。

10. 给定代数系统 $U = <N_2, +_2, \times_2>$ 和 $V = <N_3, +_3, \times_3>$。试求积代数 $U \times V, V \times U$ 的运算表。

11. 给定代数系统 $U = <X, \circ>$ 和 $V = <Y, *>$,其中 $\circ$ 和 $*$ 都是二元运算,U 和 V 的积代数是 $U \times V = <X \times Y, \otimes>$。试证明:

(1) 若 $\circ$ 和 $*$ 是可交换的,则 $\otimes$ 也是可交换的。

(2) 若 $\circ$ 和 $*$ 是可交结合的,则 $\otimes$ 也是可结合的。

12. 给定代数系统 $U_m = <N_m, +_m>$。

(1) 试证明积代数 $U_2 \times U_3$ 和 U_6 同构。

(2) 试证明积代数 $U_2 \times U_3$ 的所有同余关系。

Chapter 6   Group

The establishment of group theory expanded the research objects and applications of group theory, but also could get various inspirations from different groups. Group theory is the most fully developed part of algebra and mathematics in modern times and has been widely used in natural science. For example, in automation theory, coding theory, fast adder design and so on, the application of group has been improved.

6.1   Semigroup

Semigroup is an algebraic system, which has specific applications in formal languages, automata and other fields.

Definition 6.1.1   An algebraic system $<S, * >$, S is a non-empty set, $*$ is a binary operation on S, and if the operation $*$ is closed, the algebraic system $<S, * >$ is called a groupoid.

Definition 6.1.2   An algebraic system $<S, * >$, where S is a nonempty set, and $*$ is a binary operation on S. If: ① The operation $*$ is closed. ② The operation $*$ is combinable, that is, for any $x, y, z \in S$, satisfies $(x * y) * z = x * (y * z)$. The algebraic system $<S, * >$ is called a semigroup.

Example 6.1.1   Set $S_k = \{x \mid x \in I \wedge x \geqslant k\}$, $k \geqslant 0$, then $<S_k, +>$ is a semigroup, where $+$ is an ordinary addition operation.

Solution   Because the $+$ is closed on top, and ordinary addition operations are associative. So, $<S_k, +>$ is a semigroup.

In Example 6.1.1, the condition $k \geqslant 0$ is important, otherwise, if k is less than 0, the operation $+$ is not closed on top.

Example 6.1.2   Let $S = \{a, b, c\}$, a binary operation $\triangle$ defined in Table 6.1.1.

Table   6.1.1

$\triangle$	a	b	c
a	a	b	c
b	a	b	c
c	a	b	c

Verify that $<S, \triangle>$ is a semigroup.

Solution　Table 6.1.1 shows that the operation $\triangle$ is closed, and a, b and c are all left identity elements. So, for any x, y, $z \in S$, there is

$$x \triangle (y \triangle z) = x \triangle z = z = y \triangle z = (x \triangle y) \triangle z$$

So, $<S, \triangle>$ is a semigroup.

Obviously, algebraic systems $<I_+, ->$ and $<R, />$ are not semigroups, here, -and/are ordinary subtraction and division, respectively.

Theorem 6.1.1　If $<S, *>$ is a semigroup, $B \subseteq S$ and $*$ is closed on B, then $<B, *>$ is also a semigroup. Usually called $<B, *>$ is a subsemigroup of semigroup$<S, *>$.

Proof　Because $*$ be binding on S, and $*$ is closed on B, $*$ is also binding on B, so $<B, *>$ is a semigroup.

Example 6.1.3　Let $*$ represent ordinary multiplication, then $<[0, 1], *>$, $<[0, 1), *>$ and $<1, *>$ are all subsemigroups of $<R, *>$.

Solution　First, the operation $*$ be closed and joinable on $\mathbf{R}$, so$<\mathbf{R}, *>$ is a semigroup. Secondly, the operation $*$ be closed on $[0, 1]$, $[0, 1)$ and I, and $[0, 1] \subset \mathbf{R}$, $[0, 1) \subset \mathbf{R}$, $I \subset \mathbf{R}$, therefore, it can be known from Theorem 6.1.1 that $<[0, 1], *>$, $<[0, 1), *>$ and $<[0, 1), *>$ are all subsemigroups of $<R, *>$.

Theorem 6.1.2　Set $<S, *>$ is a semigroup, if S is a finite set, then there must be $a \in S$, such that $a * a = a$.

Proof　$<S, *>$ is a semigroup. For any $b \in S$, it can be known by the closure of $*$:

$$b * b = S, \text{ remember as } b^2 = b * b$$
$$b^2 * b = b * b^2 \in S, \text{ remember as } b^3 = b^2 * b = b * b^2$$
$$\vdots$$

Because S is a finite set, so there must be $j > i$, such that
$$b^i = b^j$$

Let $\quad p = j - i$

Get $\quad b^i = b^p * b^i$

So $\quad b^q = b^p * b^q, \quad q \geq i$

Because $p \geq 1$, so you can always find $k \geq 1$, make
$$kp \geq i$$

For the element b^{kq} in S, there is
$$b^{kq} = b^p * b^{kq} = b^p * (b^p * b^{kq}) = b^{2p} * b^{kq} = b^{2p} * (b^p * b^{kq}) = \dots = b^{kq} * b^{kq}$$

That proves that there is an element $a = b^{kq}$ in S, such that $a * a = a$.

Definition 6.1.3　A semigroup containing an identity element is called a monoid.

For example, an algebraic system $<R, +>$ is a monoid, because $<R, +>$ is a semigroup, and 0 is the identity element of the $+$ operation in R. In addition, algebraic systems $<I, \cdot>$, $<I_+, \cdot>$, $<R, \cdot>$ are all semigroups with the identity element I, $\cdot$ is normal multiplication, so they are all monoids.

However, the algebraic system $<N - \{0\}, +>$ is a semigroup, but there is no identity element about operation $+$, so this algebraic system is not a monoid.

Theorem 6.1.3 If $<S, *>$ is a monoid, then any two rows or columns in the table of operations on $*$ will not be the same.

Proof Let the identity element of the operation $*$ in S be e. Because for any a, $b \in S$ and $a \neq b$, there is always

$$e * a = a \neq b = e * b$$
$$a * e = a \neq b = b * e$$

Therefore, it is impossible for two rows or two columns to be identical in the operand table of $*$.

Example 6.1.4 Suppose I is a set of integers, m is any positive integer, Z_m is a set of congruence classes composed of congruence classes of module m. Two binary operations $+_m$ and $\times_m$ are defined as follows.

For any $[i]$, $[j] \in Z_m$, then

$$[i] +_m [j] = [(i+j)(\bmod m)]$$
$$[i] \times_m [j] = [(i \times j)(\bmod m)]$$

Try to prove that any two rows or columns in the two binary operation tables are not the same.

Proof Look at algebraic systems $<Z_m, +_m>$ and $<Z_m, \times_m>$.

(1) By the definition of operations $+_m$ and $\times_m$, we know that it is closed on Z_m.

(2) For any $[i]$, $[j]$, $[k] \in Z_m$

$$([i] +_m [j]) +_m [k] = [i] +_m ([j] +_m [k])$$
$$= [(i+j+k)(\bmod m)]$$
$$([i] \times_m [j]) \times_m [k] = [i] \times_m ([j] \times_m [k])$$
$$= [(i \times j \times k)(\bmod m)]$$

So, $+_m$ and $\times_m$ are both associative.

(3) Because $[0] +_m [i] = [i] +_m [0] = [i]$, so $[0]$ is an identity element of $<Z_m, +_m>$. Because $[1] \times_m [i] = [i] \times_m [1] = [i]$, so $[1]$ is an identity element of $<Z_m, +_m>$.

So, the algebraic system $<Z_m, +_m>$, $<Z_m, \times_m>$ are all distinct monoids. Theorem 6.1.3 shows that any two rows or columns in the operand table of these two operations are not the same.

In the above example, if $m=5$ is given, then the operation tables of $+_5$ and $\times_5$, are shown in Table 6.1.2 and Table 6.1.3.

Table 6.1.2

$+_5$	[0]	[1]	[2]	[3]	[4]
[0]	[0]	[1]	[2]	[3]	[4]
[1]	[1]	[1]	[2]	[3]	[4]
[2]	[2]	[2]	[4]	[1]	[3]
[3]	[3]	[3]	[1]	[4]	[2]
[4]	[4]	[4]	[3]	[2]	[1]

Table 6.1.3

$\times_5$	[0]	[1]	[2]	[3]	[4]
[0]	[0]	[0]	[0]	[0]	[0]
[1]	[0]	[1]	[2]	[3]	[4]
[2]	[0]	[2]	[4]	[1]	[3]
[3]	[0]	[3]	[1]	[4]	[2]
[4]	[0]	[4]	[3]	[2]	[1]

Obviously, no two rows or columns in the above table are identical.

Theorem 6.1.4   Suppose $<S, *>$ is a monoid, for any $a, b \in S$, and both a and b have an inverse element, then

(1) $(a^{-1})^{-1} = a$.

(2) $a * b$ has an inverse element, and $(a * b)^{-1} = b^{-1} * a^{-1}$.

Proof (a) Because a^{-1} is an inverse element of a, such that

$$a * a^{-1} = a^{-1} * a = e \quad (e \text{ is an identity element})$$

So $(a^{-1})^{-1} = a$

$$(a * b) * (b^{-1} * a^{-1}) = a * (b * b^{-1}) * a^{-1} = a * e * a^{-1} = a * a^{-1} = e$$

Similarly $(b^{-1} * a^{-1}) * (a * b) = e$

So $(a * b)^{-1} = b^{-1} * a^{-1}$

6.2   Group and Subgroup

In this section, we introduce the concept of subgroups and the relationship between subgroups and groups.

Definition 6.2.1   Let $<G, *>$ be an algebraic system where G is a non-empty set and $*$ is a binary operation on G, if the operation $*$ is closed, the operation $*$ is associative, there is an identity element e, and for each element $x \in G$, there exists its inverse element x^{-1}. Then $<G, *>$ is a group.

For example, $<R - \{0\}, \times>$, $< P(S), \oplus>$ and so on are groups.

Example 6.2.1   Let $R = \{0°, 60°, 120°, 180°, 240°, 300°\}$ represent six possible cases of the clockwise rotation angle of a geometric figure on the plane around the centroid. Let ★ be a binary operation on R. For any two elements A and B in R, a ★ b represents the total rotation angle obtained by the continuous rotation of a and b of a plane figure. And it says that the rotation of 360° is equal to the original state, so it's not rotated. Verify that $<R, ★>$ is a group.

Solution It can be seen from the above that the operation of binary operation ★ on R is shown in Table 6.2.1.

Table   6.2.1

★	0°	60°	120°	180°	240°	300°
0°	0°	60°	120°	180°	240°	300°
60°	60°	120°	180°	240°	300°	0°
120°	120°	180°	240°	300°	0°	60°
180°	180°	240°	300°	0°	60°	120°
240°	240°	300°	0°	60°	120°	180°
300°	300°	0°	60°	120°	180°	240°

As can be seen from Table 6.2.1, ★ is closed on R.

For any a, b, $c \in R$, $(a \bigstar b) \bigstar c$ means to rotate the graph successively a, b and c, while $a \bigstar (b \bigstar c)$ means to rotate the graph successively b, c and a, and the total rotation angle is equal to $a+b+c \pmod{360°}$, therefore, $(a \bigstar b) \bigstar c = a \bigstar (b \bigstar c)$.

$0°$ is identity element.

The inverse element of $60°$, $180°$, $120°$ is $300°$, $180°$, $240°$. So, $<R, \bigstar>$ is a group.

Definition 6.2.2 Let $<G, *>$ be a group. If G is a finite set, then $<G, *>$ is a finite group, the number of elements in G is usually called the order of the finite group, denoted as $|G|$. If G is an infinite set, then $<G, *>$ is an infinite group.

In Example 6.2.1, the $<R, \bigstar>$ is a finite group, and $|R|=6$.

Example 6.2.2 Try to prove that the algebraic system $<I, +>$ is a group, where I is the set of all integers, and $+$ is a common addition operation.

Solution Obviously, binary operations $+$ are closed and associative on I. The identity element is equal to 0. For any $a \in A$, its inverse is $-a$. So $<I, +>$ is a group, and it's an infinite group.

So far, we can say that A groupoid is simply an algebraic system with a closed binary operation on a nonempty set. A semigroup is a groupoid with an associative operation. A monoid is a semigroup with an identity element. where every element has an inverse element. That is:

$$\{group\} \subset \{monoid\} \subset \{semigroup\} \subset \{groupoid\}$$

According to the Theorem 5.2.4, the inverse element of any member of the group must be unique. Because of the uniqueness of inverse elements in a group, we can use the following theorems.

Theorem 6.2.1 There cannot be a zero element in a group.

Proof When the order of the group is 1, its only element becomes the identity element.

Let $|G|>1$ and group $<G, *>$ has zero element θ. Then any element in the group $x \in G$ has $x * \theta = \theta * x = e$, so there is no inverse element for the zero element θ, which is contradictory to the group $<G, *>$.

Theorem 6.2.2 Let $<G, *>$ be a group, for a, $b \in G$, there must be a unique $x \in G$, such that $a * x = b$.

Proof Let's say the inverse element of a is a^{-1}, let
$$x = a^{-1} * b$$
then $\qquad a * x = a * (a^{-1} * b) = (a * a^{-1}) * b = e * b = b$

If there is another solution that satisfies $a * x_1 = b$, then
$$a^{-1} * (a * x_1) = a^{-1} * b$$
Then $\qquad x_1 = a^{-1} * b$

Theorem 6.2.3 Let $<G, *>$ be a group, for any a, b, $c \in G$, if there is $a * b = a * c$ or $b * a = c * a$, then there must be $b = c$ (elimination law).

Proof If $a * b = a * c$, and the inverse element of a is a^{-1}, then

$$a^{-1} * (a * b) = a^{-1} * (a * c)$$

$$(a * a^{-1}) * b = (a * a^{-1}) * c$$

$$e * b = e * c$$

$$b = c$$

When $b * a = c * a$, we can prove that $b = c$.

Theorem 6.1.3 shows that no two rows (or two columns) in the operation table of a group are identical. To further investigate the properties of the operands of groups, the concept of permutation is introduced.

Definition 6.2.3   Assuming S is a non-empty set, a bijection from set S to S is called a permutation of S.

For example, for the set $S = \{a, b, c, d\}$, mapping a to b, mapping b to d, mapping c to a, mapping d to c is a one-to-one mapping from S to S, this permutation can be expressed as

$$\begin{pmatrix} a & b & c & d \\ b & d & a & c \end{pmatrix}$$

That is, the previous line writes out all the elements in the collection in any order, and the next line writes the image of each corresponding element.

Theorem 6.2.4   Each row or column in the operand table of a group $<G, *>$ is a permutation of an element of G.

Proof First, prove that no row or column in the operand table can contain more than one element of G. By contradiction, if a row corresponding to an element $a \in G$ has two elements both c, then there is

$$a * b_1 = a * b_2 = c, \text{ and } b_1 \neq b_2$$

$b_1 = b_2$ can be obtained by Cancellation law, which is contradictory to $b_1 \neq b_2$.

Second, prove that every element in G appears in every row and column of the operand table. Look at the row corresponding to the element $a \in G$. Let b be any element in G. Since $b = a * (a^{-1} * b)$, b must appear in the row corresponding to a.

From the fact that no two rows (or columns) are the same in the operand table, it follows that each row in the $<G, *>$ operand table is a permutation of an element of G, and each row is different. The same is true for columns.

Definition 6.2.4   In an algebraic system $<G, *>$, if $a \in G$ exists and $a * a = a$, a is called an idempotent.

Theorem 6.2.5   In the group $<A, *>$, there can be no idempotent other than the identity element e.

Proof Because $e * e = e$, so e is idempotent.

Let $a \in A$, $a \neq e$ and $a * a = a$, now $a = e * a = (a^{-1} * a) * a = a^{-1} * (a * a) = a^{-1} * a = e$, contradicting hypothesis $a \neq e$.

Now let's introduce the concept of subgroups.

Definition 6.2.5　Let $<G, *>$ be a group, S be a non-empty subset of G, if $<S, *>$ also constitute a group, then $<S, *>$ is a subgroup of $<G, *>$.

Theorem 6.2.6　If $<G, *>$ is a group, and $<S, *>$ is a subgroup of $<G, *>$, then the identity element e in $<G, *>$ must also be the identity element e in $<S, *>$.

Proof　Let e_1 be the identity element in $<S, *>$, for any $x \in S \subseteq G$, there must be $e_1 * x = x = e * x$, then $e_1 = e$.

Definition 6.2.6　Let $<G, *>$ be a group, and $<S, *>$ is a sub group of $<G, *>$, if $S = \{e\}$, or $S = G$, $<S, *>$ is called the trivial subgroup of $<G, *>$.

Example 6.2.3　$<I, +>$ is a group, so let's say that $I_E = \{x \mid x = 2n, n \in I\}$, prove $<I_E, +>$ is a subgroup of $<I, +>$.

Proof　(1) For any $x, y \in I_E$, we may assume $x = 2n_1$, $y = 2n_2$, $n_1, n_2 \in I$, then
$$x + y = 2n_1 + 2n_2 = 2(n_1 + n_2)$$
But $n_1 + n_2 \in I$, so $x + y \in I_E$, $+$ is closed on I_E.

(2) Operation $+$ remains associative on I_E.

(3) The identity element 0 in $<I, +>$ also in I_E.

(4) For any of these $x \in I_E$, you have to have n such that $x = 2n$, but $-x = -2n = 2(-n)$, $-n \in I$.

So $-x \in I_E$, but $x + (-x) = (-x) + x = 0$, so $<I_E, +>$ is a subgroup of $<I, +>$.

Theorem 6.2.7　Let $<G, *>$ be a group, B be a non-empty subset of G, and if B be a finite set, then $<B, *>$ must be a subgroup of $<G, *>$, so long as the operation $*$ are closed on B.

Proof　Let b be any member of B. If $*$ is closed on B, then the element, $b^2 = b * b$, $b^3 = b^2 * b$, ... they are all in B. Since B is a finite set, there must be positive integers i and j, let's say $i < j$, let
$$b^i = b^j, \quad b^i = b^i * b^{j-i}$$
This implies that the identity element b^{j-i} is a member of $<G, *>$, and that the identity element is also in subset B.

If $j - i > 1$, then because of $b^{j-i} = b * b^{j-i-1}$, we can know that b^{j-i-1} is an inverse element of b, $b^{j-i-1} \in B$; if $j - i = 1$, then because of $b^i = b^i * b$, we can know that b is an identity element, but the identity element is its own inverse.

So, $<B, *>$ must be a subgroup of $<G, *>$.

Theorem 6.2.8　Let $<G, \triangle>$ be a group, and S be a nonempty subset of G. If there are any elements a and b in S, we can know $a \triangle b^{-1} \in S$ then $<S, \triangle>$ is a subgroup of $<G, \triangle>$.

Proof　First of all, it is proved that the identity element e in G is also the identity element in S. Take any element a in S, $a \in S \subset G$, so $e = a \triangle a^{-1} \in S \subset G$ and $a \triangle e = e \triangle = a$, that is, e is also an identity element in S.

Second, we prove that every element in S has an inverse element. For any $a \in S$, because $e \in S$, so $e \triangle a^{-1} \in S$, such that $a^{-1} \in S$.

It turns out that $\triangle$ is closed on S. For any a, $b \in S$, it can be known from the above $b^{-1} \in S$, but $b = (b^{-1})^{-1}$, so $a \triangle b = a \triangle (b^{-1})^{-1} \in S$.

As for the operation, the associability of $\triangle$ is maintained over S. Therefore, $<S, \triangle>$ is a subgroup of $<G, \triangle>$.

6.3 Homomorphism and Isomorphism of Groups

By applying the concepts of homomorphism and isomorphism of algebraic systems to groups, the definitions of homomorphism and isomorphism of groups can be obtained.

Definition 6.3.1 Let $U = <G, *>$ and $V = <H, \odot>$ are two groups. If there is a mapping (surjective, injective, or bijective) $f: G \rightarrow H$, for any element a and b in G, we know

$$f(a * b) = f(a) \odot f(b)$$

f is said to be a group homomorphism (epimorphism or isomorphism) from U to V. In particular, the homomorphism (isomorphism) f of U to U is called the endomorphism (automorphism) of U.

Set e_G and e_H are identity element of $U = <G, *>$ and $V = <H, \odot>$, f is U to V homomorphism, so by $f(e_G) \odot f(e_G) = f(e_G * e_G) = f(e_G)$, we can get $f(e_G) = e_H$. Because $f(a^{-1}) \odot f(a) = f(a^{-1} * a) = f(e_G) = e_H$. In the same way $f(a) \odot f(a^{-1}) = e_H$, we can get $f(a^{-1}) = (f(a))^{-1}$.

Definition 6.3.2 If a group $U = <G, *>$ to $V = <H, \odot>$ has a homomorphism (isomorphism), then the U homomorphism (isomorphism) is said to V, denoted as $U \sim V$ ($U \cong V$).

Example 6.3.1 $f(a) = 2$ is not a homomorphism of the group $<Q, +>$ to $<I, +>$.

Example 6.3.2 $f(<a, b>) = a + 26$ is an epimorphism from the group $<I \times I, +>$ to $<I, +>$ (f is surjective), so $<I \times I, +> \sim <I, +>$.

Example 6.3.3 $f(a) = |a|$ is not an endomorphism of a group $<R, +>$, but it is an endomorphism of a group $<R', \times>$.

Theorem 6.3.1 Let $<G, *>$ and $<H, \odot>$ be isomorphic algebraic systems, where $<G, *>$ is a group, and if there is surjective $f: G \rightarrow H$, for any elements a and b in G, there is

$$f(a * b) = f(a) \odot f(b)$$

So $<H, \odot>$ is also a group.

6.4 Abelian Groups and Cyclic Groups

Abelian group is a group whose operation conforms to commutative law, so the Abelian group is also called commutative group. It consists of its own set G and the binary operation $*$. Cyclic groups are very important groups and have been completely solved. It is defined as follows, if every element of a group G is a power of some fixed element a of G, then G is called a cyclic group. In this section, we will introduce Abelian groups and cyclic groups.

Definition 6.4.1 If a group $<G, *>$ operation $*$ is commutative, the group is called an Abelian group, or an commutative group.

Example 6.4.1 Let $S=\{a, b, c, d\}$, define a bijective function f: $f(a)=b$, $f(b)=c$, $f(c)=d$, $f(d)=a$ on S, for any $x \in S$, construct a compound functions

$$f(x)^2 = f \circ f(x) = f(f(x))$$
$$f(x)^3 = f \circ f^2(x) = f(f^2(x))$$
$$f(x)^4 = f \circ f^3(x) = f(f^3(x))$$

If f represents the identity mapping on S, that is

$$f(x) = x, \quad x \in S$$

Obviously, there is $f^4(x) = f^0(x)$, let's call it $f^1 = f$, construct the set $F = \{f^0, f^1, f^2, f^3\}$, then $<F, \circ>$ is an Abelian group.

Solution For the composition of any two functions in F, it can be given in Table 6.4.1.

Table 6.4.1

$\circ$	f^0	f^1	f^2	f^3
f^0	f^0	f^1	f^2	f^3
f^1	f^1	f^2	f^3	f^0
f^2	f^2	f^3	f^0	f^1
f^3	f^3	f^0	f^1	f^2

So, the compound operation $\circ$ is closed about F, and it's bound.

f^0 is the identity element of the complex operation $\circ$.

The inverse element of f^0 is itself, the inverse element of f^1 and f^3 are each other, the inverse element of f^2 is itself.

According to the symmetry of Table 6.4.1, it can be seen that the composite operation $\circ$ is commutative. Therefore, $<F, \circ>$ is an Abel group.

Theorem 6.4.1 Let $<G, *>$ be a group, $<G, *>$ be an Abelian group be necessary and sufficient condition for arbitrary $a, b \in G$, have $(a * b) * (a * b) = (a * a) * (b * b)$.

Proof (1) Sufficiency.

If for any $a, b \in G$, there is

$$(a * b) * (a * b) = (a * a) * (b * b)$$

Because $\quad a * (a * b) * b = (a * a) * (b * b) = (a * b) * (a * b) = a * (b * a) * b$

So $\quad a^{-1} * (a * (a * b) * b) * b^{-1} = a^{-1} * (a * (b * a) * b) * b^{-1}$

Then
$$a * b = b * a$$
So, the group $<G, *>$ is an Abelian group.

(2) Necessity.

If $<G, *>$ is an Abelian group, then for any a, $b \in G$, there is
$$a * b = b * a$$
So $(a * a) * (b * b) = a * (a * b) * b = a * (b * a) * b = (a * b) * (b * a)$

Definition 6.4.2 Set $<G, *>$ as the group, if there is an element a in G, so that any element in G is composed of powers of a, the group is called cyclic group, and element a is called the generator of cyclic group G.

For example, in Example 6.2.1, $60°$ is the generator of the group $<\{0°, 60°, 120°, 180°, 240°, 300°\}, \bigstar>$. Therefore, the group is a cyclic group.

Theorem 6.4.2 Any cyclic group must be an Abel group.

Proof Suppose $<G, *>$ is a cyclic group, its generator is a, then, for any x, $y \in G$, there must be r, $s \in I$, such that $x = a^r$, $y = a^s$, and $x * y = a^r * a^s = a^{r+s} = a^{s+r} = a^s * a^r = y * x$. So, $<G, *>$ is an Abel group.

For finite cyclic groups, we have the following theorem.

Theorem 6.4.3 Set $<G, *>$ is a finite cyclic group generated by the element $a \in G$. If the order of G is n, that is, $|G| = n$, then, and
$$G = \{a, a^2, a^3, \ldots, a^{n-1}, a^n = e\}$$
Where e is the identity element in $<G, *>$, n is the lowest positive integer.

Proof Let's say that for some positive integer m, m less than n, there is $a^m = e$. So, since $<G, *>$ is a cyclic group, so any element in G can be written as $a^k (k \in I)$, and $k = mq + r$, where q is some integer. This is
$$a^k = a^{mq+r} = (a^m)^q * a^r = a^r$$

This leads to the fact that every element in G can be represented as $a^r (0 \leqslant r < m)$, so that there are at most m different elements in G, which contradicts $|G| = n$. So $a^m = e$ is impossible.

Further proof a, a^2, a^3, $\ldots$, a^{n-1}, a^n are not the same. By contradiction. Let $a^i = a^j$, $1 \leqslant i < j \leqslant n$, we can get a result $a^{j-i} = e$ and $1 \leqslant j - i < n$. This has been proved impossible by above. So, $a, a^2, a^3, \ldots, a^{n-1}, a^n$ are different, so
$$G = \{a, a^2, a^3, \ldots, a^{n-1}, a^n = e\}$$

Example 6.4.2 Set $G = \{\alpha, \beta, \gamma, \delta\}$ and define binary operation $*$ on G as shown in Table 6.4.2.

Table 6.4.2

$*$	α	β	γ	δ
α	α	β	γ	δ
β	β	α	δ	γ
γ	γ	δ	β	α
δ	δ	γ	α	β

Indicates that $<G, *>$ is a cyclic group.

Proof According to operation Table 6.4.2, it can be seen that operation $*$ is closed and α is an identity element. The inverse of β, γ, and δ are β, δ, and γ respectively. You can verify that the operation $*$ is associative. So $<G, *>$ is a group.

In this group, because

$$\gamma * \gamma = \gamma^2 = \beta, \quad \gamma^3 = \delta, \quad \gamma^4 = \alpha$$
$$\delta * \delta = \delta^2 = \beta, \quad \delta^3 = \gamma, \quad \delta^4 = \alpha$$

So the group $<G, *>$ is generated by γ or δ, so $<G, *>$ is a cyclic group.

As can be seen from Example 6.4.2: the generator of a cyclic group can not be unique.

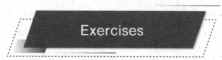

Exercises

1. For positive integer k, $N = \{0, 1, 2, ..., k-1\}$, let's say $*_k$ is a binary operation on N_k, such that $a * b$ is equal to the remainder of k divided by $a \cdot b$, here $a, b \in N_k$.

(1) When k is equal to 4, the trial-made operation $*_k$ table.

(2) For any positive integer K, prove that $<N_k, *_k>$ is a semigroup.

2. Let $<S, *>$ is a semigroup, $a \in S$, define a binary operation $\square$ on S, such that for any element x and y in S, there is

$$x \square y = x * a * y$$

Prove that binary operation $\square$ is associative.

3. Let $<R, *>$ be an algebraic system, $*$ be a binary operation on R, such that for any element a, b

$$a * b = a + b + a \cdot b$$

Prove that 0 is an identity element and $<R, *>$ is a monoid.

4. Let's say that $<A, *>$ is a semigroup, and for the elements a and b in A, if $a \neq b$ have to $a * b \neq b * a$, try to prove:

(1) For every member of A, we have $a * a = a$.

(2) For any of the elements a and b in A, we have $a * b * a = a$.

(3) For any member of a, b, and c in A, we have $a * b * c = a * c$.

5. Set $<A, *>$ is a semigroup, e is a left identity element and for each $x \in A$, $\hat{x} \in A$ exists, such that $\hat{x} * x = e$.

(1) Prove: for any $a, b, c \in A$, if $a * b = a * c$, then $b = c$.

(2) By proving that e is an identity element of A, we can prove that $<A, *>$ is a group.

6. Let $<G, *>$ be a group, for any $a \in G$, let $H = \{y \mid y * a = a * y, y \in G\}$, try to prove that $<H, *>$ is a subgroup of $<G, *>$.

7. Set $<H, \cdot>$ and $<K, \cdot>$ are subgroups of $<G, \cdot>$, let

$$HK = \{h \cdot k \mid h \in H, k \in K\}$$

Prove: $<HK, \cdot>$ is a subgroup of $<G, \cdot>$ if and only if $HK = KH$.

8. Let's say that $<A, *>$ is a group, and $|A| = 2n$. Proof: $a \neq e$ exists at least in A,

so that $a * a = e$. Where e is the identity element.

9. Given two semigroups $U = <S, \ * >$ and $V = <T, \ \odot >$, $f: S \to T$ is the isomorphism from U to V. Try to prove: If z is zero of U, then $f(z)$ is zero of V.

10. So let's say that f and g are homomorphisms of groups $U = <G_1, \ * >$ to $V = <G_2, \ \odot >$, so let's set them up $H_1 = \{x \mid x \in G_1, \ f(x) = g(x)\}$ here. Let's try to prove that $<H_1, \ * >$ is a subgroup of group U.

11. Let f be the homomorphism of the group $U = <G_1, \ * >$ to $V = <G_2, \ \odot >$. Try to prove that f is injective if $k(f) = \{e_1\}$ and only if, here e_1 is the identity element of the group U.

12. Given an element a in the group $U = <G, \ * >$, define the mapping $f: G \to G$ as $f(x) = a^{-1} * x * a$. Prove that f is an automorphism of the group U.

13. Suppose $<G, \ * >$ is a monoid, and for every element x in G, we have $x * x = e$, where e is an identity element, try to prove $<G, \ * >$ is an Abelian group.

14. It is proved that any group of orders 1, 2, 3, 4 is an Abelian group.

15. Let $<G, \ * >$ be a group, try to prove that if for any $a, b \in G$, we have $a^3 * b^3 = (a * b)^3$, $a^4 * b^4 = (a * b)^4$ and $a^5 * b^5 = (a * b)^5$, then $<G, \ * >$ is a group.

16. Prove: Any subgroup of a cyclic group must also be a cyclic group.

第6章 群

群论的建立,不仅扩大了群论研究的对象和应用,而且可以从不同的群得到多方面的启发。群论是近代数学中发展得最充分的部分,在自然科学学科中得到了广泛的应用。例如,在自动化理论、编码理论、快速加法器的设计等方面,群的应用已日趋完善。

6.1 半群

半群是一种特殊的代数系统,在形式语言、自动化等领域中都有具体的应用。

定义 6.1.1 ▶ 一个代数系统 $<S,*>$,其中 S 是非空集合,$*$ 是 S 上的一个二元运算,如果运算 $*$ 是封闭的,则称代数系统 $<S,*>$ 为广群。

定义 6.1.2 ▶ 一个代数系统 $<S,*>$,其中 S 是非空集合,$*$ 是 S 上的一个二元运算。如果:①运算 $*$ 是封闭的;②运算 $*$ 是可结合的,即对任意的 $x,y,z \in S$,满足 $(x*y)*z = x*(y*z)$。则称代数系统 $<S,*>$ 为半群。

例 6.1.1 设集合 $S_k = \{x \mid x \in I \wedge x \geqslant k\}$,$k \geqslant 0$,那么 $<S_k,+>$ 是一个半群,其中 $+$ 是普通的加法运算。

解 因为运算 $+$ 在 S_k 上是封闭的,而且普通加法运算是可结合的,所以 $<S_k,+>$ 是一个半群。

在例 6.1.1 中,$k \geqslant 0$ 这个条件是很重要的,否则,如果 $k < 0$,则运算 $+$ 在 S_k 上是不封闭的。

例 6.1.2 设 $S = \{a,b,c\}$,在 S 上的一个二元运算 $\triangle$ 定义如表 6.1.1 所示。

表 6.1.1

$\triangle$	a	b	c
a	a	b	c
b	a	b	c
c	a	b	c

验证 $<S,\triangle>$ 是一个半群。

解 从表 6.1.1 中可知运算 $\triangle$ 是封闭的,同时 a,b 和 c 都是左幺元。所以,对于任意的 $x,y,z \in S$,都有

$$x \triangle (y \triangle z) = x \triangle z = z = y \triangle z = (x \triangle y) \triangle z$$

因此,$<S,\triangle>$ 是半群。

显然,代数系统 $<I_+,->$ 和 $<R,/>$ 都不是半群,这里,$-$ 和 $/$ 分别是普通的减法和除法。

定理 6.1.1 ▶ 设$<S,*>$是一个半群，$B\subseteq S$且 $*$ 在 B 上是封闭的，那么$<B,*>$也是一个半群。通常称$<B,*>$是半群$<S,*>$的子半群。

证明　因为 $*$ 在 S 上是可结合的，而 $B\subseteq S$ 且 $*$ 在 B 上是封闭的，所以 $*$ 在 B 上也是可结合的，因此，$<B,*>$是一个半群。

例 6.1.3　设 $*$ 表示普通的乘法运算，那么$<[0,1],*>$，$<[0,1),*>$和$<1,*>$都是$<R,*>$的子半群。

解　首先，运算 $*$ 在 R 上是封闭且是可结合的，所以$<R,*>$是一个半群。其次，运算 $*$ 在$[0,1]$，$[0,1)$和 1 上都是封闭的，且$[0,1]\subset R$，$[0,1)\subset R$，$I\subset R$。因此，由定理 6.1.1 可知$<[0,1],*>$，$<[0,1),*>$和$<1,*>$都是$<R,*>$的子半群。

定理 6.1.2 ▶ 设$<S,*>$是一个半群，如果 S 是一个有限集，则必定有 $a\in S$，使 $a*a=a$。

证明　$<S,*>$是半群。对于任意的 $b\in S$，由 $*$ 的封闭性可知

$$b*b\in S,\text{记 } b^2=b*b$$
$$b^2*b=b*b^2\in S,\text{记 } b^3=b^2*b=b*b^2$$
$$\vdots$$

因为 S 是有限集，所以必定存在 $j>i$，使

$$b^i=b^j$$

令

$$p=j-i$$

便有

$$b^i=b^p*b^i$$

所以

$$b^q=b^p*b^q,\quad q\geqslant i$$

因为 $p\geqslant 1$，所以总可以找到 $k\geqslant 1$，使

$$kp\geqslant i$$

对于 S 中的元素 b^{kq}，就有

$$b^{kq}=b^p*b^{kq}=b^p*(b^p*b^{kq})=b^{2p}*b^{kq}=b^{2p}*(b^p*b^{kq})=\cdots=b^{kq}*b^{kq}$$

这就证明了在 S 中存在元素 $a=b^{kq}$，使 $a*a=a$。

定义 6.1.3 ▶ 含有幺元的半群称为独异点。

例如，代数系统$<R,+>$是一个独异点，因为，$<R,+>$是一个半群，且 0 是 R 中关于运算＋的幺元。另外，代数系统$<I,\cdot>$，$<I_+,\cdot>$，$<R,\cdot>$都是具有幺元 I 的半群，$\cdot$ 为普通乘法运算，因此它们都是独异点。

可是，代数系统$<N-\{0\},+>$虽是一个半群，但运算＋不存在幺元，所以，这个代数系统不是独异点。

定理 6.1.3 ▶ 设$<S,*>$是一个独异点，则在关于运算 $*$ 的运算表中任何两行或两列都是不相同的。

证明　设 S 中关于运算 $*$ 的幺元是 e。因为对于任意的 $a,b\in S$ 且 $a\neq b$ 时，总有

$$e*a=a\neq b=e*b$$
$$a*e=a\neq b=b*e$$

所以，在 $*$ 的运算表中不可能有两行或两列是相同的。

例 6.1.4　设 I 是整数集合，m 是任意正整数，Z_m 是由模 m 的同余类组成的同余类集，

在 Z_m 上分别定义两个二元运算 $+_m$ 和 $\times_m$。

对于任意的 $[i],[j] \in Z_m$,有

$$[i] +_m [j] = [(i+j)(\bmod m)]$$
$$[i] \times_m [j] = [(i \times j)(\bmod m)]$$

试证明在这两个二元运算的运算表中任何两行或两列都不相同。

证明 考察代数系统 $<Z_m, +_m>$ 和 $<Z_m, \times_m>$。

(1) 由运算 $+_m$ 和 $\times_m$ 的定义可知,它在 Z_m 上是封闭的。

(2) 对于任意 $[i],[j],[k] \in Z_m$,有

$$([i] +_m [j]) +_m [k] = [i] +_m ([j] +_m [k]) = [(i+j+k)(\bmod m)]$$
$$([i] \times_m [j]) \times_m [k] = [i] \times_m ([j] \times_m [k]) = [(i \times j \times k)(\bmod m)]$$

所以,$+_m$ 和 $\times_m$ 都是可结合的。

(3) 因为 $[0] +_m [i] = [i] +_m [0] = [i]$,所以 $[0]$ 是 $<Z_m, +m>$ 中的幺元。因为 $[1] \times_m [i] = [i] \times_m [1] = [i]$,所以 $[1]$ 是 $<Z_m, \times_m>$ 的幺元。

因此,代数系统 $<Z_m, +_m>$,$<Z_m, \times_m>$ 都是独异点。由定理 6.1.3 可知,这两个运算的运算表中任何两行或者两列都不相同。

上例中,如果给定 $m=5$,$+_5$,$\times_5$ 的运算表分别如表 6.1.2 和表 6.1.3 所示。

表 6.1.2

$+_5$	$[0]$	$[1]$	$[2]$	$[3]$	$[4]$
$[0]$	$[0]$	$[1]$	$[2]$	$[3]$	$[4]$
$[1]$	$[1]$	$[2]$	$[3]$	$[4]$	$[0]$
$[2]$	$[2]$	$[3]$	$[4]$	$[0]$	$[1]$
$[3]$	$[3]$	$[4]$	$[0]$	$[1]$	$[2]$
$[4]$	$[4]$	$[0]$	$[1]$	$[2]$	$[3]$

表 6.1.3

$\times_5$	$[0]$	$[1]$	$[2]$	$[3]$	$[4]$
$[0]$	$[0]$	$[0]$	$[0]$	$[0]$	$[0]$
$[1]$	$[0]$	$[1]$	$[2]$	$[3]$	$[4]$
$[2]$	$[0]$	$[2]$	$[4]$	$[1]$	$[3]$
$[3]$	$[0]$	$[3]$	$[1]$	$[4]$	$[2]$
$[4]$	$[0]$	$[4]$	$[3]$	$[2]$	$[1]$

显然,上述运算表中没有两行或两列是相同的。

定理 6.1.4 ▶ 设 $<S, *>$ 是独异点,对于任意 $a,b \in S$,且 a,b 均有逆元,则

(1) $(a^{-1})^{-1} = a$。

(2) $a * b$ 有逆元,且 $(a * b)^{-1} = b^{-1} * a^{-1}$。

证明 (1) 因为 a^{-1} 是 a 的逆元,即

$$a * a^{-1} = a^{-1} * a = e \quad (e \text{ 为幺元})$$

所以

$$(a^{-1})^{-1} = a$$
$$(a * b) * (b^{-1} * a^{-1}) = a * (b * b^{-1}) * a^{-1} = a * e * a^{-1} = a * a^{-1} = e$$

同理可证

$$(b^{-1} * a^{-1}) * (a * b) = e$$

所以

$$(a * b)^{-1} = b^{-1} * a^{-1}$$

6.2 群与子群

在这一节中,我们将介绍子群的概念,以及子群和群的关系。

定义 6.2.1 ▶ 设 $<G, *>$ 是一个代数系统,其中 G 是非空集合,$*$ 是 G 上的二元运算,如果运算 $*$ 是封闭的;运算 $*$ 是可结合的;存在幺元 e;对于每一个元素 $x \in G$,存在它的逆元

x^{-1}。则称$<G,*>$是一个群。

例如，$<R-\{0\},\times>,<P(S),\oplus>$等都是群。

例 6.2.1　设 $R=\{0°,60°,120°,180°,240°,300°\}$ 表示在平面上几何图形绕质心顺时针旋转角度的六种可能情况，设★是 R 上的二元运算，对于 R 中任意两个元素 a 和 b，$a★b$ 表示平面图形连续旋转 a 和 b 得到的总旋转角度，并规定旋转 $360°$ 等于原来的状态，视为没有经过旋转。试验证$<R,★>$是一个群。

解　由题意可知，R 上二元运算★的运算如表 6.2.1 所示。

表　6.2.1

★	0°	60°	120°	180°	240°	300°
0°	0°	60°	120°	180°	240°	300°
60°	60°	120°	180°	240°	300°	0°
120°	120°	180°	240°	300°	0°	60°
180°	180°	240°	300°	0°	60°	120°
240°	240°	300°	0°	60°	120°	180°
300°	300°	0°	60°	120°	180°	240°

由表 6.2.1 可知，运算★在 R 上是封闭的。

对于任意的 $a,b,c\in R$，$(a★b)★c$ 表示将图形依次旋转 a,b 和 c，而 $a★(b★c)$ 表示将图形依次旋转 b,c 和 a，而总的旋转角度等于 $a+b+c(\bmod 360°)$，因此，$(a★b)★c=a★(b★c)$。

$0°$是幺元。

$60°$、$180°$、$120°$的逆元分别是 $300°$、$180°$、$240°$。因此，$<R,★>$是一个群。

定义 6.2.2 ▶设$<G,*>$是一个群。如果 G 是有限集，那么称$<G,*>$为有限群，G 中元素的个数通常称为该有限群的阶数，记为$|G|$；如果 G 是无限集，则称$<G,*>$为无限群。

例 6.2.1 中的$<R,★>$就是一个有限群，且$|R|=6$。

例 6.2.2　试验证代数系统$<I,+>$是一个群，这里 I 是所有整数的集合，$+$是普通加法运算。

解　显然，二元运算$+$在 I 上是封闭的且是可结合的。幺元为 0。对于任意 $a\in A$，它的逆元是$-a$，所以$<I,+>$是一个群，且是一个无限群。

至此，我们可以概括地说，广群仅仅是一个非空集合上具有封闭二元运算的代数系统；半群是一个具有结合运算的广群；独异点是具有幺元的半群；群是每个元素都有逆元的独异点。即

$$\{群\}\subset\{独异点\}\subset\{半群\}\subset\{广群\}$$

由定理 5.2.4 可知，群中任何一个元素的逆元必定是唯一的。由群中逆元的唯一性，可以有以下几个定理。

定理 6.2.1 ▶群中不可能有零元。

证明　当群的阶为 1 时，它的唯一元素成为幺元。

设$|G|>1$且群$<G,*>$有零元 θ，那么群中任何元素 $x\in G$，都有 $x*\theta=\theta*x=\theta\neq e$。所以，零元 θ 不存在逆元，这与$<G,*>$是群相矛盾。

定理 6.2.2 ▶设$<G,*>$是一个群，对于 $a,b\in G$，必定存在唯一的 $x\in G$，使 $a*x=b$。

证明 设 a 的逆元是 a^{-1},令

$$x=a^{-1}*b$$

则 $\qquad a*x=a*(a^{-1}*b)=(a*a^{-1})*b=e*b=b$

若另有一个解 x_1,满足 $a*x_1=b$,则

$$a^{-1}*(a*x_1)=a^{-1}*b$$

即 $\qquad\qquad\qquad\qquad x_1=a^{-1}*b$

定理 6.2.3 ▶ 设 $<G,*>$ 是一个群,对于任意的 $a,b,c\in G$,如果有 $a*b=a*c$ 或者 $b*a=c*a$,那么必有 $b=c$(消去律)。

证明 设 $a*b=a*c$,且 a 的逆元是 a^{-1},则有

$$a^{-1}*(a*b)=a^{-1}*(a*c)$$
$$(a*a^{-1})*b=(a*a^{-1})*c$$
$$e*b=e*c$$
$$b=c$$

当 $b*a=c*a$ 时,可同样证得 $b=c$。

由定理 6.1.3 可知,群的运算表中没有两行(或两列)是相同的。为进一步考察群的运算表所具有的性质,现在引进置换的概念。

定义 6.2.3 ▶ 设 S 是一个非空集合,从集合 S 到 S 的一个双射称为 S 的一个置换。

例如,对于集合 $S=\{a,b,c,d\}$,将 a 映射到 b,b 映射到 d,c 映射到 a,d 映射到 c,这是一个从 S 到 S 上的一对一映射,这个置换可以表示为

$$\begin{pmatrix} a & b & c & d \\ b & d & a & c \end{pmatrix}$$

即在上一行中按任意顺序写出集合中的全部元素,而在下一行中写出每个对应元素的像。

定理 6.2.4 ▶ 群 $<G,*>$ 的运算表中的每一行或每一列都是 G 的元素的一个置换。

证明 首先,证明运算表中的任一行或任一列所含 G 中的一个元素不可能多于一次。用反证法,如果对应元素 $a\in G$ 的那一行中有两个元素都是 c,即有

$$a*b_1=a*b_2=c,\text{且 }b_1\neq b_2$$

由消去律可得 $b_1=b_2$,这与 $b_1\neq b_2$ 矛盾。

其次,要证明 G 中的每一个元素都在运算表的每一行和每一列中出现。考察对应元素 $a\in G$ 的那一行,设 b 是 G 中的任一元素,由于 $b=a*(a^{-1}*b)$,所以 b 必定出现在对应于 a 的那一行中。

再由运算表中没有两行(或两列)相同的事实便可得出:$<G,*>$ 的运算表中每一行都是 G 的元素的一个置换,且每一行都是不相同的。同样的结论对于列也是成立的。

定义 6.2.4 ▶ 在代数系统 $<G,*>$ 中,如果存在 $a\in G$,有 $a*a=a$,则称 a 为等幂元。

定理 6.2.5 ▶ 在群 $<A,*>$ 中,除幺元 e 外,不可能有任何别的等幂元。

证明 因为 $e*e=e$,所以 e 是等幂元。

现设 $a\in A$,$a\neq e$ 且 $a*a=a$,则有 $a=e*a=(a^{-1}*a)*a=a^{-1}*(a*a)=a^{-1}*a=e$,与假设 $a\neq e$ 相矛盾。

下面再介绍子群的概念。

定义 6.2.5 ▶ 设$<G,*>$是一个群,S是G的非空子集,如果$<S,*>$也构成群,则称$<S,*>$是$<G,*>$的一个子群。

定理 6.2.6 ▶ 设$<G,*>$是一个群,$<S,*>$是$<G,*>$的一个子群,那么,$<G,*>$中的幺元e必定也是$<S,*>$中的幺元。

证明 设$<S,*>$中的幺元为e_1,对于任何一个$x\in S\subseteq G$,必有$e_1*x=x=e*x$,故$e_1=e$。

定义 6.2.6 ▶ 设$<G,*>$是一个群,$<S,*>$是$<G,*>$的子群,如果$S=\{e\}$,或者$S=G$,则称$<S,*>$为$<G,*>$的平凡子群。

例 6.2.3 $<I,+>$是一个群,设$I_E=\{x\mid x=2n,n\in I\}$,证明$<I_E,+>$是$<I,+>$的一个子群。

证明 (1) 对于任意的$x,y\in I_E$,我们不妨假设$x=2n_1,y=2n_2,n_1,n_2\in I$,则
$$x+y=2n_1+2n_2=2(n_1+n_2)$$
而$n_1+n_2\in I$,所以$x+y\in I_E$,即$+$在I_E上是封闭的。

(2) 运算$+$在I_E上保持可结合性。

(3) $<I,+>$中的幺元0也在I_E中。

(4) 对于任意的$x\in I_E$,必有n使$x=2n$,而$-x=-2n=2(-n)$,$-n\in I$。

所以$-x\in I_E$,而$x+(-x)=(-x)+x=0$,因此,$<I_E,+>$是$<I,+>$的一个子群。

定理 6.2.7 ▶ 设$<G,*>$是一个群,B是G的非空子集,如果B是一个有限集,那么,只要运算$*$在B上封闭,$<B,*>$必定是$<G,*>$的子群。

证明 设b是B中的任一个元素。若$*$在B上封闭,则元素$b^2=b*b,b^3=b^2*b,\cdots$都在B中。由于B是有限集,所以必存在正整数i和j,不妨假设$i<j$,使
$$b^i=b^j,\ \text{即}\ b^i=b^i*b^{j-i}$$
这就说明b^{j-i}是$<G,*>$中的幺元,且这个幺元也在子集B中。

如果$j-i>1$,那么由$b^{j-i}=b*b^{j-i-1}$可知b^{j-i-1}是b的逆元,$b^{j-i-1}\in B$;如果$j-i=1$,那么由$b^i=b^i*b$可知b就是幺元,而幺元是以自身为逆元的。

因此,$<B,*>$必定是$<G,*>$的子群。

定理 6.2.8 ▶ 设$<G,\triangle>$是群,S是G的非空子集,如果对于S中的任意元素a和b有$a\triangle b^{-1}\in S$,则$<S,\triangle>$是$<G,\triangle>$的子群。

证明 首先证明,G中的幺元e也是S中的幺元。任取S中的元素$a,a\in S\subset G$,所以$e=a\triangle a^{-1}\in S\subset G$且$a\triangle e=e\triangle a=a$,即$e$也是$S$中的元素。

其次证明,S中的每一元素都有逆元。对任意$a\in S$,因为$e\in S$,所以,$e\triangle a^{-1}\in S$,即$a^{-1}\in S$。

最后证明,$\triangle$在S上是封闭的。对于任意$a,b\in S$,由上可知$b^{-1}\in S$,而$b=(b^{-1})^{-1}$,所以$a\triangle b=a\triangle(b^{-1})^{-1}\in S$。

由于,运算$\triangle$在S上的可结合性是保持的,所以$<S,\triangle>$是$<G,\triangle>$的子群。

6.3 群的同态与同构

在第5章中我们介绍过代数系统的同构与同态,现在将代数系统同态与同构的概念应用到群,便可得出群的同态与同构的定义。

定义 6.3.1 ▶设 $U=<G,*>$ 和 $V=<H,\odot>$ 是两个群。若存在映射(满射、单射或双射)$f:G{\rightarrow}H$,对 G 中任意元素 a 和 b,有

$$f(a*b)=f(a)\odot f(b)$$

则称 f 是 U 到 V 的一个群同态(满同态、单同态或同构),特别地,U 到 U 上的同态(同构)f 称为 U 的自同态(自同构)。

设 e_G 和 e_H 分别是群 $U=<G,*>$ 和 $V=<H,\odot>$ 的幺元,f 是 U 到 V 的同态,故由 $f(e_G)\odot f(e_G)=f(e_G*e_G)=f(e_G)$ 可得 $f(e_G)=e_H$。因 $f(a^{-1})\odot f(a)=f(a^{-1}*a)=f(e_G)=e_H$,同理 $f(a)\odot f(a^{-1})=e_H$,可得 $f(a^{-1})=(f(a))^{-1}$。

定义 6.3.2 ▶若群 $U=<G,*>$ 到 $V=<H,\odot>$ 存在一个同态(同构),则称 U 同态(同构)于 V,记作 $U{\sim}V(U{\cong}V)$。

例 6.3.1 $f(a)=2$ 不是群 $<Q,+>$ 到 $<I,+>$ 的同态。

例 6.3.2 $f(<a,b>)=a+26$ 是群 $<I{\times}I,\mp>$ 到 $<I,+>$ 的满同态(f 是满射的),故 $<I{\times}I,+>{\sim}<I,+>$。

例 6.3.3 $f(a)=|a|$ 不是群 $<R,+>$ 的自同态,但是群 $<R',{\times}>$ 的自同态。

定理 6.3.1 ▶设 $<G,*>$ 和 $<H,\odot>$ 是同型的代数系统,其中 $<G,*>$ 是群,若存在满射 $f:G{\rightarrow}H$,对 G 中任意元素 a 和 b,有

$$f(a*b)=f(a)\odot f(b)$$

则 $<H,\odot>$ 也是群。

6.4 阿贝尔群与循环群

阿贝尔群是群运算符合交换律性质的群,因此阿贝尔群也被称为交换群。它由自身的集合 G 和二元运算 $*$ 构成。循环群是一种很重要的群,也是已被完全解决了的一类群。其定义为若一个群 G 的每一个元素都是 G 的某一个固定元 a 的乘方,则称 G 为循环群。这节我们将介绍阿贝尔群与循环群。

定义 6.4.1 ▶如果群 $<G,*>$ 中的运算 $*$ 是可交换的,则称该群为阿贝尔群,或称交换群。

例 6.4.1 设 $S=\{a,b,c,d\}$,在 S 上定义一个双射函数 $f:f(a)=b,f(b)=c,f(c)=d,f(d)=a$,对于任意 $x{\in}S$,构造复合函数

$$f(x)^2=f{\circ}f(x)=f(f(x))$$
$$f(x)^3=f{\circ}f^2(x)=f(f^2(x))$$
$$f(x)^4=f{\circ}f^3(x)=f(f^3(x))$$

如果用 f 表示 S 上的恒等映射,即

$$f(x)=x,\quad x{\in}S$$

显然,有 $f^4(x)=f^0(x)$,记作 $f^1=f$,构造集合 $F=\{f^0,f^1,f^2,f^3\}$,那么 $<F,{\circ}>$ 是一个阿贝尔群。

解 对于 F 中任意两个函数的复合,可以由表 6.4.1 给出。

表　6.4.1

∘	f^0	f^1	f^2	f^3
f^0	f^0	f^1	f^2	f^3
f^1	f^1	f^2	f^3	f^0
f^2	f^2	f^3	f^0	f^1
f^3	f^3	f^0	f^1	f^2

可见,复合运算∘关于 F 是封闭的,并且是结合的。

f^0 是关于复合运算∘的幺元。

f^0 的逆元是它本身,f^1 和 f^3 互为逆元,f^2 的逆元也是它本身。

由表 6.4.1 的对称性可知,复合运算∘是可交换的。因此,$<F,∘>$ 是一个阿贝尔群。

定理 6.4.1 ▶设 $<G,*>$ 是一个群,$<G,*>$ 是阿贝尔群的充要条件是对任意 $a,b\in G$,有 $(a*b)*(a*b)=(a*a)*(b*b)$。

证明 (1) 充分性。

设对任意 $a,b\in G$,有
$$(a*b)*(a*b)=(a*a)*(b*b)$$

因为　$a*(a*b)*b=(a*a)*(b*b)=(a*b)*(a*b)=a*(b*a)*b$

所以　　　$a^{-1}*(a*(a*b)*b)*b^{-1}=a^{-1}*(a*(b*a)*b)*b^{-1}$

即得　　　　　　　　　$a*b=b*a$

因此,群 $<G,*>$ 是阿贝尔群。

(2) 必要性。

设 $<G,*>$ 是阿贝尔群,则对任意 $a,b\in G$,有
$$a*b=b*a$$

因此　$(a*a)*(b*b)=a*(a*b)*b=a*(b*a)*b=(a*b)*(b*a)$

定义 6.4.2 ▶设 $<G,*>$ 为群,若在 G 中存在一个元素 a,使 G 中的任意元素都由 a 的幂组成,则称该群为循环群,元素 a 称为循环群 G 的生成元。

例如,在例 6.2.1 中,60° 就是群 $<\{0°,60°,120°,180°,240°,300°\},★>$ 的生成元,因此,该群是循环群。

定理 6.4.2 ▶任何一个循环群必定是阿贝尔群。

证明 设 $<G,*>$ 是一个循环群,它的生成元是 a,那么,对于任意 $x,y\in G$,必有 $r,s\in I$,使 $x=a^r$, $y=a^s$,而且 $x*y=a^r*a^s=a^{r+s}=a^{s+r}=a^s*a^r=y*x$,因此,$<G,*>$ 是一个阿贝尔群。

对于有限循环群,有下面的定理。

定理 6.4.3 ▶设 $<G,*>$ 是一个由元素 $a\in G$ 生成的有限循环群。如果 G 的阶数是 n,即 $|G|=n$,则 $a^n=e$,且
$$G=\{a,a^2,a^3,\cdots,a^{n-1},a^n=e\}$$

其中,e 是 $<G,*>$ 中的幺元,n 是使 $a^n=e$ 的最小正整数。

证明 假设对于某个正整数 $m,m<n$,有 $a^m=e$。那么,由于 $<G,*>$ 是一个循环群,所以 G 中的任何元素都能写为 $a^k(k\in I)$,而且 $k=mq+r$,其中,q 是某个整数,$0\leqslant r<m$。这就有

$$a^k = a^{mq+r} = (a^m)q * a^r = a^r$$

这就导致 G 中每一个元素都可以表示成 $a^r (0 \leqslant r < m)$，这样，G 中最多有 m 个不同的元素，与 $|G| = n$ 相矛盾。所以 $a^m = e$ 是不可能的。

进一步证明 $a, a^2, a^3, \cdots, a^{n-1}, a^n$ 都不相同。用反证法。假设 $a^i = a^j$，其中，$1 \leqslant i < j \leqslant n$，就有 $a^{j-i} = e$，而且 $1 \leqslant j - i < n$，这已经由上面证明是不可能的。所以，$a, a^2, a^3, \cdots, a^{n-1}$，$a^n$ 是不相同的，因此

$$G = \{a, a^2, a^3, \cdots, a^{n-1}, a^n = e\}$$

例 6.4.2 设 $G = \{\alpha, \beta, \gamma, \delta\}$，在 G 上定义二元运算 $*$ 如表 6.4.2 所示。

表 6.4.2

$*$	α	β	γ	δ
α	α	β	γ	δ
β	β	α	δ	γ
γ	γ	δ	β	α
δ	δ	γ	α	β

证明 $<G, *>$ 是一个循环群。

证明 由运算表 6.4.2 可知运算 $*$ 是封闭的，α 是幺元。β, γ 和 δ 的逆元分别是 β, δ 和 γ。可以验证运算 $*$ 是可结合的。所以 $<G, *>$ 是一个群。

在这个群中，由于

$$\gamma * \gamma = \gamma^2 = \beta, \quad \gamma^3 = \delta, \quad \gamma^4 = \alpha$$
$$\delta * \delta = \delta^2 = \beta, \quad \delta^3 = \gamma, \quad \delta^4 = \alpha$$

故群 $<G, *>$ 是由 γ 或 δ 生成的，因此 $<G, *>$ 是一个循环群。

从例 6.4.2 中可以看出：一个循环群的生成元可以不是唯一的。

习 题

1. 对于正整数 k，$N = \{0, 1, 2, \cdots, k-1\}$，设 $*_k$ 是 N_k 上的一个二元运算，使 $a * b$ 等于 k 除 $a \cdot b$ 所得的余数，这里 $a, b \in N_k$。

(1) 当 $k = 4$ 时，试画出 $*_k$ 的运算表。

(2) 对于任意正整数 K，证明 $<N_k, *_k>$ 是一个半群。

2. 设 $<S, *>$ 是一个半群，$a \in s$，在 S 上定义一个二元运算 $\square$，使对于 S 中的任意元素 x 和 y，都有

$$x \square y = x * a * y$$

证明：二元运算 $\square$ 是可结合的。

3. 设 $<R, *>$ 是一个代数系统，$*$ 是 R 上的一个二元运算，使对于任意元素 a, b 都有

$$a * b = a + b + a \cdot b$$

证明：0 是幺元且 $<R, *>$ 是独异点。

4. 设 $<A, *>$ 是一个半群，而且对于 A 中的元素 a 和 b，如果 $a \neq b$ 必有 $a * b \neq b * a$，试证明：

(1) 对于 A 中每个元素 a，有 $a * a = a$。

（2）对于 A 中任何元素 a 和 b，有 $a*b*a=a$。

（3）对于 A 中任何元素 a,b 和 c，有 $a*b*c=a*c$。

5. 设 $<A,*>$ 是半群，e 是左幺元且对每一个 $x\in A$，存在 $\hat{x}\in A$，使 $\hat{x}*x=e$。

（1）证明：对于任意 $a,b,c\in A$，如果 $a*b=a*c$，则 $b=c$。

（2）通过证明 e 是 A 中的幺元，证明 $<A,*>$ 是群。

6. 设 $<G,*>$ 是群，对任意 $a\in G$，令 $H=\{y\,|\,y*a=a*y,y\in G\}$，试证明 $<H,*>$ 是 $<G,*>$ 的子群。

7. 设 $<H,\cdot>$ 和 $<K,\cdot>$ 都是群 $<G,\cdot>$ 的子群，令

$$HK=\{h\cdot k\,|\,h\in H,k\in K\}$$

证明：$<HK,\cdot>$ 是 $<G,\cdot>$ 的子群的充要条件是 $HK=KH$。

8. 设 $<A,*>$ 是群，且 $|A|=2n$。证明：在 A 中至少存在 $a\neq e$，使 $a*a=e$。其中 e 是幺元。

9. 给定两个半群 $U=<S,*>$ 和 $V=<T,\odot>$，$f:S\rightarrow T$ 是 U 到 V 的同构。试证：若 z 是 U 的零元，则 $f(z)$ 是 V 的零元。

10. 设 f 和 g 都是群 $U=<G_1,*>$ 到 $V=<G_2,\odot>$ 的同态，令 $H_1=\{x\,|\,x\in G_1,f(x)=g(x)\}$。试证：$<H_1,*>$ 是群 U 的子群。

11. 设 f 是群 $U=<G_1,*>$ 到 $V=<G_2,\odot>$ 的同态。试证：f 为单射的充要条件条件为 $k(f)=\{e_1\}$，这里 e_1 是群 U 的幺元。

12. 在群 $U=<G,*>$ 中给定元素 a，定义映射 $f:G\rightarrow G$ 为 $f(x)=a^{-1}*x*a$。试证：f 是群 U 的自同构。

13. 设 $<G,*>$ 是一个独异点，并且对于 G 中的每一个元素 x 都有 $x*x=e$，其中 e 是幺元，证明 $<G,*>$ 是一个阿贝尔群。

14. 证明任何阶数分别为 1、2、3、4 的群都是阿贝尔群。

15. 设 $<G,*>$ 是一个群，证明：如果对任意 $a,b\in G$ 都有 $a^3*b^3=(a*b)^3$，$a^4*b^4=(a*b)^4$ 和 $a^5*b^5=(a*b)^5$，则 $<G,*>$ 是一个群。

16. 证明：循环群的任何子群必定也是循环群。

Chapter 7　Lattice and Boolean Algebra

This chapter introduces another kind of algebraic system, lattice. Lattice theory was formed around 1935. It is not only a branch of algebra but also plays an important role in modern analytic geometry, semi-ordered space and so on.

7.1　The Concept and Properties of Lattice

Definition 7.1.1　Let $<X, \leqslant>$ be a partially ordered set (for short poset), if $\forall x$, $y \in X$, the set $\{x, y\}$ has a minimum upper bound and a maximum lower bound, then $<X, \leqslant>$ is called a lattice.

Example 7.1.1　Let $S_{12} = \{1, 2, 3, 4, 6, 12\}$ be the set of factors of 12. The above divisible relation $R = \{<x, y> \mid x \in S_{12} \land y \in S_{12} \land x$ divisible $y\}$ is a partial order on S_{12}, then R is the partial order relation on S_{12}, and $<S_{12}, R>$ is poset. Write the covering relation on S_{12}, COV S_{12}, and draw the Hasse diagram, then verify that the poset $<S_{12}, R>$ is a lattice.

Solution　The covering relation on S_{12} COV $S_{12} = \{<1,2>, <1,3>,$
$<2,4>, <2,6>, <3,6>, <4,12>, <6,12>\}$, the Hasse graph is shown in Figure 7.1.1. From this graph, we can see that any two elements of the set S_{12} have a minimum upper bound and a maximum lower bound, so the poset $<S_{12}, R>$ is a lattice.

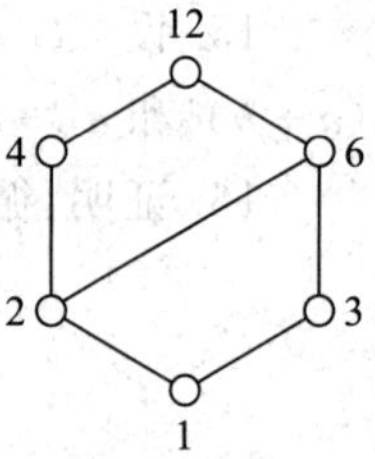

Figure　7.1.1

Example 7.1.2　In Figure 7.1.2, Hasse diagrams of some posets are given to determine whether they constitute lattices.

Solution　None of them are lattices. In (a), $\{1, 2\}$ has no lower bound and therefore no

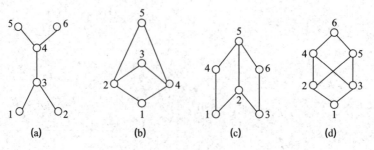

(a)　　　(b)　　　(c)　　　(d)

Figure　7.1.2

maximum lower bound. In (b), $\{2, 4\}$ has two upper bounds, but no minimum upper bound. In (c), $\{1, 3\}$ has no lower bound, and therefore no maximum lower bound. In (d), $\{2, 3\}$ has three upper bounds, but no minimum upper bound.

Let $<X, \leqslant>$ be a lattice, $\forall x, y \in X$, and later use $x \vee y$ represent the minimum upper bound of the set $\{x, y\}$, binary operations $\vee$ called the join operation; Similarly, $x \wedge y$ represents the greatest lower bound of the set $\{x, y\}$, and the binary operation $\wedge$ is called the meet operation.

Definition 7.1.2   Let $<X, \leqslant>$ be a lattice, $\vee$ is a join operation on X, $\wedge$ is an meet operation on X. Then we call $<X, \vee, \wedge>$ the algebraic system, which is derived from the lattice $<X, \leqslant>$.

In Example 7.1.1, according to Figure 7.1.1, the minimum upper bound of the set $\{4, 6\}$ is 12, that is, $4 \vee 6 = 12$ is the least common multiple of 4 and 6; the largest lower bound is 2, it also is the greatest common divisor of $4 \wedge 6 = 2 = 4$ and 6. Furthermore, this result can be generalized to the general case. In the algebraic system of a lattice, $<S_{12}, \vee, \wedge>$, binary operations $\vee$ is the least common multiple; The binary operation $\wedge$ is to find the greatest common divisor.

Now let's introduce the duality principle of the lattice.

Set $<X, \leqslant>$ as poset, and define a binary relation $\geqslant = \{<a, b> | <b, a> \in \leqslant\}$ on X, and it can prove that $<X, \geqslant>$ is also a poset.

Definition 7.1.3   Let f be a proposition that contains the elements and symbols of lattices, namely $=, \leqslant, \geqslant, \vee$ and $\wedge$. To obtain $\leqslant$ of f with $\geqslant$, replace $\geqslant$ with $\leqslant$, replace $\vee$ with $\wedge$, replace $\wedge$ with $\vee$, and get a new proposition, it is called the dual propositions of f, also called $f *$.

For example, in the lattice, f is $a \wedge (b \vee c) \leqslant a$, then $f *$, the dual propositions of f, is $a \vee (b \wedge c) \geqslant a$. Proposition f and its dual proposition $f *$ follow the following rule, which is called the duality principle of lattices.

Let f be a proposition that contains the elements or symbols of lattices, these are $=, \leqslant, \geqslant, \vee$ and $\wedge$. If f is true for all lattices, then $f *$, the dual proposition of f, is also true for all lattices.

Many properties of lattice are dual propositions. With the duality principle of lattices, it is only necessary to prove one of them when proving the properties of lattices.

Theorem 7.1.1   Set $<X, \leqslant>$ as a lattice, $<X, \vee, \wedge>$ is the algebraic system derived from $<X, \leqslant>$. If $\forall a, b, c \in X$, then:

(1) $a \vee b = b \vee a$, $a \wedge b = b \wedge a$ (Commutative law)

(2) $(a \vee b) \vee c = a \vee (b \vee c)$

$(a \wedge b) \wedge c = a \wedge (b \wedge c)$ (Associative law)

(3) $a \vee a = a$, $a \wedge a = a$ (Idempotent law)

(4) $a \vee (a \wedge b) = a$

$a \wedge (a \vee b) = a$ (Absorption law)

Proof

(1) $\forall a, b \in X$, $\{a, b\} = \{b, a\}$, so they have the same minimum upper bound, namely $a \vee b = b \vee a$, by the same token, we can prove $a \wedge b = b \wedge a$.

(2) The largest lower bound of a and b must be the lower bound, namely $a \wedge b \leqslant a$, by the same token, $(a \wedge b) \wedge c \leqslant a \wedge b$, so, $(a \wedge b) \wedge c \leqslant a \wedge b \leqslant a$. By the same token, $(a \wedge b) \wedge c \leqslant a \wedge b \leqslant a$.

From the above 3 formulas, we can get $(a \wedge b) \wedge c \leqslant b \wedge c$ and $(a \wedge b) \wedge c \leqslant a \wedge (b \wedge c)$, by the same token, $a \wedge (b \wedge c) \leqslant (a \wedge b) \wedge c$.

According to the antisymmetry of partial order relation, there is $(a \wedge b) \wedge c = a \wedge (b \wedge c)$, due to the duality principle, we can get $(a \vee b) \vee c = a \vee (b \vee c)$.

(3) Obviously, $a \leqslant a \vee a$, and get $a \leqslant a$ from the reflexivity of $\leqslant$, then inference to $a \vee a \leqslant a$, According to the antisymmetry of partial order relation, there is $a \vee a = a$, due to duality principle, we can get $a \wedge a = a$.

(4) Obviously, $a \leqslant a \vee (a \wedge b)$, and also due to $a \leqslant a$, $a \wedge b \leqslant a$, we can get $a \vee (a \wedge b) \leqslant a$, then we get $a \vee (a \wedge b) = a$.

Due to the duality principle, we get $a \wedge (a \vee b) = a$.

Theorem 7.1.2 Let $<X, \vee, \wedge>$ be an algebraic system, $\vee, \wedge$ are all binary operations. If $\vee$ and $\wedge$ meet absorption law, $\vee$ and $\wedge$ meet idempotent law.

Proof $a \vee a = a \vee (a \wedge (a \vee b)) = a$, similarly to $a \wedge a = a$.

Theorem 7.1.3 Let $<X, \vee, \wedge>$ be an algebraic system, including $\vee, \wedge$ are binary operations, meet the commutative law, associative law and absorption law, can properly defined $X \leqslant$ partial order relation, make $<X, \vee, \wedge>$ form a lattice.

Definition 7.1.4 Let $<X, *, \circ>$ be the algebraic system where $*$ and $\circ$ are dual operations. If $*$ and $\circ$ are enclosed on X and it satisfies the commutative, associative and absorbing laws, we call $<X, *, \circ>$ a lattice.

According to Definition 7.1.4 and Theorem 7.1.1, the algebraic system $<X, \vee, \wedge>$ derived from lattice $<X, \leqslant>$ is a lattice, and the lattices defined by partial order set and the lattices are defined by algebraic system will no longer be distinguished, which are collectively called lattices.

7.2 Distributive Lattice

For any member of lattice $<X, \leqslant>$, a, b, c, must hold a $a \vee (b \wedge c) \leqslant (a \vee b) \wedge (a \vee c)$ and $a \wedge (b \vee c) \leqslant (a \wedge b) \vee (a \wedge c)$. When the equal sign in the two formulas above holds, we get a special kind of lattice.

Definition 7.2.1 Let $<X, \leqslant>$ be the lattice, $<X, \vee, \wedge>$ be the algebraic system derived from $<X, \leqslant>$, if $\forall a, b, c \in X$ has

$$a \vee (b \wedge c) = (a \vee b) \wedge (a \vee c)$$

(union operations to make intersection operations can be allocated)

$$a \wedge (b \vee c) = (a \wedge b) \vee (a \wedge c)$$

(intersection operations of union operations can be allocated)

Then $<X, \vee, \wedge>$ is called the distributive lattice.

Because $a \vee (b \wedge c) = (a \vee b) \wedge (a \vee c)$ and $a \wedge (b \vee c) = (a \wedge b) \vee (a \wedge c)$ are dual propositions, according to the principle of duality, Definition 7.2.1 can also be rewritten as if intersection operations to make union operations can be allocated or union operations to make intersection operations can be allocated, has described the lattice as a distributive lattice.

Example　7.2.1　Set $A = \{a, b, c\}$, $P(A) = \{\varnothing, \{a\}, \{b\}, \{c\}, \{a, b\}, \{a, c\}, \{b, c\}, \{a, b, c\}\}$ as the set of powers of A, $R_{\subseteq} = \{<x, y> | x \in P(A) \wedge y \in P(A) \wedge x \subseteq y\}$ is partial order relation of $P(A)$. $<P(A), R_{\subseteq}>$ is poset, the covering relationship of $P(A)$.

$$\text{COV } P(A) = \{<\varnothing, \{a\}>, <\varnothing, \{b\}>, <\varnothing, \{c\}>, <\{a\}, \{a, b\}>, <\{b\},$$
$$\{a, b\}>, <\{a\}, \{a, c\}>, <\{c\}, \{a, c\}>, <\{b\}, \{b, c\}>,$$
$$<\{c\}, \{b, c\}>, <\{a, b\}, \{a, b, c\}>, <\{a, c\}, \{a, b, c\}>,$$
$$<\{b, c\}, \{a, b, c\}>\}$$

Its hasse graph is shown in Figure 7.2.1. $<P(A), \vee, \wedge>$ is the algebraic system derived from $<P(A), R_{\subseteq}>$, prove that $<P(A), \vee, \wedge>$ is a distributive lattice.

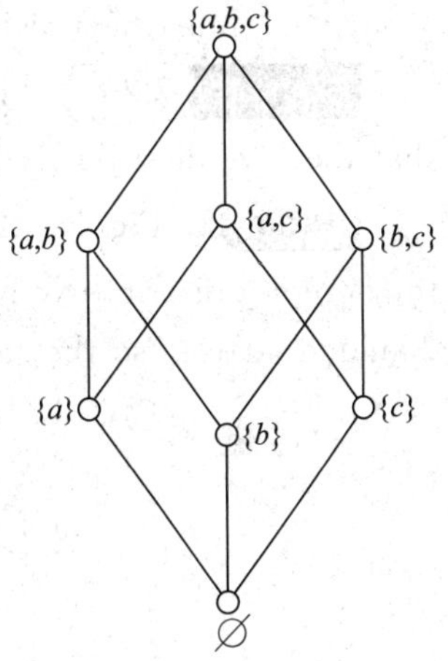

Proof　It has been proved that the algebraic system $<P(A), \vee, \wedge>$ derived from lattice $<P(A), R_{\subseteq}>$ is actually an algebraic system $<P(A), \cup, \cap>$, where $\cup$ is the union of sets and $\cap$ is the intersection operation of sets. And the union and intersection operations of sets satisfy the distributive property:

$$\forall P, Q, R \in P(A)$$
$$P \cup (Q \cap R) = (P \cup Q) \cap (P \cup R)$$
$$P \cap (Q \cup R) = (P \cap Q) \cup (P \cap R)$$

So, $<P(A), \vee, \wedge>$ is distributive lattice.

Example　7.2.2　$A = \{a, b, c, d, e\}$, $<A, \leqslant>$ is a lattice, its hasse graph is shown in Figure 7.2.2, please prove that $<A, \leqslant>$ is not a distributive lattice.

Proof

$$b \vee (c \wedge d) = b \vee e = b$$
$$(b \vee c) \wedge (b \vee d) = a \wedge a = a$$
$$b \vee (c \wedge d) \neq (b \vee c) \wedge (b \vee d)$$

So, $<A, \leqslant>$ is not a distributive lattice.

The lattice in this example is called a diamond lattice, which is not a distributive lattice.

Example　7.2.3　Let $A = \{a, b, c, d, e\}$, $<A, \leqslant>$ be a lattice, its hasse graph is

Figure　7.2.1

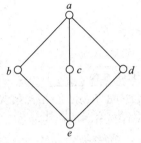

Figure　7.2.2

shown in Figure 7.2.3, please prove that $<A, \leqslant>$ is not a distributive lattice.

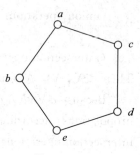

Figure 7.2.3

Proof
$$d \vee (b \wedge c) = d \vee e = d$$
$$(d \vee b) \wedge (d \vee c) = a \wedge c = c$$
$$d \vee (b \wedge c) \neq (d \vee b) \wedge (d \vee c)$$

So, $<A, \leqslant>$ is not a distributive lattice.

In this example, the lattice is referred to as a pentagon lattice, and it is not a distributive lattice. The diamond lattice and the pentagon lattice are two important types of lattices.

Theorem 7.2.1　The sufficient and necessary condition for a lattice to be distributive is that there are no sublattices isomorphic(objects are completely equivalent) to diamond lattice or pentagonal lattice.

The proof of the theorem is beyond the scope of this book, so it is omitted.

Corollary 1　Let $<A, \leqslant>$ be a lattice, if $|A| < 5$, then $<A, \leqslant>$ must be a distributive lattice.

Corollary 2　Let $<A, \leqslant>$ be a lattice, and if $<A, \leqslant>$ is a fully ordered set, then $<A, \leqslant>$ must be a distributive lattice.

Example　7.2.4　Figure 7.2.4 shows the hasse diagram of two lattices. Try to prove that they are distributive lattices.

Proof In Figure 7.2.4 (a), there are sublattices isomorphic to pentagonal lattices, so they are not distributive lattices; In Figure 7.2.4 (b), there are sublattices isomorphic to the diamond lattice, so they are not distributive.

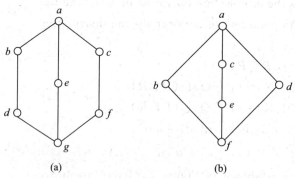

(a)　　　　　　　(b)

Figure　7.2.4

Theorem 7.2.2　Let $<X, \leqslant>$ be lattice and $<X, \vee, \wedge>$ be the algebraic system derived from lattice $<X, \leqslant>$. $<X, \leqslant>$ is the necessary and sufficient condition of distributive lattice is $\forall a, b, c \in X$, when $a \wedge b = a \wedge c$ and $a \vee b = a \vee c$, then $b = c$.

Proof Let $<X, \leqslant>$ be a distributive lattice, $\forall a, b, c \in X$, $a \wedge b = a \wedge c$ and $a \vee b = a \vee c$

$$b = b \vee (b \wedge a) = b \vee (a \wedge b) \text{ (absorption law and commutative law)}$$
$$= b \vee (a \wedge c) = (b \vee a) \wedge (b \vee c)$$

$$=(a \lor c) \land (b \lor c)$$
$$=(a \land b) \lor c \text{ (distributive law)}$$
$$=(a \land c) \lor c$$
$$=c \text{ (commutative law and absorption law)}$$

Set $\forall a, b, c \in X$, when $a \land b = a \land c$ and $a \lor b = a \lor c$, there is $b = c$, but $<X, \lor, \land>$ is not a distributive lattice. It is known from Theorem 7.2.1 that there must be a sublattice isomorphic to a diamond lattice or pentagonal lattice in $<X, \lor, \land>$. Assume that $<X, \lor, \land>$ contains a sublattice isomorphic to the diamond lattice, and this sublattice is $<S, \lor, \land>$, where $S = \{u, v, x, y, z\}$, u is its maximum element, and v is its minimum element. Then $x \lor y = u = x \lor z$, $x \land y = v = x \land z$, but $y \neq z$, contrary to what is known.

The pentagonal case can be similarly proved.

For example, in Figure 7.2.4(a), $b \land c = g = b \land f$ and $b \lor c = a = b \lor f$, but $c \neq f$; in Figure 7.2.4(b), $b \land c = f = b \land e$ and $b \lor c = a = b \lor e$, but $c \neq e$. According to Theorem 7.2.3 they are not distributive lattices.

7.3 Complemented Lattice

Before we introduce the complementary lattice, we first introduce the bounded lattice.

Definition 7.3.1 Set $<X, \leqslant>$ as a lattice, if $\exists a \in X$, $\forall x \in X$ and $a \leqslant x$ ($x \leqslant a$), then call a as the all lower (upper) bounds of lattice $<X, \leqslant>$, denoted by 0(1).

Theorem 7.3.1 Set $<X, \leqslant>$ as lattice, if the lattice $<X, \leqslant>$ has all lower or upper bounds, then they must be unique.

Proof With the reduction to absurdity.

If there are two lower bounds a and b, $a, b \in X$. Because a is lower, so $a \leqslant b$; And because b is lower, so $b \leqslant a$. Again by the antisymmetry $\leqslant$ have $a = b$.

Similarly, the uniqueness of all upper bounds can be proved.

Definition 7.3.2 Let $<X, \leqslant>$ be a lattice, $<X, \lor, \land>$ be the algebraic system derived from $<X, \leqslant>$. If lattice $<X, \lor, \land>$ has all lower bound 0 and all upper bound 1, then $<X, \lor, \land>$ is called bounded lattice and denoted as $<X, \lor, \land, 0, 1>$.

In Example 7.2.1, $<P(A), R_\subseteq>$ is lattice. $<P(A), \lor, \land>$ is the algebraic system derived from $<P(A), R_\subseteq>$. Empty set $\varnothing$ is the lower bound of the lattice, and set A is the upper bound of the lattice. So $<P(A), \lor, \land>$ is bounded, so we can denoted it by $<P(A)\lor, \land, \varnothing, A>$. In Example 7.2.3, it can be seen from the Hasse diagram (Figure 7.2.3) that a is fully upper bound and e is fully lower bound, so $<A, \leqslant>$ is bounded lattice.

Theorem 7.3.2 If $<X, \lor, \land, 0, 1>$ is a bounded lattice, then $\forall a \in X$ the following results hold:

$$a \land 0 = 0, a \lor 0 = a$$
$$a \land 1 = a, a \lor 1 = 1$$

Proof Since 0 is lower and $a \wedge 0 \in X$, it follows that $0 \leqslant a \wedge 0$, and for $a \wedge 0 = 0$, and because of the antisymmetry of $\leqslant$, then $a \wedge 0 = 0$. Obviously, $a \leqslant a$, because 1 is the upper bound, there's $a \leqslant 1$, which launched $a \leqslant a \wedge 1$, again for $a \wedge 1 \leqslant a$, again because of the antisymmetry of $\leqslant$, then $a \wedge 1 = a$.

$a \vee 0 = a$ and $a \vee 1 = 1$ can be similarly proved.

Set $<X, \vee, \wedge, 0, 1>$ as a bounded lattice, $a \in X$ has $a \wedge 0 = 0$, and $0 \wedge a = 0$ because it satisfies the law of exchange. This tells us that 0 is the zero element of the meet operation. In the same way, 0 is the identity element of the join operation, and 1 is the identity element of intersection operation and the zero element of union operation.

Definition 7.3.3 Set $<X, \vee, \wedge, 0, 1>$ as a bounded lattice, If for $a \in X$, $\exists b \in X$, such that $a \vee b = 1$ and $a \wedge b = 0$, then b is a complement element of a.

If b is the complement element of a, it can be seen from Definition 7.3.3 that a is also the complement element of b. So we could say that a and b are complementary, or that a and b complement to each other.

For example, in Example 7.2.3, it can be seen from the Hasse graph (Figure 7.2.3) that b and c are complementary elements, and b and d are also complementary elements, and b has two complementary elements c and d. So the complement of an element in a lattice is not unique.

Example 7.3.1 Figure 7.3.1 is a Hasse graph of a bounded lattice. Please find the complement elements for a, b, c, d and e.

Solution It can be seen from Figure 7.3.1 that the complement element of a is e; b has no complement element; c's complement element is d; the complement element of d are c and e; the complement element of e are a and d, 0 and 1 complement each other.

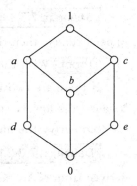

Figure 7.3.1

Obviously, in bounded lattices, the only complement of all upper bound 1 is all lower bound 0, and the only complement of all lower bound 0 is all upper bound 1. Other than 1 and 0, some elements have a complement, and some have no complement. If a complement exists for an element, there may be one complement or more. But in a bounded distributive lattice, if the complement of an element exists, it must be unique.

Theorem 7.3.3 Let $<X, \vee, \wedge, 0, 1>$ be a bounded distributive lattice. If there is a complement b for $a \in X$, then b is the unique complement of a.

Proof Because b is a complement to a, so we have $a \vee b = 1$, $a \wedge b = 0$. Let $c \in X$, c be another complement of a, which is also $a \vee c = 1$, $a \wedge c = 0$. So we have $a \vee b = a \vee c$, $a \wedge b = a \wedge c$.

Since $<X, \vee, \wedge, 0, 1>$ is a distributive lattice, according to Theorem 7.2.3, there must be $b = c$, namely the complement of a is unique.

Definition 7.3.4　Let $<X,\vee,\wedge,0,1>$ be a bounded lattice, and if $\forall a \in X$ has a complement in X, then $<X,\vee,\wedge,0,1>$ is a complementary lattice.

For example, Figure 7.3.2 is a Hasse graph of three bounded lattices. Since each element has at least one complement, they are both complement lattices.

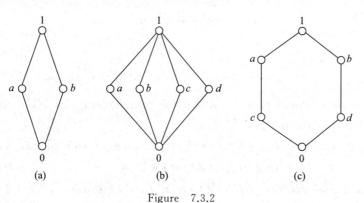

Figure　7.3.2

7.4　Boolean Algebra

Boolean algebra is the foundation of computers. Without it, there would be no computer. In this section, we're going to introduce Boolean algebra.

Definition 7.4.1　Complemented distributive lattice is called a Boolean lattice.

Every element in a Boolean lattice has a complement element. Because a complemented lattice must be a bounded distributive lattice, so a Boolean lattice must be a bounded distributive lattice. According to Theorem 7.3.3, the complement element of each element in a Boolean lattice exists and is unique. So we can think of the complement as a unary operation and write the complement of a as a'.

Definition 7.4.2　Let $<X,\leqslant>$ be a Boolean lattice and $<X,\vee,\wedge,'>$ be the algebraic system derived from lattice $<X,\leqslant>$. Algebraic system $<X,\vee,\wedge,'>$ is called Boolean algebra.

In Section 7.2 it is proved that $<P(A),R_{\subseteq}>$ is a distributive lattice, where $A=\{a,b,c\}$. $<P(A),\cup,\cap>$ is an algebraic system derived from $<P(A),R_{\subseteq}>$, where $\cup$ and $\cap$ are set union operation and intersection operation. It is further explained that $<P(A),\cup,\cap>$ is distributive lattice and bounded lattice, $\varnothing$ is full lower bound and A is full upper bound. So $<P(A),\cup,\cap>$ is a bounded distributive lattice. Let A be the complete set, $\forall T \in P(A)$, the complement of T, $\sim T=A-T \in P(A)$, satisfy $T \cup \sim T=A$ and $T \cap \sim T=\varnothing$, so the complement of $T'=\sim T$. By Definition 7.4.2, $<P(A),\cup,\cap,\sim>$ is a Boolean algebra.

It can be proved that when A is any set, $<P(A),\cup,\cap,\sim>$ is also a Boolean algebra. It's called set algebra. Set algebra is a concrete model of Boolean algebra.

Boolean algebra must be a lattice. According to Theorem 7.1.1, the two binary operations in Boolean algebra satisfy the commutative law, associative law, idempotent law and absorption law. In addition, Boolean algebra has the following properties.

Theorem 7.4.1 Let $<X, \vee, \wedge, '>$ be Boolean algebra, $\forall a, b \in X$, there must be:

(1) $(a')'=a$

(2) $(a \vee b)'=a' \wedge b'$

(3) $(a \wedge b)'=a' \vee b'$

Proof

(1) a' is the complement of a, a is also the complement of a'. The uniqueness of the complement in Boolean algebra is $(a')'=a$.

(2) $(a \vee b) \vee (a' \wedge b')=((a \vee b) \vee a') \wedge ((a \vee b) \vee b')=(b \vee (a \vee a')) \wedge (a \vee (b \vee b'))$
$$=(b \vee 1) \wedge (a \vee 1)=1 \wedge 1=1$$

$(a \vee b) \wedge (a' \wedge b')=(a \wedge (a' \wedge b')) \vee (b \wedge (a' \wedge b'))=((a \wedge a') \wedge b') \vee ((b \wedge b') \wedge a')$
$$=(0 \wedge b') \vee (0 \wedge a')=0 \vee 0=0$$

So, $(a \vee b)'=a' \wedge b'$.

(3) Similarly, $(a \wedge b)'=a' \vee b'$.

The (1) in Theorem 7.4.1 is called the law of double negation, the (2) and (3) are called the Law of DE Morgan. Boolean algebra satisfies the law of double negation and the Law of DE Morgan.

Theorem 7.4.2 Let $<X, *, \circ, '>$ be the algebraic system, where $*$ and $\circ$ are both binary operations, and $'$ are unary operations. $<X, *, \circ, '>$ is a sufficient and necessary condition for Boolean algebra:

(1) The $*$ and $\circ$ are enclosed on X and satisfy the commutative law.

(2) The $*$ and $\circ$ satisfying the distribution law.

(3) In the set X, there exist identity elements for two operation, denoted as $*$ and $\circ$. Let the identity element of operation $*$ be 0 and the identity element of the $\circ$ be 1. Namely, $\forall a \in X$, having $a * 0=a$, $a \circ 1=a$.

(4) $\forall a \in X$, $\exists a' \in X$, making $a * a'=1$, $a \circ a'=0$.

The four conditions given in Theorem 7.4.2 can be used as equivalent definitions of Boolean algebra.

Boolean algebra $<X, *, \circ, '>$ can also be expressed as $<X, *, \circ, ', 0, 1>$, where 0 is the identity element of operation $*$ and 1 is the identity element of operation $\circ$.

Definition 7.4.3 Set $<X, \vee, \wedge, ', 0, 1>$ as Boolean algebra, B is a nonempty subset of X, if $0, 1 \in B$ and $<B, \vee, \wedge, ', 0, 1>$ is also Boolean algebra, then call $<B, \vee, \wedge, ', 0, 1>$ is a subboolean algebra of $<X, \vee, \wedge, ', 0, 1>$.

Theorem 7.4.3 Set $<X, \vee, \wedge, ', 0, 1>$ as Boolean algebra, B is a nonempty subset of X, if $0, 1 \in B$ and operation $\vee, \wedge, '$ are closed on B, then call $<B, \vee, \wedge, ', 0, 1>$ is a subboolean algebra of $<X, \vee, \wedge, ', 0, 1>$.

Proof (1) $\forall a$, $b \in B$, because $B \subseteq X$, a, $b \in X$. And $<X$, $\vee$, $\wedge$, $'$, 0, $1>$ is also a Boolean algebra, so $a \vee b = b \vee a$, $a \wedge b = b \wedge a$.

(2) Similar to (1) can prove $*$ and $\circ$ satisfy the distributive law.

(3) We know 0, $1 \in B$, $\forall a \in B \subseteq X$, $a \in X$, so we know $a \vee 0 = a$ 和 $a \wedge 1 = a$.

(4) $\forall a \in B$, by the analysis on the B closed $a' \in B$, makes a $\vee a' = 1$, $a \wedge a' = 0$.

According to Theorem 7.4.2, $<B$, $\vee$, $\wedge$, $'$, 0, $1>$ is a Boolean algebra, and it is a subboolean algebra of $<X$, $\vee$, $\wedge$, $'$, 0, $1>$.

For convenience, the following will $x \leqslant y$ and $x \neq y$ indicate $x < y$.

Definition 7.4.4   Set $<X_1$, $\vee_1$, $\wedge_1$, $'$, 0, $1>$ and $<X_2$, $\vee_2$, $\wedge_2$, $''$, θ, $E>$ as two Boolean algebra, including $\vee_1$, $\wedge_1$, $\vee_2$ and $\wedge_2$ are binary operation, $'$ and $''$ is the unary, 0 and 1 is upper bound and lower bound of the X_1. f is a mapping from X_1 to X_2, for any a, $b \in X_1$, having:

$$f(a \vee_1 b) = f(a) \vee_2 f(b)$$
$$f(a \wedge_1 b) = f(a) \wedge_2 f(b)$$
$$(f(a))' = (f(a))''$$

f is a homomorphism of Boolean algebra $<X_1$, $\vee_1$, $\wedge_1$, $'$, 0, $1>$ through $<X_2$, $\vee_2$, $\wedge_2$, $''$, θ, $E>$, or simply Boolean algebra homomorphism. If f is injective, surjective, and bijective, it is called Boolean algebra homomorphism, Boolean algebra full homomorphism, and Boolean algebra isomorphism, respectively. Call $<f(X_1)$, $\vee_2$, $\wedge_2$, $''$, θ, $E>$ a Boolean algebraic homomorphism of $<X_1$, $\vee_1$, $\wedge_1$, $'$, 0, $1>$.

7.5   Boolean Expression

Boolean algebra can be used in the design of logic circuits. A combined circuit with several inputs and some logical function can be represented by a circuit function defined in circuit algebra, while a circuit function can be represented by a Boolean expression.

Definition 7.5.1   Set $<S$, $\oplus$, $\odot$, $'$, 0, $1>$ as a Boolean algebra, the element in S is called a Boolean constant; Arguments in S are called Boolean variables.

Definition 7.5.2   Set $<S$, $\oplus$, $\odot$, $'$, 0, $1>$ as a Boolean algebra, x_1, x_2, ..., x_n as Boolean constant, then the Boolean expression generated by n Boolean arguments can be recursively defined as follows.

(1) Any element and argument in S is a Boolean expression.

(2) If F and G are Boolean, then F', $F \oplus G$ and $F \odot G$ are also Boolean.

(3) A symbol string constructed only with a finite number of uses of (1) or (2) is a Boolean.

(4) For simplicity, stipulate the operation priority of $\oplus$ to be lower than $\odot$.

Example 7.5.1   In Boolean algebra $<\{0, 1, \alpha, \beta\}$, $\oplus$, $\odot$, $'$, 0, $1>$, the Boolean is $0 \odot 1'$, $1 \oplus (\alpha \odot x_1) \oplus (x_2' \odot x_3)$, $(\beta' \oplus x_1 \oplus x_3) \odot 1$.

Any n-element Boolean can be defined as a function from S_n to S.

Example 7.5.2 In Boolean algebra $<\{0, 1, \alpha, \beta\}, \oplus, \odot, ', 0, 1>$, $f(x, y) = (\beta \odot x' \odot y) \oplus (\alpha \odot x \odot (x \oplus y'))$ is a binary Boolean function.

The two Booleans are equal.

Boolean expression $f(x_1, x_2, \ldots, x_n)$ is to take the element in S as $x_i (i = 1, 2, \ldots, n)$ is substituted into the expression;

Two Boolean expressions are said to be equal or equivalent if the values of both Boolean expressions are equal for any assignment of n Boolean arguments (i.e., taking an element in S for each argument).

Exercises

1. **N** is the set of natural Numbers, $\leqslant$ is the relation of less than or equal to, then $<\mathbf{N}, \leqslant>$ is ( ).

    A. bounded lattice   B. complemented lattice

    C. distributive lattice   D. complemented distributive lattice

2. In a bounded lattice, if only one element has a complement, then the complement( ).

    A. must be unique   B. not unique

    C. not necessarily unique   D. may be the only

3. The following is the Hasse diagram corresponding to 4 lattices, ( ) is a distributive lattice.

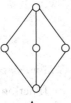 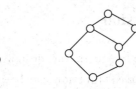 

    A.   B.   C.   D.

4. A lattice with only finite elements becomes a finite lattice, and the finite lattice must be ( ).

    A. distributive lattice   B. complemented lattice

    C. Boolean lattice   D. bounded lattice

5. Suppose $<L, \leqslant>$ is a chain, among them $|L| \geqslant 3$, then $<L, \leqslant>$ ( ).

    A. is not a lattice   B. is a complemented lattice

    C. is a distributive lattice   D. is a Boolean lattice

6. Suppose A is a set, $<P(A), \subseteq>$ is a complemented lattice, and the complement of each element in $P(A)$ ( ).

    A. exists and only   B. does not exist

    C. exists but not unique   D. may exist

7. The number of elements in finite Boolean algebra must be equal to ( ).

    A. $2n$ B. n^2 C. 2^n D. $4n$

8. In Boolean $<A, \leqslant>$, there are 3 atoms a_1, a_2, a_3, then $\overline{a_1} = ($ $)$.

    A. $a_2 \wedge a_3$ B. $a_2 \vee a_3$ C. $\overline{a_2} \wedge \overline{a_3}$ D. $\overline{a_2} \vee \overline{a_3}$

9. In Boolean $<A, \leqslant>$, $A = \{X \mid X$ is an integer multiple of 5 and a positive factor of 210}, and | is a divisible relationship. Then the complement of 30 is (   ).

    A. 15   B. 30   C. 35   D. 70

10. Partially ordered set $<L, \leqslant>$ formed by the following set L, where $\leqslant$ is defined as For $n_1, n_2 \in L$, $n_1 \leqslant n_2$ if and only if n_1 is a factor of n_2. Ask which of the partially ordered sets are lattices (explain the reason).

    (1) $L = \{1, 2, 3, 4, 5, 6, 12\}$

    (2) $L = \{1, 2, 3, 4, 6, 8, 12, 14\}$

    (3) $L = \{1, 2, 3, 4, 5, 6, 7, 8, 9, 10, 11, 12\}$

11. S_n is a set consisting of all factors of a positive integer n, and $m \mid n$ means m divides n. For lattice $<D_{30}, \mid>$

    (1) Prove that $<D_{30}, \mid>$ is a Boolean.

    (2) Make the Hasse diagram of the corresponding partially ordered set.

    (3) Find all the atoms of D_{30}.

12. Let $<B, \vee, \wedge, ^->$ be a Boolean algebra, if the two binary operations $+$ and o on B are defined as:

$$a + b = (a \wedge \overline{b}) \vee (\overline{a} \wedge b)$$

$$a \circ b = a \wedge b$$

Prove that $<B, +, \circ>$ is a ring with 1 as the identity element.

13. $<B, \vee, \wedge, ^->$ is a Boolean algebra, $\forall a, b \in B$, proof: $a = b$ if and only if $(a \wedge \overline{b}) \vee (\overline{a} \wedge b) = 0$.

第 7 章　格与布尔代数

这一章介绍的是另一类代数系统——格。格论是在 1935 年左右形成的,它不仅是代数学的一个分支,而且在近代解析几何、半序空间等方面也都有重要的作用。

7.1　格的概念与性质

定义 7.1.1 ▶ 设 $<X,\leqslant>$ 是偏序集,如果 $\forall x,y\in X$,集合 $\{x,y\}$ 都有最小上界和最大下界,则称 $<X,\leqslant>$ 是格。

例 7.1.1　设 $S_{12}=\{1,2,3,4,6,12\}$ 是 12 的因子构成的集合。其上的整除关系 $R=\{<x,y>|x\in S_{12}\wedge y\in S_{12}\wedge x$ 整除 $y\}$,R 是 S_{12} 上的偏序关系,$<S_{12},R>$ 是偏序集。写出 S_{12} 上的盖住关系 COV S_{12},画出哈斯图,验证偏序集 $<S_{12},R>$ 是格。

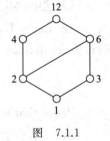

图　7.1.1

解　S_{12} 上的盖住关系 COV $S_{12}=\{<1,2>,\ <1,3>,\ <2,4>,\ <2,6>,$ $<3,6>,\ <4,12>,\ <6,12>\}$,其哈斯图如图 7.1.1 所示。从哈斯图中可看出,集合 S_{12} 的任意两个元素都有最小上界和最大下界,故偏序集 $<S_{12},R>$ 是格。

例 7.1.2　图 7.1.2 中给出了一些偏序集的哈斯图,判断它们是否构成格。

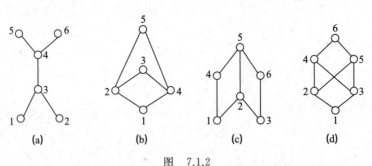

图　7.1.2

解　它们都不是格。在(a)中,$\{1,2\}$ 没有下界,因而没有最大下界。在(b)中,$\{2,4\}$ 虽有两个上界,但没有最小上界。在(c)中,$\{1,3\}$ 没有下界,因而没有最大下界。在(d)中,$\{2,3\}$ 虽有三个上界,但没有最小上界。

设 $<X,\leqslant>$ 是格,$\forall x,y\in X$,用 $x\vee y$ 表示集合 $\{x,y\}$ 的最小上界,二元运算 $\vee$ 称为并运算;用 $x\wedge y$ 表示集合 $\{x,y\}$ 的最大下界,二元运算 $\wedge$ 称为交运算。

定义 7.1.2 ▶ 设 $<X,\leqslant>$ 是格,$\vee$ 是 X 上的并运算,$\wedge$ 是 X 上的交运算,则称

$<X,\vee,\wedge>$ 是格 $<X,\leqslant>$ 导出的代数系统。

在例 7.1.1 中，根据图 7.1.1，集合 $\{4,6\}$ 的最小上界为 12，即 $4\vee6=12$ 为 4 和 6 的最小公倍数；它的最大下界为 2，即 $4\wedge6=2$ 为 4 和 6 的最大公约数。同样，这个结果也可以推广到一般情况。即在格 $<S_{12},R>$ 导出的代数系统 $<S_{12},\vee,\wedge>$ 中，二元运算 $\vee$ 是求最小公倍数；二元运算 $\wedge$ 是求最大公约数。

下面介绍格的对偶原理。

设 $<X,\leqslant>$ 是偏序集，在 X 上定义二元关系 $\geqslant=\{<a,b>|<b,a>\in\leqslant\}$，可以证明 $<X,\geqslant>$ 也是偏序集。

定义 7.1.3 ▶ 设 f 是含有格中元素及符号 $=,\leqslant,\geqslant,\vee$ 和 $\wedge$ 的命题。将 f 中的 $\leqslant$ 替换成 $\geqslant$，$\geqslant$ 替换成 $\leqslant$，$\vee$ 替换成 $\wedge$，$\wedge$ 替换成 $\vee$，得到一个新命题，叫作 f 的对偶命题，记为 $f*$。

例如，在格中，f 为 $a\wedge(b\vee c)\leqslant a$，则 f 的对偶命题 $f*$ 为 $a\vee(b\wedge c)\geqslant a$。命题 f 和它的对偶命题 $f*$ 遵循格的对偶原理。

设 f 是含有格中元素及符号 $=,\leqslant,\geqslant,\vee$ 和 $\wedge$ 的命题。若 f 对一切格为真，则 f 的对偶命题 $f*$ 也对一切格为真。

格的许多性质都是互为对偶命题的。有了格的对偶原理，在证明格的性质时，只需证明其中的一个就可以了。

定理 7.1.1 ▶ 设 $<X,\leqslant>$ 是格，$<X,\vee,\wedge>$ 是格 $<X,\leqslant>$ 导出的代数系统，则 $\forall a,b,c\in X$ 有：

(1) $a\vee b=b\vee a,a\wedge b=b\wedge a$　（交换律）

(2) $(a\vee b)\vee c=a\vee(b\vee c)$

　　$(a\wedge b)\wedge c=a\wedge(b\wedge c)$　（结合律）

(3) $a\vee a=a,a\wedge a=a$　（幂等律）

(4) $a\vee(a\wedge b)=a$

　　$a\wedge(a\vee b)=a$　（吸收律）

证明

(1) $\forall a,b\in X$，$\{a,b\}=\{b,a\}$，所以它们的最小上界相等，即 $a\vee b=b\vee a$。同理可证 $a\wedge b=b\wedge a$。

(2) a 和 b 的最大下界一定是 a、b 的下界，即 $a\wedge b\leqslant a$。同理，$(a\wedge b)\wedge c\leqslant a\wedge b$，所以，$(a\wedge b)\wedge c\leqslant a\wedge b\leqslant a$。同理，有 $(a\wedge b)\wedge c\leqslant a\wedge b\leqslant b$ 和 $(a\wedge b)\wedge c\leqslant c$。

由以上 3 式得 $(a\wedge b)\wedge c\leqslant b\wedge c$ 和 $(a\wedge b)\wedge c\leqslant a\wedge(b\wedge c)$，类似可证 $a\wedge(b\wedge c)\leqslant(a\wedge b)\wedge c$。

根据偏序关系的反对称性有 $(a\wedge b)\wedge c=a\wedge(b\wedge c)$，由对偶原理得 $(a\vee b)\vee c=a\vee(b\vee c)$。

(3) 显然 $a\leqslant a\vee a$，又由 $\leqslant$ 的自反性得 $a\leqslant a$，从而推出 $a\vee a\leqslant a$，根据偏序关系的反对称性有 $a\vee a=a$，由对偶原理得 $a\wedge a=a$。

(4) 显然，$a\leqslant a\vee(a\wedge b)$。又由 $a\leqslant a,a\wedge b\leqslant a$ 得 $a\vee(a\wedge b)\leqslant a$，从而得 $a\vee(a\wedge b)=a$。

由对偶原理得 $a\wedge(a\vee b)=a$。

定理 7.1.2 ▶ 设 $<X,\vee,\wedge>$ 是代数系统，其中 $\vee$、$\wedge$ 都是二元运算。如果 $\vee$ 和 $\wedge$ 满足吸收律，则 $\vee$ 和 $\wedge$ 满足幂等律。

证明 $a \lor a = a \lor (a \land (a \lor b)) = a$，同理可证 $a \land a = a$。

定理 7.1.3 ▶ 设 $<X, \lor, \land>$ 是代数系统，其中 $\lor$，$\land$ 都是二元运算，满足交换律、结合律和吸收律，则可适当定义 X 的偏序关系 $\leqslant$，使 $<X, \leqslant>$ 构成一个格。

定义 7.1.4 ▶ 设 $<X, *, \circ>$ 是代数系统，其中 $*$ 和 $\circ$ 都是二元运算，如果 $*$ 和 $\circ$ 在 X 上封闭且满足交换律、结合律和吸收律，则称 $<X, *, \circ>$ 为格。

根据定义 7.1.4 和定理 7.1.1，格 $<X, \leqslant>$ 导出的代数系统 $<X, \lor, \land>$ 是格，以后不再区分偏序集定义的格和代数系统定义的格，统称为格。

7.2 分配格

对格 $<X, \leqslant>$ 中的任意元素 a, b, c，必有 $a \lor (b \land c) \leqslant (a \lor b) \land (a \lor c)$ 和 $a \land (b \lor c) \leqslant (a \land b) \lor (a \land c)$ 成立。当上述两个式子中的等号成立时，我们就得到一类特殊的格。

定义 7.2.1 ▶ 设 $<X, \leqslant>$ 是格，$<X, \lor, \land>$ 是 $<X, \leqslant>$ 导出的代数系统，如果 $\forall a, b, c \in X$ 有

$$a \lor (b \land c) = (a \lor b) \land (a \lor c) \quad (\text{并运算对交运算可分配})$$
$$a \land (b \lor c) = (a \land b) \lor (a \land c) \quad (\text{交运算对并运算可分配})$$

则称 $<X, \lor, \land>$ 为分配格。

因为 $a \lor (b \land c) = (a \lor b) \land (a \lor c)$ 和 $a \land (b \lor c) = (a \land b) \lor (a \land c)$ 互为对偶命题，根据对偶原理，定义 7.2.1 还可以改写：一个格如果交运算对并运算可分配或并运算对交运算可分配，则称该格为分配格。

例 7.2.1 设 $A = \{a, b, c\}$，$P(A) = \{\varnothing, \{a\}, \{b\}, \{c\}, \{a,b\}, \{a,c\}, \{b,c\}, \{a,b,c\}\}$ 是 A 的幂集合，$P(A)$ 上的包含关系 $R_{\subseteq} = \{<x,y> | x \in P(A) \land y \in P(A) \land x \subseteq y\}$ 是 $P(A)$ 上的偏序关系。$<P(A), R_{\subseteq}>$ 是偏序集，$P(A)$ 上的盖住关系

$$\begin{aligned} \text{COV } P(A) = \{&<\varnothing, \{a\}>, <\varnothing, \{b\}>, <\varnothing, \{c\}>, <\{a\}, \{a,b\}>, <\{b\}, \{a,b\}>, \\ &<\{a\}, \{a,c\}>, <\{c\}, \{a,c\}>, <\{b\}, \{b,c\}>, <\{c\}, \{b,c\}>, \\ &<\{a,b\}, \{a,b,c\}>, <\{a,c\}, \{a,b,c\}>, <\{b,c\}, \{a,b,c\}>\} \end{aligned}$$

其哈斯图如图 7.2.1 所示。$<P(A), \lor, \land>$ 是 $<P(A), R_{\subseteq}>$ 导出的代数系统，证明 $<P(A), \lor, \land>$ 是分配格。

证明 前面已经证明了格 $<P(A), R_{\subseteq}>$ 导出的代数系统 $<P(A), \lor, \land>$ 实际就是代数系统 $<P(A), \cup, \cap>$，其中 $\cup$ 是集合的并运算，$\cap$ 是集合的交运算。而集合的并、交运算满足分配律：

$$\forall P, Q, R \in P(A)$$
$$P \cup (Q \cap R) = (P \cup Q) \cap (P \cup R)$$
$$P \cap (Q \cup R) = (P \cap Q) \cup (P \cap R)$$

所以，$<P(A), \lor, \land>$ 是分配格。

例 7.2.2 $A = \{a, b, c, d, e\}$，$<A, \leqslant>$ 是格，其哈斯图如图 7.2.2 所示，证明 $<A, \leqslant>$ 不是分配格。

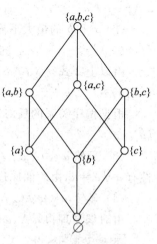

图 7.2.1

证明

$$b \lor (c \land d) = b \lor e = b$$
$$(b \lor c) \land (b \lor d) = a \land a = a$$
$$b \lor (c \land d) \neq (b \lor c) \land (b \lor d)$$

所以，$<A, \leqslant>$不是分配格。

本例中的格叫作钻石格，钻石格不是分配格。

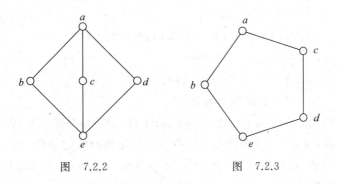

图　7.2.2　　　　　　　图　7.2.3

例 7.2.3　设 $A = \{a, b, c, d, e\}$，$<A, \leqslant>$是格，其哈斯图如图 7.2.3 所示，证明$<A,$ $\leqslant>$不是分配格。

证明

$$d \lor (b \land c) = d \lor e = d$$
$$(d \lor b) \land (d \lor c) = a \land c = c$$
$$d \lor (b \land c) \neq (d \lor b) \land (d \lor c)$$

所以，$<A, \leqslant>$不是分配格。

本例中的格叫作五角格，五角格也不是分配格。钻石格和五角格是两种很重要的格。

定理 7.2.1 ▶ 一个格是分配格的充分必要条件是该格中不含有与钻石格或五角格同构（即对象是完全等价）的子格。

这个定理的证明已经超过了本书的范围，故略去。

推论 1　设$<A, \leqslant>$是格，如果$|A| < 5$，则$<A, \leqslant>$一定是分配格。

推论 2　设$<A, \leqslant>$是格，如果$<A, \leqslant>$是全序集，则$<A, \leqslant>$一定是分配格。

例 7.2.4　图 7.2.4 给出了两个格的哈斯图。试证明它们都不是分配格。

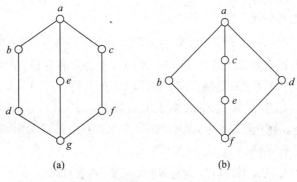

(a)　　　　　　　(b)

图　7.2.4

证明　图 7.2.4（a）中含有与五角格同构的子格，所以不是分配格；图 7.2.4（b）中含有与钻石格同构的子格,所以不是分配格。

定理 7.2.2 ► 设$<X,\leqslant>$是格,$<X,\vee,\wedge>$是格$<X,\leqslant>$导出的代数系统。$<X,\leqslant>$是分配格的充分必要条件是$\forall a,b,c\in X$,当$a\wedge b=a\wedge c$且$a\vee b=a\vee c$时,必有$b=c$。

证明 设$<X,\leqslant>$是分配格,$\forall a,b,c\in X$,$a\wedge b=a\wedge c$且$a\vee b=a\vee c$

$$b=b\vee(b\wedge a)=b\vee(a\wedge b) \quad (吸收律和交换律)$$
$$=b\vee(a\wedge c)=(b\vee a)\wedge(b\vee c) \quad (已知代入和分配律)$$
$$=(a\vee c)\wedge(b\vee c) \quad (交换律和已知代入)$$
$$=(a\wedge b)\vee c \quad (分配律)$$
$$=(a\wedge c)\vee c \quad (已知条件代入)$$
$$=c \quad (交换律和吸收律)$$

设$\forall a,b,c\in X$,当$a\wedge b=a\wedge c$且$a\vee b=a\vee c$时,有$b=c$,但$<X,\vee,\wedge>$不是分配格。由定理7.2.1知,$<X,\vee,\wedge>$中必含有与钻石格或五角格同构的子格。假设$<X,\vee,\wedge>$含有与钻石格同构的子格,且此子格为$<S,\vee,\wedge>$,其中$S-\{u,v,x,y,z\}$,u为它的最大元,v为它的最小元。从而$x\vee y=u=x\vee z$,$x\wedge y=v=x\wedge z$,但$y\neq z$,与已知矛盾。

对五角格的情况,可类似证明。

例如,在图7.2.4(a)中,$b\wedge c=g=b\wedge f$且$b\vee c=a=b\vee f$,但$c\neq f$;在图7.2.4(b)中,$b\wedge c=f=b\wedge e$且$b\vee c=a=b\vee e$,但$c\neq e$。根据定理7.2.3,它们不是分配格。

7.3 有补格

在介绍有补格之前,先要介绍有界格。

定义 7.3.1 ► 设$<X,\leqslant>$是格,如果$\exists a\in X$,$\forall x\in X$都有$a\leqslant x$ $(x\leqslant a)$,则称a为格$<X,\leqslant>$的全下(上)界,记为0(1)。

定理 7.3.1 ► 设$<X,\leqslant>$是格,若格$<X,\leqslant>$有全下界或全上界,则它们一定是唯一的。

证明 用反证法。

如果有两个全下界a和b,$a,b\in X$。因为a是全下界,所以$a\leqslant b$;又因为b是全下界,所以$b\leqslant a$。再由$\leqslant$的反对称性有$a=b$。

类似地可证明全上界的唯一性。

定义 7.3.2 ► 设$<X,\leqslant>$是格,$<X,\vee,\wedge>$是格$<X,\leqslant>$导出的代数系统。若格$<X,\vee,\wedge>$存在全下界0和全上界1,则称$<X,\vee,\wedge>$为有界格,记为$<X,\vee,\wedge,0,1>$。

在例7.2.1中,$<P(A),R_\subseteq>$是格。$<P(A),\vee,\wedge>$是$<P(A),R_\subseteq>$导出的代数系统。空集$\varnothing$是格的全下界,而集合A是格的全上界。因此$<P(A),\vee,\wedge>$是有界格,可记为$<P(A)\vee,\wedge,\varnothing,A>$。在例7.2.3中,从哈斯图(图7.2.3)中可以看出,a是全上界,而e是全下界,所以$<A,\leqslant>$是有界格。

定理 7.3.2 ► 设$<X,\vee,\wedge,0,1>$为有界格,则$\forall a\in X$有

$$a\wedge 0=0, \ a\vee 0=a$$
$$a\wedge 1=a, \ a\vee 1=1$$

证明　因为 0 是全下界且 $a \wedge 0 \in X$，所以 $0 \leqslant a \wedge 0$。又因为 $a \wedge 0 \leqslant 0$，由 $\leqslant$ 的反对称性有 $a \wedge 0 = 0$。显然 $a \leqslant a$，由于 1 是全上界，所以有 $a \leqslant 1$，从而推出 $a \leqslant a \wedge 1$。又因为 $a \wedge 1 \leqslant a$，再由 $\leqslant$ 的反对称性有 $a \wedge 1 = a$。

$a \vee 0 = a$ 和 $a \vee 1 = 1$ 可以类似地证明。

设 $<X, \vee, \wedge, 0, 1>$ 为有界格，$a \in X$ 有 $a \wedge 0 = 0$，因为格满足交换律，所以 $0 \wedge a = 0$，这说明 0 是交运算的零元；同样的道理，0 是并运算的幺元，而 1 是交运算的幺元和并运算的零元。

定义 7.3.3 ▶ 设 $<X, \vee, \wedge, 0, 1>$ 为有界格，如果对于 $a \in X$，$\exists b \in X$，使 $a \vee b = 1$ 且 $a \wedge b = 0$，则称 b 是 a 的补元。

如果 b 是 a 的补元，从定义 7.3.3 可以看出，a 也是 b 的补元。因此，可以说 a 和 b 是互补的，或者说 a 和 b 互为补元。

例如，在例 7.2.3 中，从哈斯图(图 7.2.3)中可以看出，b 和 c 互为补元，b 和 d 也互为补元，b 有两个补元 c 和 d。所以格中元素的补元并不唯一。

例 7.3.1　图 7.3.1 是一个有界格的哈斯图。找出 a, b, c, d, e 的补元。

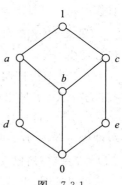

图　7.3.1

解　从图 7.3.1 中可以看出，a 的补元是 e；b 没有补元；c 的补元是 d；d 的补元是 c 和 e，e 的补元是 a 和 d，0 和 1 互为补元。

显然，在有界格中，全上界 1 的唯一补元是全下界 0，而全下界 0 的唯一补元是全上界 1。除 1 和 0 外，其他元素有的有补元，有的没有补元。如果某个元素的补元存在，补元可能有一个，也可能有多个。但在有界分配格中，如果元素的补元存在，则一定唯一。

定理 7.3.3 ▶ 设 $<X, \vee, \wedge, 0, 1>$ 为有界分配格，如果对于 $a \in X$，a 存在补元 b，则 b 是 a 的唯一补元。

证明　因为 b 是 a 的补元，所以有 $a \vee b = 1$，$a \wedge b = 0$。设 $c \in X$，c 是 a 的另一个补元，同样也有 $a \vee c = 1$，$a \wedge c = 0$，从而有 $a \vee b = a \vee c$，$a \wedge b = a \wedge c$。

由于 $<X, \vee, \wedge, 0, 1>$ 为分配格，根据定理 7.2.3，必有 $b = c$，即 a 的补元唯一。

定义 7.3.4 ▶ 设 $<X, \vee, \wedge, 0, 1>$ 为有界格，如果 $\forall a \in X$，在 X 中都有 a 的补元存在，则称 $<X, \vee, \wedge, 0, 1>$ 为有补格。

例如，图 7.3.2 是三个有界格的哈斯图，由于每一个元素至少有一个补元，所以它们都是有补格。

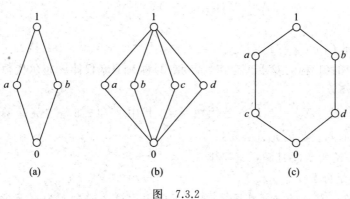

图　7.3.2

7.4 布尔代数

布尔代数是计算机的基础。没有它,就不会有计算机。这一节我们来介绍布尔代数的相关知识。

定义 7.4.1 ▶ 有补分配格称为布尔格。

在布尔格中每一个元素都有补元。因为有补格一定是有界格,所以布尔格一定是有界分配格。根据定理 7.3.3,布尔格中的每一个元素的补元存在且唯一。于是可以将求补元的运算看作一元运算且把 a 的补元记为 a'。

定义 7.4.2 ▶ 设 $<X,\leqslant>$ 是布尔格,$<X,\vee,\wedge,'>$ 是格 $<X,\leqslant>$ 导出的代数系统。称代数系统 $<X,\vee,\wedge,'>$ 为布尔代数。

在 7.2 节中证明了 $<P(A),R_\subseteq>$ 是分配格,其中 $A=\{a,b,c\}$。$<P(A),\cup,\cap>$ 是 $<P(A),R_\subseteq>$ 导出的代数系统,其中 $\cup$ 和 $\cap$ 是集合并运算和交运算。之后又进一步说明了 $<P(A),\cup,\cap>$ 是分配格和有界格,$\varnothing$ 是全下界,A 是全上界,从而 $<P(A),\cup,\cap>$ 是有界分配格。令 A 为全集,$\forall T\in P(A)$,T 的补集 $\sim T=A-T\in P(A)$,满足 $T\cup\sim T=A$ 和 $T\cap\sim T=\varnothing$,所以 T 的补元 $T'=\sim T$。根据定义 7.4.2,$<P(A),\cup,\cap,\sim>$ 是布尔代数。

可以证明,当 A 为任意集合时,$<P(A),\cup,\cap,\sim>$ 也是布尔代数,称为集合代数。集合代数是布尔代数的一个具体模型。

布尔代数一定是格。根据定理 7.1.1,布尔代数中的两个二元运算满足交换律、结合律、幂等律和吸收律。此外,布尔代数还有下面的性质。

定理 7.4.1 ▶ 设 $<X,\vee,\wedge,'>$ 为布尔代数,$\forall a,b\in X$,必有

(1) $(a')'=a$

(2) $(a\vee b)'=a'\wedge b'$

(3) $(a\wedge b)'=a'\vee b'$

证明

(1) a' 是 a 的补元,a 也是 a' 的补元,由布尔代数中补元的唯一性有 $(a')'=a$。

(2) $(a\vee b)\vee(a'\wedge b')=((a\vee b)\vee a')\wedge((a\vee b)\vee b')=(b\vee(a\vee a'))\wedge(a\vee(b\vee b'))$
$$=(b\vee 1)\wedge(a\vee 1)=1\wedge 1=1$$
$(a\vee b)\wedge(a'\wedge b')=(a\wedge(a'\wedge b'))\vee(b\wedge(a'\wedge b'))=((a\wedge a')\wedge b')\vee((b\wedge b')\wedge a')$
$$=(0\wedge b')\vee(0\wedge a')=0\vee 0=0$$

所以,$(a\vee b)'=a'\wedge b'$。

(3) 同理可证 $(a\wedge b)'=a'\vee b'$。

定理 7.4.1 中的(1)称为双重否定律,(2)和(3)称为德摩根律。布尔代数满足双重否定律和德摩根律。

定理 7.4.2 ▶ 设 $<X,*,\circ,'>$ 是代数系统,其中 $*$ 和 $\circ$ 都是二元运算,$'$ 是一元运算。$<X,*,\circ,'>$ 是布尔代数的充分必要条件:

(1) $*$ 和 $\circ$ 在 X 上封闭且满足交换律。

(2) $*$ 和 $\circ$ 满足分配律。

(3) X 中存在运算 $*$ 的幺元和运算 $\circ$ 的幺元。设运算 $*$ 的幺元为 0,运算 $\circ$ 的幺元为 1,即

$\forall a \in X$,有 $a * 0 = a, a \circ 1 = a$。

(4) $\forall a \in X, \exists a' \in X$,使 $a * a' = 1, a \circ a' = 0$。

定理 7.4.2 给出的四个条件可以作为布尔代数的等价定义。

布尔代数 $<X, *, \circ, '>$ 也可以表示为 $<X, *, \circ, ', 0, 1>$,其中 0 是运算 $*$ 的幺元,1 是运算 $\circ$ 的幺元。

定义 7.4.3 ▶ 设 $<X, \vee, \wedge, ', 0, 1>$ 是布尔代数,B 是 X 的非空子集,若 $0, 1 \in B$ 且 $<B, \vee, \wedge, ', 0, 1>$ 也是布尔代数,则称 $<B, \vee, \wedge, ', 0, 1>$ 是 $<X, \vee, \wedge, ', 0, 1>$ 的子布尔代数。

定理 7.4.3 ▶ 设 $<X, \vee, \wedge, ', 0, 1>$ 是布尔代数,B 是 X 的非空子集,若 $0, 1 \in B$ 且运算 $\vee, \wedge, '$ 在 B 上封闭,则 $<B, \vee, \wedge, ', 0, 1>$ 是 $<X, \vee, \wedge, ', 0, 1>$ 的子布尔代数。

证明 (1) $\forall a, b \in B$,因为 $B \subseteq X$,所以 $a, b \in X$。又因为 $<X, \vee, \wedge, ', 0, 1>$ 是布尔代数,故 $a \vee b = b \vee a, a \wedge b = b \wedge a$。

(2) 类似(1)可以证明 $*$ 和 $\circ$ 满足分配律。

(3) 已知 $0, 1 \in B, \forall a \in B \subseteq X, a \in X$,有 $a \vee 0 = a$ 和 $a \wedge 1 = a$。

(4) $\forall a \in B$,由 $'$ 在 B 上封闭可知 $a' \in B$,使得 $a \vee a' = 1, a \wedge a' = 0$。

根据定理 7.4.2,$<B, \vee, \wedge, ', 0, 1>$ 是布尔代数,它是 $<X, \vee, \wedge, ', 0, 1>$ 的子布尔代数。

为了方便,以下将 $x \leqslant y$ 且 $x \neq y$ 记为 $x \prec y$。

定义 7.4.4 ▶ 设 $<X_1, \vee_1, \wedge_1, ', 0, 1>$ 和 $<X_2, \vee_2, \wedge_2, '', \theta, E>$ 是两个布尔代数,其中 $\vee_1, \wedge_1, \vee_2$ 和 $\wedge_2$ 都是二元运算,$'$ 和 $''$ 是一元运算,0 和 1 是 X_1 的全下界和全上界,θ 和 E 是 X_2 的全下界和全上界。f 是从 X_1 到 X_2 的一个映射,对任意 $a, b \in X_1$ 有

$$f(a \vee_1 b) = f(a) \vee_2 f(b)$$
$$f(a \wedge_1 b) = f(a) \wedge_2 f(b)$$
$$(f(a))' = (f(a))''$$

则称 f 是布尔代数 $<X_1, \vee_1, \wedge_1, ', 0, 1>$ 到 $<X_2, \vee_2, \wedge_2, '', \theta, E>$ 的同态,简称布尔代数同态。如果 f 是单射的、满射的和双射的,分别称 f 是布尔代数单同态、布尔代数满同态和布尔代数同构。称 $<f(X_1), \vee_2, \wedge_2, '', \theta, E>$ 是 $<X_1, \vee_1, \wedge_1, ', 0, 1>$ 的布尔代数同态像。

7.5 布尔表达式

布尔代数可用于逻辑电路的设计。具有若干输入和某种逻辑功能的组合线路可以用一个定义在电路代数上的电路函数表示,而一个电路函数则可以用布尔表达式来表示。

定义 7.5.1 ▶ 设 $<S, \oplus, \odot, ', 0, 1>$ 为布尔代数,则 S 中的元素称为布尔常元;取值于 S 中的变元称为布尔变量。

定义 7.5.2 ▶ 设 $<S, \oplus, \odot, ', 0, 1>$ 为布尔代数,$x_1, x_2, \cdots, x_n$ 为布尔变元,则由这 n 个布尔变元产生的布尔表达式可递归定义如下。

(1) S 中的任何元素和变元为一个布尔表达式。

(2) 若 F 和 G 都是布尔式,则 $F', F \oplus G, F \odot G$ 也是布尔式。

(3) 只有有限次使用(1)或(2)构造而成的符号串才是一个布尔式。

(4) 为简便起见,规定 $\oplus$ 的运算优先级低于 $\odot$。

例 7.5.1 在布尔代数 $<\{0, 1, \alpha, \beta\}, \oplus, \odot, ', 0, 1>$ 中,布尔式有 $0 \odot 1', 1 \oplus (\alpha \odot x_1) \oplus$

$(x_2' \odot x_3),(\beta' \oplus x_1 \oplus x_3) \odot 1$。

任一 n 元布尔式都可定义为是一个从 S_n 到 S 的一个函数。

例 7.5.2 在布尔代数 $<\{0,1,\alpha,\beta\},\oplus,\odot,',0,1>$ 中,$f(x,y)=(\beta \odot x' \odot y) \oplus (\alpha \odot x \odot (x \oplus y'))$ 是一个二元布尔函数。

两个布尔式相等。

布尔表达式 $f(x_1,x_2,\cdots,x_n)$ 的值是将 S 中的元素作为 $x_i(i=1,2,\cdots,n)$ 的值代入表达式以后计算出来的表达式的值。

若对 n 个布尔变元任意指派(即给每个变元取上 S 中的元素),两个布尔表达式的值均相等,则称这两个布尔表达式是相等或等价的。

习 题

1. **N** 是自然数集,$\leqslant$ 是小于等于关系,则 $<\mathbf{N},\leqslant>$ 是(   )。

    A. 有界格   B. 有补格   C. 分配格   D. 有补分配格

2. 在有界格中,若只有一个元素有补元,则补元(   )。

    A. 必唯一   B. 不唯一   C. 不一定唯一   D. 可能唯一

3. 以下为 4 个格对应的哈斯图,(   )是分配格。

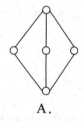 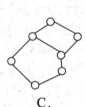 

    A.   B.   C.   D.

4. 只有有限个元素的格称为有限格,有限格必是(   )。

    A. 分配格   B. 有补格   C. 布尔格   D. 有界格

5. 设 $<L,\leqslant>$ 是一条链,其中 $|L| \geqslant 3$,则 $<L,\leqslant>$(   )。

    A. 不是格   B. 是有补格   C. 是分配格   D. 是布尔格

6. 设 A 为一个集合,$<P(A),\subseteq>$ 为有补格,$P(A)$ 中每个元素的补元(   )。

    A. 存在且唯一   B. 不存在   C. 存在但不唯一   D. 可能存在

7. 有限布尔代数的元素的个数必定等于(   )。

    A. $2n$ B. n^2 C. 2^n D. $4n$

8. 在布尔格 $<A,\leqslant>$ 中有 3 个原子 a_1,a_2,a_3,则 $\overline{a_1}=$(   )。

    A. $a_2 \wedge a_3$ B. $a_2 \vee a_3$ C. $\overline{a_2} \wedge \overline{a_3}$ D. $\overline{a_2} \vee \overline{a_3}$

9. 在布尔格 $<A,\leqslant>$ 中,$A=\{X \mid X$ 是 5 的整数倍且是 210 的正因子$\}$,$|$ 为整除关系,则 30 的补元为(   )。

    A. 15   B. 30   C. 35   D. 70

10. 由下列集合 L 构成的偏序集 $<L,\leqslant>$,其中 $\leqslant$ 定义:对于 $n_1,n_2 \in L$,$n_1 \leqslant n_2$ 当且仅当 n_1 是 n_2 的因子。判断其中哪几个偏序集是格(说明理由)。

    (1) $L=\{1,2,3,4,5,6,12\}$

    (2) $L=\{1,2,3,4,6,8,12,14\}$

(3) $L = \{1,2,3,4,5,6,7,8,9,10,11,12\}$

11. S_n 是由正整数 n 的所有因子构成的集合，$m \mid n$ 表示 m 整除 n。对于格 $<D_{30}, \mid >$

(1) 证明 $<D_{30}, \mid >$ 是布尔格。

(2) 作出其对应偏序集的哈斯图。

(3) 找出 D_{30} 的所有原子。

12. 设 $<B, \vee, \wedge, ^{-} >$ 是一个布尔代数，如果在 B 上的两个二元运算 $+$ 和 $\circ$ 定义为

$$a + b = (a \wedge \bar{b}) \vee (\bar{a} \wedge b)$$

$$a \circ b = a \wedge b$$

证明 $<B, +, \circ >$ 是以 1 为幺元的环。

13. $<B, \vee, \wedge, ^{-} >$ 是布尔代数，$\forall a, b \in B$，求证：$a = b$ 当且仅当 $(a \wedge \bar{b}) \vee (\bar{a} \wedge b) = 0$。

第4篇
Part IV

图 论

Graph Theory

Chapter 8　Basic Concepts of Graphs

In daily life, production activities, and scientific research, people often use points to represent things, and use whether there is a line between points to indicate a certain relationship between things. The graph formed in this way is the graph in graph theory. In fact, the graphs of binary relations in set theory are graphs in graph theory. In these graphs, people only care about whether there is a line between the points, but not the position of the points, and the curve of the line, which is the essential difference between graphs in graph theory and graphs in geometry.

8.1　Concept of Graph

The disordered sequence of two individuals x, y is called a disordered pair, denoted as (x, y). In the disordered pair (x, y), x and y are disordered, and their order can be reversed, that is $(x, y)=(y, x)$.

$\boxed{\text{Definition 8.1.1}}$　Figure G is a triple recombination $<V(G), E(G), \varphi_G>$

Among them: $V(G)$ is a non-empty node set, $E(G)$ is the edge set, φ_G is a function of the set of ordered pairs or unordered pairs from the edge set to the node.

$\boxed{\text{Example　8.1.1}}$　$G=<V(G), E(G), \varphi_G>$, among them: $V(G)=\{a,b,c,d\}$, φ_G: $\varphi_G(e_1)=(a,b)$, $\varphi_G(e_2)=(b,c)$, $\varphi_G(e_3)=(a,c)$, $\varphi_G(e_4)=(a,a)$.

Try to draw a graph of G.

$\boxed{\text{Solution}}$　The graph of G is shown in Figure 8.1.1.

Edges of a graph can be represented directly as ordered pairs or unordered pairs without causing confusion.

Therefore, the graph can be simply represented as:

$$G=<V, E>$$

Among them: V is a nonnull set of nodes; E is the set of ordered or unordered pairs of edges.

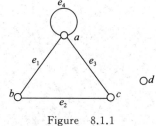

Figure　8.1.1

In this notation, the figure in Example 8.1.1 can be abbreviated as:

$$G=<V, E>$$

Among them: $V=\{a, b, c, d\}$, $E=\{(a, b), (b, c), (a, c), (a, a)\}$.

$\boxed{\text{Definition 8.1.2}}$　If the graph G has n nodes, it is called an n-th order graph.

Definition 8.1.3   Let G be a graph. If all edges of G are directed, then G is a directed graph. If all sides of G are undirected, then G is called an undirected graph. If G has both directional and undirectional edges, then G is called a mixed graph.

In Figure 8.1.2, (a) is an undirected graph, (b) is a directed graph and (c) is a mixed graph.

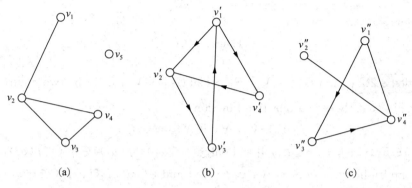

Figure   8.1.2

In a graph, if two nodes are related by an edge (directed or undirected), then one of them is said to be the adjacent node of the other node. And they say that these two nodes are adjacent to each other.

A node that is not adjacent to any node in a graph is called a isolated node.

In a graph, if two edges are related to the same node, one of the edges is said to be an adjacent edge to the other. They say these two sides are adjacent to each other. An edge associated with the same node is called a loop or self-loop. The directions in the center of the directed graph can be clockwise or counterclockwise, they are equivalent.

Definition 8.1.4   A graph composed of isolated nodes is called a zero graph. A graph consisting of a isolated node is called a trivial graph.

By Definition 8.1.4, a trivial graph must be a zero graph.

8.2   Subgraph and Isomorphic Graph

Subgraph is one of the basic concepts of graph theory. Node set and edge set are the graphs of the subset of node set and edge set of a graph respectively. In this section, besides introducing the concept of subgraphs, we also introduce the isomorphism of graphs.

Definition 8.2.1   Let $G = <V, E>$, $G' = <V', E'>$ be two graphs (the same undirected graph or the supergraph), if $V' \subseteq V$ and $E' \subseteq E$, then we call G' is the subgraph of G, G is the supergraph of G', denoted as $G' \subseteq G$, and if $V' \subset V$ and $E' \subset E$, then G' is said to be the true subgraph of G, if $V' = V$, it is said that G' is the generated subgraph of G.

In Figure 8.2.1, (b) is the subgraph, true subgraph and generated subgraph of (a).

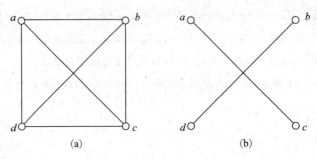

Figure 8.2.1

Definition 8.2.2 Let $G_1 = <V_1, E_1>$ and $G_2 = <V_2, E_2>$ be two undirected graphs (directed graphs), if there is a bijective function

$$f: V_1 \rightarrow V_2, \ \forall v_1 \in V_1, \ \forall v_2 \in V_1$$

$(v_1, v_2) \in E_1 (<v_1, v_2> \in E_1)$ If and only if $(f(v_1), f(v_2)) \in E_2 (<f(v_1), f(v_2)> \in E_2)$ and the multiplicity of $(v_1, v_2)(<v_1, v_2>)$ and $(f(v_1), f(v_2))(<f(v_1), f(v_2)>)$ is the same, it is said that graph G_1 is isomorphic with graph G_2. It is written as $G_1 \cong G_2$. The bijective function f is called the isomorphic function of graph G_1 and graph G_2.

The isomorphism of two graphs must satisfy the following conditions: ① The same number of nodes. ②The same number of edges. ③The nodes with the same degree have the same number of nodes.

These three conditions are necessary but not sufficient for the isomorphism of two graphs. In general, the above three necessary conditions are used to determine that the two graphs are not isomorphic.

Two graphs are isomorphic, and their isomorphic functions must realize the same degree node corresponding to the same degree node.

8.3　Path and Loop

In the real world, it is often necessary to consider such a problem: how to start from a given node in a graph G and continuously move along some edges to reach another specified node. This sequence composed of points and edges, thus forms the concept of the path.

Definition 8.3.1 Let $G = <V, E>$ be a graph. The alternating sequence of nodes and edges in G $L: v_0 e_1 v_1 e_2 v_2 ... e_n v_n$ is called the path from v_0 to v_n. Where v_{i-1} and v_i are the endpoints of e_i, $i = 1, ..., n$. v_0 and v_n are called the start and end points of path L, respectively. The number of sides in path L is called the length of the path.

For example, in Figure 8.3.1, $L_1: v_5 e_8 v_4 e_5 v_2 e_6 v_5 e_7 v_3$ is the path from v_5 to v_3, v_5 is the start point, v_3 is the end point, and the length is 4. $L_2: v_1 e_1 v_2 e_3 v_3$ is the path from v_1 to v_3, v_1 is the start point, v_3 is the end point, and the length is 2.

There are no parallel edges (graphs with neither parallel edges nor self rings) and loops (parallel edges only exist in a multigraph, that is, there are at least two edges in a node pair, which are parallel to each other) in a simple graph. The path L: $v_0 e_1 v_1 e_2 v_2 \ldots e_n v_n$ is completely determined by its node sequence $v_0 v_1 v_2 \ldots v_n$. Therefore, the path in a simple graph can be represented by a sequence of nodes.

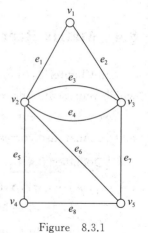

Figure 8.3.1

Definition 8.3.2　Let $G = <V, E>$ be a graph, and L be the path from v_0 to v_n. If $v_0 = v_n$, then L is called a loop. If all edges in L are different, then L is called a simple path. If $v_0 = v_n$ again at this time, then L is called a simple loop. If all the nodes in L are different, then L is called the basic path. If $v_0 = v_n$ again at this time, then L is called the basic loop. If $v_0 = v_n$ again at this time, then L is called the primary loop.

In Figure 8.3.1, L_1 is a simple path. L_2 is a simple path, a basic path and a primary path.

Theorem 8.3.1　In n order graph G, if there is a path from node v_i to v_j ($v_i \neq v_j$), there must be a path with a length less than or equal to $n-1$.

Proof　Suppose L: $v_i e_1 v_1 e_2 v_2 \ldots e_j v_j$ is a path with length l from node v_i to node v_j in G, there are $l+1$ nodes on the path.

If $l \leqslant n-1$, the theorem is proved.

Otherwise, $l > n-1$, at this time, $l+1 > n$, that is, the number of nodes $l+1$ on the path L is greater than the number of nodes n in the graph G, and there must be the same node on this path. Let v_k and v_s be the same. Therefore, this path passes through the same node $v_k (v_s)$ twice, so there is a loop C_{ks} from v_s to itself on L. Delete all edges on C_{ks} and all nodes except v_s on L, and get L_1: $v_i e_1 v_1 \ldots e_k v_s \ldots e_j v_j$. L_1 is still the path from node v_i to v_j, and the length is at least 1 less than L. If the length of the path L_1 is less than or equal to $n-1$, the theorem has been proved. Otherwise, repeat the above process. Since G has n nodes, after finite steps, a path from v_i to v_j whose length is less than or equal to $n-1$ must be obtained.

Inference　In an n-order graph G, if there is a path from node v_i to v_j ($v_i \neq v_j$), there must be a basic path with a length less than or equal to $n-1$.

According to the proof process of Theorem 8.3.1, this corollary is valid. Similarly, the following theorems and corollaries can be proved.

Theorem 8.3.2　In a graph G of order n, if there is a loop from node v_i to itself, there must be a loop from v_i to itself whose length is less than or equal to n.

Inference　In the n-order graph G, if there is a simple loop from node v_i to itself, there must be a basic loop from v_i to itself whose length is less than or equal to n.

8.4　Matrix Representation of Graph

In Chapter 3, the relation R on a given set A can be represented by a directed graph, which represents the relation between the elements in set A and the adjacency relation between the elements in the set. For the diagram, it can be represented by a matrix, and a matrix must also correspond to a diagram with the ordinal number of the node.

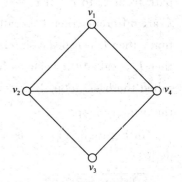

Definition 8.4.1　Let $G=<V, E>$ be a simple graph, $V=\{v_1, v_2, \cdots, v_n\}$, $A(G)=(a_{ij})_{n\times n}$, among them:

$$a_{ij}=\begin{cases}1, & v_i \text{ to } v_j \text{ have edges} \\ 0, & v_i \text{ to } v_j \text{ have no edges or } i=j\end{cases} \quad i,j=1,\cdots,n$$

Call $A(G)$ the adjacency matrix of G, abbreviated as A.

For example, the adjacency matrix in Figure 8.4.1 is

$$A(G)=\begin{pmatrix} 0 & 1 & 0 & 1 \\ 1 & 0 & 1 & 1 \\ 0 & 1 & 0 & 1 \\ 1 & 1 & 1 & 0 \end{pmatrix}$$

Figure　8.4.1

For example, the adjacency matrix in Figure 8.4.2(a) is

$$A(G)=\begin{pmatrix} 0 & 1 & 0 & 0 \\ 0 & 0 & 1 & 1 \\ 1 & 1 & 0 & 1 \\ 1 & 0 & 0 & 0 \end{pmatrix}$$

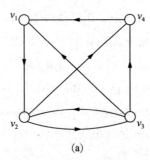

(a)

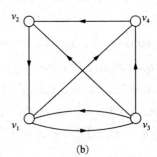

(b)

Figure　8.4.2

The adjacency matrix has the following properties.

(1) The elements of the adjacency matrix are either 0 or 1. Such a matrix is called a Boolean matrix. The adjacency matrix is a Boolean matrix.

(2) The adjacency matrix of an undirected graph is a symmetric matrix, but the adjacency matrix of a directed graph is not necessarily a symmetric matrix.

(3) The adjacency matrix is related to the order in which nodes are demarcated in the graph. For example, the adjacency matrix of Figure 8.4.2(a) is $A(G)$. If the calibration order of the contacts v_1 and v_2 in Figure 8.4.2(a) is reversed, the adjacency matrix of Figure 8.4.2 (b) is $A'(G)$.

$$\boldsymbol{A}'(G)=\begin{pmatrix}0&0&1&1\\1&0&0&0\\1&1&0&1\\0&1&0&0\end{pmatrix}$$

When we examine $\boldsymbol{A}(G)$ and $\boldsymbol{A}'(G)$, we find that the first row of $\boldsymbol{A}(G)$ is switched with the second row, and then the first column is switched with the second column to obtain $\boldsymbol{A}'(G)$. To say that $\boldsymbol{A}'(G)$ and $\boldsymbol{A}(G)$ are permutation equivalent.

Generally speaking, $\boldsymbol{A}'$ is a permutation equivalent to $\boldsymbol{A}$ if some rows of the n-th square matrix $\boldsymbol{A}$ are reversed, and the corresponding columns are reversed to obtain $\boldsymbol{A}$ new n-th square matrix $\boldsymbol{A}'$. It can be proved that the substitution equivalence is an equivalence relation on the set of n-th order Boolean square matrices.

Although different adjacency matrices can be obtained for the same graph due to the different demarcating order of nodes, these adjacency matrices are permutation equivalent. In the future, the randomness of node calibration order is omitted and any adjacency matrix is chosen to represent the graph.

(4) For a directed graph, the number of the i-th row 1 of the adjacency matrix $A(G)$ is outdegree of v_i, and the number of the j-th column 1 is indegree of v_j.

(5) The adjacency matrix of a zero graph has all zero elements, which is called the zero matrices of the zero graphs. Conversely, a graph must be a zero graph if its adjacency matrix is the zero matrices.

Theorem 8.4.1　Let $\boldsymbol{A}(G)$ be the adjacency matrix of graph G, $\boldsymbol{A}(G)^k=\boldsymbol{A}(G)\boldsymbol{A}(G)^{k-1}$, row i of $\boldsymbol{A}(G)^k$, column j of the element a_{ij}^k is equal to the number of paths from v_i to v_j of length k. Where a_{ii}^k is the loop from v_i to itself with a length of k.

Inference　$G=<V,\ E>$ is n order simple directed graph, adjacency matrix $\boldsymbol{A}$ is directed graph G, $\boldsymbol{B}^k=\boldsymbol{A}+\boldsymbol{A}^2+...+\boldsymbol{A}^k$, $\boldsymbol{B}^k=(b_{ij}^k)_{n\times n}$, then b_{ij}^k is the number of paths in G whose length from v_i to v_j is less than or equal to k. $\sum\limits_{i=1}^{n}\sum\limits_{j=1}^{n}b_{ij}^k$ is the total number of paths in G of length less than or equal to k. $\sum\limits_{i=1}^{n}b_{ii}^k$ is the number of loops in G where length is less than or equal to k.

Example 8.4.1　$G=<V,\ E>$ for simple directed graph, the graph as shown in Figure 8.4.3, write G adjacency matrix $\boldsymbol{A}$, calculate the $\boldsymbol{A}^2$, $\boldsymbol{A}^3$, $\boldsymbol{A}^4$ and determine how many paths of length 3 there are from v_1 to v_2. How many paths are there of length two from v_1 to v_3? How many loops are there from v_2 to itself of length 3 and length 4?

Solution　Adjacency matrix $\boldsymbol{A}$ and $\boldsymbol{A}^2$, $\boldsymbol{A}^3$, $\boldsymbol{A}^4$ are as follows:

$$A = \begin{bmatrix} 0 & 1 & 0 & 0 & 0 \\ 1 & 0 & 1 & 0 & 0 \\ 0 & 1 & 0 & 0 & 0 \\ 0 & 0 & 0 & 0 & 1 \\ 0 & 0 & 0 & 1 & 0 \end{bmatrix} \quad A^2 = \begin{bmatrix} 1 & 0 & 1 & 0 & 0 \\ 0 & 2 & 0 & 0 & 0 \\ 1 & 0 & 1 & 0 & 0 \\ 0 & 0 & 0 & 1 & 0 \\ 0 & 0 & 0 & 0 & 1 \end{bmatrix}$$

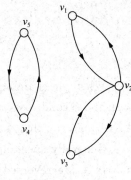

$$A^3 = \begin{bmatrix} 0 & 2 & 0 & 0 & 0 \\ 2 & 0 & 2 & 0 & 0 \\ 0 & 2 & 0 & 0 & 0 \\ 0 & 0 & 0 & 0 & 1 \\ 0 & 0 & 0 & 1 & 0 \end{bmatrix} \quad A^4 = \begin{bmatrix} 2 & 0 & 2 & 0 & 0 \\ 0 & 4 & 0 & 0 & 0 \\ 2 & 0 & 2 & 0 & 0 \\ 0 & 0 & 0 & 1 & 0 \\ 0 & 0 & 0 & 0 & 1 \end{bmatrix}$$

Figure 8.4.3

$a_{12}^{3} = 2$, therefore, there are two paths with a length of 3 from v_1 to v_2, which are respectively $v_1 v_2 v_1 v_2$ and $v_1 v_2 v_3 v_2$.

$a_{13}^{2} = 1$, therefore, there is one path with a length of 2 from v_1 to v_3, which is $v_1 v_2 v_3$.

$a_{22}^{3} = 0$, v_2 to itself has no loop of length 3.

$a_{22}^{4} = 4$, v_2 to itself has four loops of length 4, and they are $v_2 v_1 v_2 v_1 v_2$、$v_2 v_3 v_2 v_3 v_2$、$v_2 v_3 v_2 v_1 v_2$ 和 $v_2 v_1 v_2 v_3 v_2$.

Definition 8.4.2 $G = <V, E>$ is simple directed graph, $V = \{v_1, v_2, ..., v_n\}$, $P(G) = (p_{ij})_{n \times n}$, among them:

$$p_{ij} = \begin{cases} 1, & v_i \text{ to } v_j \\ 0, & \text{Unreachable from } v_i \text{ to } v_j \end{cases} \quad i,j = 1, ..., n$$

Call $P(G)$ the reachability matrix of G. Shorthand for P.

The main diagonal elements of the accessibility matrix $P(G)$ are all 1.

$G = <V, E>$ is n order simple directed graph, $V = \{v_1, v_2, ..., v_n\}$, by the definition of reachability matrix is known, when $i \neq j$, if there is a path v_i to v_j, $p_{ij} = 1$; if v_i has no way to v_j, then $p_{ij} = 0$; if there is a path from v_i to v_j, there must be a path of length less than or equal to $n-1$. According to the corollary of Theorem 8.4.1, the reachability matrix P of figure G is calculated as follows.

First to calculate $B_{n-1} = A + A^2 + ... + A^{n-1}$, suppose that $B_{n-1} = (b_{ij}^{n-1})_{n \times n}$. If $b_{ij}^{n-1} \neq 0$, then set $p_{ij} = 1$, if $b_{ij}^{n-1} = 0$, then set $p_{ij} = 0$, $i, j = 1, ..., n$. Then set $p_{ii} = 1$, $i = 1, ..., n$. Then we can get the reachability matrix P of figure G.

If A^0 is n order identity matrix, the above algorithm can also be improved as:

Calculate $C_{n-1} = A^0 + B_{n-1} = A^0 + A + A^2 + ... + A^{n-1}$, set $C_{n-1} = (c_{ij}^{n-1})_{n \times n}$. If $c_{ij}^{n-1} \neq 0$, then set $p_{ij} = 1$, if $c_{ij}^{n-1} = 0$, then set $p_{ij} = 0$, $i, j = 1, ..., n$.

Using the above method, the reachability matrix of figure G in Example 8.4.3 is calculated,

$$C_4 = A^0 + A + A^2 + A^3 + A^4 = \begin{bmatrix} 4 & 3 & 3 & 0 & 0 \\ 3 & 7 & 3 & 0 & 0 \\ 3 & 3 & 4 & 0 & 0 \\ 0 & 0 & 0 & 3 & 1 \\ 0 & 0 & 0 & 1 & 3 \end{bmatrix} \quad P = \begin{bmatrix} 1 & 1 & 1 & 0 & 0 \\ 1 & 1 & 1 & 0 & 0 \\ 1 & 1 & 1 & 0 & 0 \\ 0 & 0 & 0 & 1 & 1 \\ 0 & 0 & 0 & 1 & 1 \end{bmatrix}$$

To calculate the reachability matrix P of simple directed graph G, the following methods can also be used.

Let A be the adjacency matrix for G, set $A=(a_{ij})_{n\times n}$, $A^{(k)}=(a_{ij}^{(k)})_{n\times n}$, A^0 as n order identity matrix.

$A^{(2)}=A\circ A$, among them $a_{ij}^{(2)}=(a_{i1}\wedge a_{1j})\vee(a_{i2}\wedge a_{2j})\vee\ldots\vee(a_{in}\wedge a_{nj})$, $i,j=1,\ldots,n$.

$A^{(3)}=A\circ A^{(2)}$, among them $a_{ij}^{(3)}=(a_{i1}\wedge a_{1j}^{(2)})\vee\ldots\vee(a_{in}\wedge a_{nj}^{(2)})$, $i,j=1,\ldots,n$.

$P=A^0\vee A\vee A^{(2)}\vee A^{(3)}\vee\ldots\vee A^{(n-1)}$. Among them, the operations $\vee$ are the disjunction matrix corresponding element.

The reachability matrix is used to describe whether a node of a directed graph has a way to another node, that is, whether it is reachable. An undirected graph can also be used as a matrix to describe whether there is a path from one node to another. In an undirected graph, if there is a path between nodes, the two nodes are said to be connected and not reachable. So the matrix that describes whether there is a way from one node to another is called the connected matrix, not the reachability matrix.

Definition 8.4.3　$G=<V,E>$ is a simple undirected graph, $V=\{v_1,v_2,\ldots,v_n\}$, $P(G)=(p_{ij})_{n\times n}$, among them:

$$p_{ij}=\begin{cases}1, & v_i \text{ is connected to } v_j \\ 0, & v_i \text{ is disconnected to } v_j\end{cases}\quad i,j=1,\ldots,n$$

Call $P(G)$ the connected matrix of G. Shorthand for P.

The adjacency matrix of an undirected graph is a symmetric matrix, and the connected matrix of an undirected graph is also a symmetric matrix. The method of finding the connected matrix is similar to the reachability matrix.

Definition 8.4.4　$G=<V,E>$ is a undirected graph, $V=\{v_1,v_2,\ldots,v_p\}$, $E=\{e_1,e_2,\ldots,e_q\}$, $M(G)=(m_{ij})_{p\times q}$, among them:

$$P_{ij}=\begin{cases}1, & v_i \text{ is associated with } e_j \\ 0, & \text{otherwise}\end{cases}\quad i=1,\ldots,p,\ j=1,\ldots,q$$

$M(G)$ is called the complete incidence matrix of the undirected graph G. Referred to as M.

For example, the complete incidence matrix in Figure 8.4.4 is

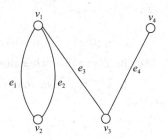

Figure　8.4.4

$$M(G)=\begin{bmatrix}1 & 1 & 1 & 0 \\ 1 & 1 & 0 & 0 \\ 0 & 0 & 1 & 1 \\ 0 & 0 & 0 & 1\end{bmatrix}$$

$G=<V,E>$ is an undirected graph, the complete incidence matrix G $M(G)$ has the following properties.

(1) The sum of the elements in each column is 2. That means that each edge is associated with two nodes.

(2) The sum of the elements in each row is the degree of the corresponding node.

(3) The sum of all the elements is the sum of the degrees of the nodes in the graph,

which is also twice the number of edges.

(4) If two columns are the same, the corresponding two sides are parallel.

(5) If all elements in a row are zero, the corresponding node is a isolated node.

Definition 8.4.5 $G=<V, E>$ is a directed graph, $V=\{v_1, v_2, ..., v_p\}$, $E=\{e_1, e_2, ..., e_q\}$, $\mathbf{M}(G)=(m_{ij})_{p\times q}$, among them:

$$m_{ij}=\begin{cases} 1, & v_i \text{ is the starting point of } e_j \\ -1, & v_i \text{ is the ending point of } e_j \\ 0, & v_i \text{ is disassociated with } e_j \end{cases} \quad i=1, ..., p, j=1, ..., q$$

$\mathbf{M}(G)$ is called the complete incidence matrix of the directed graph G. Referred to as $\mathbf{M}$.

The complete incidence matrix in Figure 8.4.5 is

$$\mathbf{M}(G)=\begin{bmatrix} -1 & 1 & 0 & 0 & 0 \\ 1 & -1 & 1 & 0 & 0 \\ 0 & 0 & 0 & 1 & 1 \\ 0 & 0 & -1 & -1 & -1 \end{bmatrix}$$

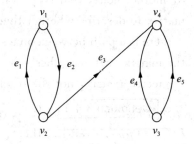

Figure 8.4.5

$G=<V, E>$ is a directed graph, the complete incidence matrix of G, $\mathbf{M}(G)$ has the following properties.

(1) Each column has a 1 and a -1, which means that each directed edge has a starting point and an ending point.

(2) The number of 1 in each row is the outdegree of the corresponding node, and the number of -1 is the indegree of the corresponding node.

(3) The sum of all the elements is equal to 0, which means that the sum of the outdegrees of all the nodes is equal to the sum of the indegrees of all the nodes.

(4) If two columns are the same, then the corresponding sides are parallel.

Exercises

1. Suppose the undirected graph G has 16 edges, 3 4-degree nodes, 4 3-degree nodes, and the degree of other nodes is less than 3. Ask: How many nodes are there at least in G?

2. Try to prove that (a) and (b) in Exercise 2 Figure are isomorphic.

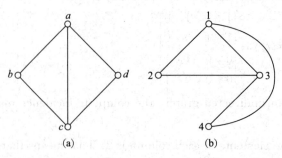

Exercise 2 Figure

3. Let $G_1=<V_1, E_1>$ and $G_2=<V_2, E_2>$ be two undirected graphs, Among them: $V_1=\{a,b,c,d,e\}$, $E_1=\{(a,b),(a,c),(a,c),(b,c),(b,d),(d,e),(c,e),(e,e)\}$

$V_2 = \{1,2,3,4,5\}, E_2 = \{(1,2),(1,3),(1,3),(2,3),(2,4),(4,5),(3,5),(4,4)\}$

The disordered pairs that repeat in E_1 and E_2 are parallel edges in the graph, such as (a, c) in E_1 and $(1, 3)$ in E_2.

(1) Plot G_1 and G_2.

(2) Prove that G_1 and G_2 are not isomorphic.

4. Suppose G is an n-order self-complement graph, try to prove that $n = 4k$ or $n = 4k + 1$, where k is a positive integer. Draw a self-complement diagram of 5 nodes. Is there a self-complementing diagram with 3 nodes or 6 nodes?

5. If G is a connected graph with at least three nodes, the following statements are equivalent.

(1) G is not a bridge.

(2) Every two nodes of G are on a common simple loop.

(3) Every node and every edge of G is on a common simple loop.

(4) Each of the two sides of G is on a common simple loop.

(5) For every pair of nodes and every edge of G, there's a simple path that connects these two nodes and contains this edge.

(6) For every pair of nodes and every edge of G, there's a basic path that connects these two nodes without this edge.

(7) For every three nodes, there is a simple path that connects any two nodes with a third node.

6. $G = <V, E>$ is a simple directed graph, $V = \{v_1, v_2, v_3, v_4\}$, the adjacency matrix is as follows:

$$A(G) = \begin{bmatrix} 0 & 1 & 0 & 0 \\ 0 & 0 & 1 & 1 \\ 1 & 1 & 0 & 1 \\ 1 & 1 & 0 & 0 \end{bmatrix}$$

(1) Figure out the outdegree of v_1 $\deg^+(v_1)$.

(2) Figure out the indegree of v_4 $\deg^-(v_4)$.

(3) How many paths are there from v_1 to v_4 of length 2?

7. Directed graph G is shown in Exercise 7 Figure.

(1) Write the adjacency matrix for G.

(2) To calculate the outdegree and indegree of each node according to the adjacency matrix.

(3) Find the total number of paths of length 3 in G, and how many loops are there?

(4) Find the reachability matrix for G.

(5) Find the complete incidence matrix for G.

(6) To calculate the outdegree and indegree of each node by the complete incidence matrix.

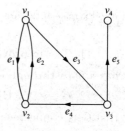

Exercise 7 Figure

8. Undirected graph G is shown in Exercise 8 Figure.

(1) Write the adjacency matrix for G.

(2) Calculate the degree of each node according to the adjacency matrix.

(3) Find the total number of paths of length 3 in G, and how many loops are there?

(4) Find the connected matrix for G.

(5) Find the complete incidence matrix for G.

(6) To calculate the degree of each node.

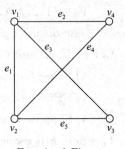

Exercise 8 Figure

9. $G = <V, E>$ is a simple directed graph, $V = \{v_1, v_2, \ldots, v_n\}$, $\mathbf{P} = (p_{ij})_{n \times n}$ is the reachability matrix of G, and $\mathbf{P}^T = (p'_{ij})_{n \times n}$ is the transposable matrix of $\mathbf{P}$. Well, $p_{ij} = 1$ means that v_i to v_j is reachable; $p'_{ij} = p_{ji} = 1$ means v_j to v_i is reachable. Therefore, when $p_{ij} \wedge p'_{ij} = 1$, v_i and v_j are mutually reachable. Thus, a strong connectedgraph of G can be obtained. For example, the reachability matrix $\mathbf{P}$ of figure G is:

$$\mathbf{P} = \begin{pmatrix} 1 & 0 & 1 & 1 & 1 \\ 0 & 1 & 1 & 1 & 1 \\ 0 & 0 & 1 & 1 & 1 \\ 0 & 0 & 1 & 1 & 1 \\ 0 & 0 & 1 & 1 & 1 \end{pmatrix} \quad \mathbf{P}^T = \begin{pmatrix} 1 & 0 & 0 & 0 & 0 \\ 0 & 1 & 0 & 0 & 0 \\ 1 & 1 & 1 & 1 & 1 \\ 1 & 1 & 1 & 1 & 1 \\ 1 & 1 & 1 & 1 & 1 \end{pmatrix} \quad \mathbf{P} \wedge \mathbf{P}^T = \begin{pmatrix} 1 & 0 & 0 & 0 & 0 \\ 0 & 1 & 0 & 0 & 0 \\ 0 & 0 & 1 & 1 & 1 \\ 0 & 0 & 1 & 1 & 1 \\ 0 & 0 & 1 & 1 & 1 \end{pmatrix}$$

Among them: $\mathbf{P} \wedge \mathbf{P}^T$ is defined as the union of the corresponding elements of matrix $\mathbf{P}$ and matrix $\mathbf{P}^T$.

It can be seen that the subgraphs derived from $\{v_1\}$, $\{v_2\}$, $\{v_3, v_4, v_5\}$ are strong connectedgraphs of G.

Try this method to find all strong subgraphs of the graph in problem 2.

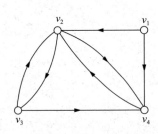

Exercise 10 Figure

10. $G = <V, E>$ is a simple directed graph, $V = \{v_1, v_2, \ldots, v_n\}$, $\mathbf{A}$ is the adjacency matrix of G, and the distance matrix $\mathbf{D} = (d_{ij})_{n \times n}$ is defined as follows:

$d_{ij} = \infty$ if $d <v_i, v_j> = \infty$.

$d_{ii} = 0$ $i = 1, \ldots, n$.

$d_{ij} = k$ k is the minimum positive integer for $a_{ij}^k \neq 0$

Try to write the distance matrix $\mathbf{D}$ in Exercise 10 Figure. And what does $d_{ij} = 1$ mean?

11. Try to prove that a directed graph G is a one-way connected graph if and only if it has a path through each node.

12. $G = <V, E>$ is a simple diagram, $|V| = n$, $|E| = m$, $m > 1/2(n-1)(n-2)$. Prove that G is connected.

第8章 图的基本概念

在日常生活、生产活动及科学研究中，人们常用点表示事物，用点与点之间是否有连线表示事物之间是否有某种关系，这样构成的图形就是图论中的图。其实，集合论中二元关系的关系图都是图论中的图。在这些图中，人们只关心点之间是否有连线，而不关心点的位置，以及连线的曲直，这就是图论中的图和几何学中的图形的本质区别。

8.1 图的概念

两个个体 x,y 的无序序列称为无序对，记为 (x,y)。在无序对 (x,y) 中，x,y 是无序的，它们的顺序可以颠倒，即 $(x,y)=(y,x)$。

定义 8.1.1 ▶ 图 G 是一个三重组 $<V(G),E(G),\varphi_G>$，其中：$V(G)$ 是非空结点集，$E(G)$ 是边集，φ_G 是边集到结点的有序对或无序对集合的函数。

例 8.1.1 $G=<V(G),E(G),\varphi_G>$，其中：$V(G)=\{a,b,c,d\}$，φ_G：$\varphi_G(e_1)=(a,b)$，$\varphi_G(e_2)=(b,c)$，$\varphi_G(e_3)=(a,c)$，$\varphi_G(e_4)=(a,a)$。

试画出 G 的图形。

解 G 的图形如图 8.1.1 所示。

由于在不引起混淆的情况下，图的边可以用有序对或无序对直接表示，因此图可以简单地表示为

$$G=<V,E>$$

其中：V 是非空的结点集；E 是由边的有序对或无序对组成的集合。

按照这种表示法，例 8.1.1 中的图可以简记为

$$G=<V,E>$$

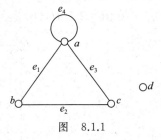

图 8.1.1

其中：$V=\{a,b,c,d\}$，$E=\{(a,b),(b,c),(a,c),(a,a)\}$。

定义 8.1.2 ▶ 若图 G 有 n 个结点，则称该图为 n 阶图。

定义 8.1.3 ▶ 设 G 为图，如果 G 的所有边都是有向边，则称 G 为有向图。如果 G 的所有边都是无向边，则称 G 为无向图。如果 G 中既有有向边，又有无向边，则称 G 为混合图。

图 8.1.2 的(a)是无向图，(b)是有向图，(c)是混合图。

在一个图中，若两个结点由一条边(有向边或无向边)关联，则称其中的一个结点是另一个结点的邻接点，并称这两个结点相互邻接。

在一个图中，不与任何结点相邻接的结点称为孤立点。

在一个图中，如果两条边关联于同一个结点，则称其中的一条边是另一条边的邻接边，并

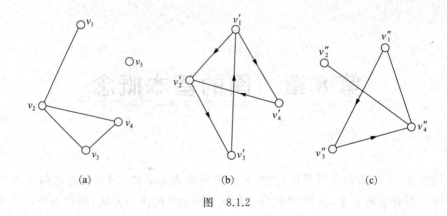

图　8.1.2

称这两条边相互邻接。关联于同一个结点的一条边叫作环或自回路。在有向图中,环的方向可以是顺时针,也可以是逆时针,它们是等效的。

定义 8.1.4 ▶ 由孤立点组成的图叫作零图。由一个孤立点组成的图叫作平凡图。

根据定义 8.1.4,平凡图一定是零图。

8.2　子图与图的同构

子图是图论的基本概念之一。结点集和边集分别是某一图的结点集的子集和边集的子集的图。本节除介绍子图的概念外,还介绍图的同构。

定义 8.2.1 ▶ 设 $G=<V,E>$,$G'=<V',E'>$为两个图(同为无向图或同为有向图),若 $V'\subseteq V$ 且 $E'\subseteq E$,则称 G' 是 G 的子图,G 是 G' 的母图,记作 $G'\subseteq G$。若 $V'\subset V$ 且 $E'\subset E$,则 G' 称是 G 的真子图。若 $V'=V$,则称 G' 是 G 的生成子图。

在图 8.2.1 中,(b)是(a)的子图、真子图、生成子图。

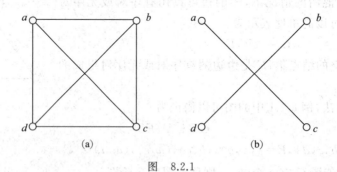

图　8.2.1

定义 8.2.2 ▶ 设 $G_1=<V_1,E_1>$ 与 $G_2=<V_2,E_2>$ 是两个无向图(有向图),若存在双射函数

$$f:V_1 \to V_2, \forall v_1 \in V_1, \forall v_2 \in V_1$$

$(v_1,v_2)\in E_1(<v_1,v_2>\in E_1)$ 当且仅当 $(f(v_1),f(v_2))\in E_2(<f(v_1),f(v_2)>\in E_2)$,并且 $(v_1,v_2)(<v_1,v_2>)$ 与 $(f(v_1),f(v_2))(<f(v_1),f(v_2)>)$ 的重数相同,则称图 G_1 与图 G_2 同构,记为 $G_1\cong G_2$。双射函数 f 称为图 G_1 与图 G_2 的同构函数。

两个图同构必须满足下列条件:①结点数相同;②边数相同;③度数相同的结点数相同。

这三个条件是两个图同构的必要条件,不是充分条件。一般地,上述三个必要条件可用于判断两个图是不同构的。

两个图同构,它们的同构函数必须实现同度结点对应同度结点。

8.3　路与回路

在现实世界中,常常要考虑这样的问题:如何从一个图 G 中的给定结点出发,沿着一些边连续移动后到达另一指定结点,这种依次由点和边组成的序列,就形成了路的概念。

定义 8.3.1 ▶ 设 $G=<V,E>$ 是图,G 中的结点与边的交替序列 $L:v_0e_1v_1e_2v_2\cdots e_nv_n$ 叫作 v_0 到 v_n 的路。其中 v_{i-1} 与 v_i 是 e_i 的端点,$i=1,\cdots,n$。v_0 和 v_n 分别叫作路 L 的始点和终点。路 L 中边的条数叫作该路的长度。

例如,在图 8.3.1 中,$L_1:v_5e_8v_4e_5v_2e_6v_5e_7v_3$ 是从 v_5 到 v_3 的路,v_5 是始点,v_3 是终点,长度为 4。$L_2:v_1e_1v_2e_3v_3$ 是从 v_1 到 v_3 的路,v_1 是始点,v_3 是终点,长度为 2。

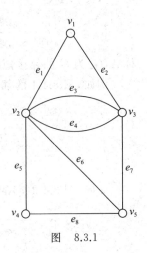

图　8.3.1

在简单图(既不含平行边也不包含自环的图)中没有平行边(平行边只存在于多重图中,也就是存在一个结点对有至少 2 条边,这些边互为平行边)和环,路 $L:v_0e_1v_1e_2v_2\cdots e_nv_n$ 完全由它的结点序列 $v_0v_1v_2\cdots v_n$ 确定。所以,在简单图中的路可以用结点序列表示。

定义 8.3.2 ▶ 设 $G=<V,E>$ 是图,L 是从 v_0 到 v_n 的路。若 $v_0=v_n$,则称 L 为回路。若 L 中所有边各异,则称 L 为简单路。若此时又有 $v_0=v_n$,则称 L 为简单回路。若 L 中所有结点各异,则称 L 为基本路。若此时又有 $v_0=v_n$,则称 L 为基本回路。若此时又有 $v_0=v_n$,则称 L 为初级回路。

在图 8.3.1 中,L_1 是一条简单路。L_2 是一条简单路、基本路、初级路。

定理 8.3.1 ▶ 在 n 阶图 G 中,若从结点 v_i 到 $v_j(v_i\neq v_j)$ 存在一条路,则必存在长度小于或等于 $n-1$ 的路。

证明 设 $L:v_ie_1v_1e_2v_2\cdots e_jv_j$ 是 G 中一条从结点 v_i 到 v_j 长度为 l 的路,路上有 $l+1$ 个结点。

若 $l\leqslant n-1$,则定理已证。

否则,$l>n-1$,此时,$l+1>n$,即路 L 上的结点数 $l+1$ 大于图 G 中的结点数 n,此路上必有相同结点。设 v_k 和 v_s 相同。于是,此路两次通过同一个结点 $v_k(v_s)$,所以在 L 上存在 v_s 到自身的回路 C_{ks}。在 L 上删除 C_{ks} 上的一切边和除 v_s 以外的一切结点,得路 $L_1:v_ie_1v_1e_kv_s\cdots e_jv_j$。$L_1$ 仍为从结点 v_i 到 v_j 的路,且长度至少比 L 减少 1。若路 L_1 的长度小于或等于 $n-1$,则定理已证。否则,重复上述过程。由于 G 有 n 个结点,经过有限步后,必得到从 v_i 到 v_j 长度小于或等于 $n-1$ 的路。

推论 在 n 阶图 G 中,若从结点 v_i 到 $v_j(v_i\neq v_j)$ 存在路,则必存在长度小于或等于 $n-1$ 的基本路。

由定理 8.3.1 的证明过程知,本推论成立。类似地可证明下列定理和推论。

定理 8.3.2 ▶ 在 n 阶图 G 中,若存在结点 v_i 到自身的回路,则必存在 v_i 到自身长度小于

或等于 n 的回路。

推论 在 n 阶图 G 中,若存在结点 v_i 到自身的简单回路,则必存在 v_i 到自身长度小于或等于 n 的基本回路。

8.4 图的矩阵表示

在第 3 章中,对于给定集合 A 上的关系 R,可用一个有向图表示,这种图形表示了集合 A 上元素之间的关系,关系图也表示了集合中元素间的邻接关系。对于关系图,可用一个矩阵表示,一个矩阵也必对应于一个标定结点序号的关系图。

定义 8.4.1 ▶ 设 $G = <V, E>$ 是一个简单图,$V = \{v_1,$ $v_2, \cdots, v_n\}$,$\boldsymbol{A}(G) = (a_{ij})_{n \times n}$,其中:

$$a_{ij} = \begin{cases} 1, & v_i \text{ 到 } v_j \text{ 有边} \\ 0, & v_i \text{ 到 } v_j \text{ 无边或 } i-j \end{cases} \quad i,j = 1, \cdots, n$$

称 $\boldsymbol{A}(G)$ 为 G 的邻接矩阵,简记为 $\boldsymbol{A}$。

例如,图 8.4.1 的邻接矩阵为

$$\boldsymbol{A}(G) = \begin{bmatrix} 0 & 1 & 0 & 1 \\ 1 & 0 & 1 & 1 \\ 0 & 1 & 0 & 1 \\ 1 & 1 & 1 & 0 \end{bmatrix}$$

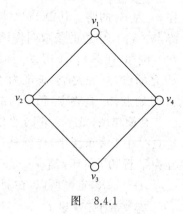

图 8.4.1

又如,图 8.4.2(a)的邻接矩阵为

$$\boldsymbol{A}(G) = \begin{bmatrix} 0 & 1 & 0 & 0 \\ 0 & 0 & 1 & 1 \\ 1 & 1 & 0 & 1 \\ 1 & 0 & 0 & 0 \end{bmatrix}$$

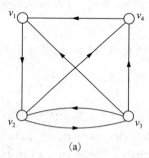 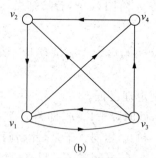

（a）   （b）

图 8.4.2

邻接矩阵具有以下性质。

(1) 邻接矩阵的元素或是 0 或是 1。这样的矩阵叫布尔矩阵。邻接矩阵是布尔矩阵。

(2) 无向图的邻接矩阵是对称矩阵,有向图的邻接矩阵不一定是对称矩阵。

(3) 邻接矩阵与结点在图中标定次序有关。例如,图 8.4.2(a)的邻接矩阵是 $\boldsymbol{A}(G)$,若将图 8.4.2(a)中的接点 v_1 和 v_2 的标定次序调换,得到图 8.4.2(b),图 8.4.2(b)的邻接矩阵是 $\boldsymbol{A}'(G)$。

$$A'(G) = \begin{pmatrix} 0 & 0 & 1 & 1 \\ 1 & 0 & 0 & 0 \\ 1 & 1 & 0 & 1 \\ 0 & 1 & 0 & 0 \end{pmatrix}$$

考察 $A(G)$ 和 $A'(G)$ 发现,先将 $A(G)$ 的第一行与第二行对调,再将第一列与第二列对调可得到 $A'(G)$。称 $A'(G)$ 与 $A(G)$ 是置换等价的。

一般地说,把 n 阶方阵 A 的某些行对调,再把相应的列做同样的对调,得到一个新的 n 阶方阵 A',则称 A' 与 A 是置换等价的。可以证明置换等价是 n 阶布尔方阵集合上的等价关系。

虽然,对于同一个图,由于结点的标定次序不同而得到不同的邻接矩阵,但是这些邻接矩阵是置换等价的。今后略去结点标定次序的任意性,取任意一个邻接矩阵表示该图。

(4) 对有向图来说,邻接矩阵 $A(G)$ 的第 i 行中 1 的个数是 v_i 的出度,第 j 列中 1 的个数是 v_j 的入度。

(5) 零图的邻接矩阵的元素全为零,叫作零矩阵。反过来,如果一个图的邻接矩阵是零矩阵,则此图一定是零图。

定理 8.4.1 ▶ 设 $A(G)$ 是图 G 的邻接矩阵,$A(G)^k = A(G)A(G)^{k-1}$,$A(G)^k$ 的第 i 行第 j 列元素 a_{ij}^k 等于从 v_i 到 v_j 长度为 k 的路的条数。其中 a_{ii}^k 为 v_i 到自身长度为 k 的回路数。

推论 设 $G = <V,E>$ 是 n 阶简单有向图,A 是有向图 G 的邻接矩阵,$B^k = A + A^2 + \cdots + A^k$,$B^k = (b_{ij}^k)_{n \times n}$,则 b_{ij}^k 是 G 中由 v_i 到 v_j 长度小于或等于 k 的路的条数。$\sum\limits_{i=1}^{n}\sum\limits_{j=1}^{n} b_{ij}^k$ 是 G 中长度小于或等于 k 的路的总条数。$\sum\limits_{i=1}^{n} b_{ii}^k$ 是 G 中长度小于或等于 k 的回路数。

例 8.4.1 设 $G = <V,E>$ 为简单有向图,如图 8.4.3 所示,写出 G 的邻接矩阵 A,算出 A^2, A^3, A^4,确定 v_1 到 v_2 有多少条长度为 3 的路。v_1 到 v_3 有多少条长度为 2 的路? v_2 到自身长度为 3 和长度为 4 的回路各有多少条?

解 邻接矩阵 A 和 A^2, A^3, A^4 如下。

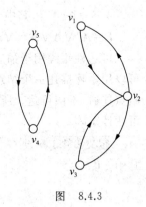

图　8.4.3

$$A = \begin{pmatrix} 0 & 1 & 0 & 0 & 0 \\ 1 & 0 & 1 & 0 & 0 \\ 0 & 1 & 0 & 0 & 0 \\ 0 & 0 & 0 & 0 & 1 \\ 0 & 0 & 0 & 1 & 0 \end{pmatrix} \quad A^2 = \begin{pmatrix} 1 & 0 & 1 & 0 & 0 \\ 0 & 2 & 0 & 0 & 0 \\ 1 & 0 & 1 & 0 & 0 \\ 0 & 0 & 0 & 1 & 0 \\ 0 & 0 & 0 & 0 & 1 \end{pmatrix}$$

$$A^3 = \begin{pmatrix} 0 & 2 & 0 & 0 & 0 \\ 2 & 0 & 2 & 0 & 0 \\ 0 & 2 & 0 & 0 & 0 \\ 0 & 0 & 0 & 0 & 1 \\ 0 & 0 & 0 & 1 & 0 \end{pmatrix} \quad A^4 = \begin{pmatrix} 2 & 0 & 2 & 0 & 0 \\ 0 & 4 & 0 & 0 & 0 \\ 2 & 0 & 2 & 0 & 0 \\ 0 & 0 & 0 & 1 & 0 \\ 0 & 0 & 0 & 0 & 1 \end{pmatrix}$$

$a_{12}^3 = 2$,所以 v_1 到 v_2 长度为 3 的路有 2 条,它们分别是 $v_1v_2v_1v_2$ 和 $v_1v_2v_3v_2$。

$a_{13}^2 = 1$,所以 v_1 到 v_3 长度为 2 的路有 1 条,即 $v_1v_2v_3$。

$a_{22}^3 = 0$,v_2 到自身无长度为 3 的回路。

$a_{22}^4 = 4$,v_2 到自身有 4 条长度为 4 的回路,它们分别是 $v_2v_1v_2v_1v_2$、$v_2v_3v_2v_3v_2$、

$v_2v_3v_2v_1v_2$ 和 $v_2v_1v_2v_3v_2$。

定义 8.4.2 ▶设 $G-<V,E>$ 是简单有向图，$V=\{v_1,v_2,\cdots,v_n\}$，$\boldsymbol{P}(G)=(p_{ij})_{n\times n}$，其中：

$$p_{ij}=\begin{cases}1, & v_i \text{到} v_j \text{可达} \\ 0, & v_i \text{到} v_j \text{不可达}\end{cases} \quad i,j=1,\cdots,n$$

称 $\boldsymbol{P}(G)$ 为 G 的可达性矩阵，简记为 $\boldsymbol{P}$。

可达性矩阵 $\boldsymbol{P}(G)$ 的主对角线元素全为 1。

设 $G=<V,E>$ 是 n 阶简单有向图，$V=\{v_1,v_2,\cdots,v_n\}$，由可达性矩阵的定义知，当 $i\neq j$ 时，如果 v_i 到 v_j 有路，则 $p_{ij}=1$；如果 v_i 到 v_j 无路，则 $p_{ij}=0$；如果 v_i 到 v_j 有路，则必存在长度小于或等于 $n-1$ 的路。依据定理 8.4.1 的推论，计算图 G 的可达性矩阵 $\boldsymbol{P}$。

先计算 $\boldsymbol{B}_{n-1}=\boldsymbol{A}+\boldsymbol{A}^2+\cdots+\boldsymbol{A}^{n-1}$，设 $\boldsymbol{B}_{n-1}=(b_{ij}^{n-1})_{n\times n}$。若 $b_{ij}^{n-1}\neq 0$，则令 $p_{ij}=1$。若 $b_{ij}^{n-1}=0$，则令 $p_{ij}=0$，$i,j=1,\cdots,n$。再令 $p_{ii}=1$，$i=1,\cdots,n$，就得到了图 G 的可达性矩阵 $\boldsymbol{P}$。

令 $\boldsymbol{A}^0$ 为 n 阶单位阵，则上述算法也可做如下改进。

计算 $\boldsymbol{C}_{n-1}=\boldsymbol{A}^0+\boldsymbol{B}_{n-1}=\boldsymbol{A}^0+\boldsymbol{A}+\boldsymbol{A}^2+\cdots+\boldsymbol{A}^{n-1}$，设 $\boldsymbol{C}_{n-1}=(c_{ij}^{n-1})_{n\times n}$。若 $c_{ij}^{n-1}\neq 0$，则令 $p_{ij}=1$。若 $c_{ij}^{n-1}=0$，则令 $p_{ij}=0$，$i,j=1,\cdots,n$。

使用上述方法，计算例 8.4.3 中图 G 的可达性矩阵，

$$\boldsymbol{C}_4=\boldsymbol{A}^0+\boldsymbol{A}+\boldsymbol{A}^2+\boldsymbol{A}^3+\boldsymbol{A}^4=\begin{pmatrix}4&3&3&0&0\\3&7&3&0&0\\3&3&4&0&0\\0&0&0&3&1\\0&0&0&1&3\end{pmatrix} \quad \boldsymbol{P}=\begin{pmatrix}1&1&1&0&0\\1&1&1&0&0\\1&1&1&0&0\\0&0&0&1&1\\0&0&0&1&1\end{pmatrix}$$

计算简单有向图 G 的可达性矩阵 $\boldsymbol{P}$，还可以用下述方法。

设 $\boldsymbol{A}$ 是 G 的邻接矩阵，令 $\boldsymbol{A}=(a_{ij})_{n\times n}$，$\boldsymbol{A}^{(k)}=(a_{ij}^{(k)})_{n\times n}$，$\boldsymbol{A}^0$ 为 n 阶单位阵。

$\boldsymbol{A}^{(2)}=\boldsymbol{A}\circ\boldsymbol{A}$，其中 $a_{ij}^{(2)}=(a_{i1}\wedge a_{1j})\vee(a_{i2}\wedge a_{2j})\vee\cdots\vee(a_{in}\wedge a_{nj})$，$i,j=1,\cdots,n$。

$\boldsymbol{A}^{(3)}=\boldsymbol{A}\circ\boldsymbol{A}^{(2)}$，其中 $a_{ij}^{(3)}=(a_{i1}\wedge a_{1j}^{(2)})\vee\cdots\vee(a_{in}\wedge a_{nj}^{(2)})$，$i,j=1,\cdots,n$。

$\boldsymbol{P}=\boldsymbol{A}^0\vee\boldsymbol{A}\vee\boldsymbol{A}^{(2)}\vee\boldsymbol{A}^{(3)}\vee\cdots\vee\boldsymbol{A}^{(n-1)}$。其中，运算 $\vee$ 是矩阵对应元素的析取。

可达性矩阵用于描述有向图的一个结点到另一个结点是否有路，即是否可达。无向图也可以用矩阵描述一个结点到另一个结点是否有路。在无向图中，如果结点之间有路，称这两个结点连通，不叫可达。所以把描述一个结点到另一个结点是否有路的矩阵叫连通矩阵，而不叫可达性矩阵。

定义 8.4.3 ▶设 $G=<V,E>$ 是简单无向图，$V=\{v_1,v_2,\cdots,v_n\}$，$\boldsymbol{P}(G)=(p_{ij})_{n\times n}$，其中：

$$p_{ij}=\begin{cases}1, & v_i \text{与} v_j \text{连通} \\ 0, & v_i \text{与} v_j \text{不连通}\end{cases} \quad i,j=1,\cdots,n$$

称 $\boldsymbol{P}(G)$ 为 G 的连通矩阵，简记为 $\boldsymbol{P}$。

无向图的邻接矩阵是对称矩阵，无向图的连通矩阵也是对称矩阵。求连通矩阵的方法与求可达性矩阵的方法类似。

定义 8.4.4 ▶设 $G=<V,E>$ 是无向图，$V=\{v_1,v_2,\cdots,v_p\}$，$E=\{e_1,e_2,\cdots,e_q\}$，$\boldsymbol{M}(G)=(m_{ij})_{p\times q}$。其中：

$$P_{ij}=\begin{cases}1, & v_i \text{ 与 } e_j \text{ 关联} \\ 0, & v_i \text{ 与 } e_j \text{ 无关联}\end{cases} \quad i=1,\cdots,p,\ j=1,\cdots,q$$

称 $\boldsymbol{M}(G)$ 为无向图 G 的完全关联矩阵,简记为 $\boldsymbol{M}$。

例如,图 8.4.4 的完全关联矩阵为

$$\boldsymbol{M}(G)=\begin{pmatrix}1 & 1 & 1 & 0 \\ 1 & 1 & 0 & 0 \\ 0 & 0 & 1 & 1 \\ 0 & 0 & 0 & 1\end{pmatrix}$$

设 $G=<V,E>$ 是无向图,G 的完全关联矩阵 $\boldsymbol{M}(G)$ 有以下性质。

(1) 每列元素之和均为 2。这说明每条边关联两个结点。

(2) 每行元素之和是对应结点的度数。

(3) 所有元素之和是图各结点度数的和,也是边数的 2 倍。

(4) 两列相同,则对应的两条边是平行边。

(5) 某行元素全为零,则对应结点为孤立点。

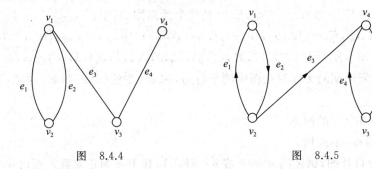

图　8.4.4　　　　　　　　图　8.4.5

定义 8.4.5 ▶ 设 $G=<V,E>$ 是有向图,$V=\{v_1,v_2,\cdots,v_p\}$,$E=\{e_1,e_2,\cdots,e_q\}$,$\boldsymbol{M}(G)=(m_{ij})_{p\times q}$。其中:

$$m_{ij}=\begin{cases}1, & v_i \text{ 是 } e_j \text{ 的始点} \\ -1, & v_i \text{ 是 } e_j \text{ 的终点} \quad i=1,\cdots,p,\ j=1,\cdots,q \\ 0, & v_i \text{ 与 } e_j \text{ 不关联}\end{cases}$$

称 $\boldsymbol{M}(G)$ 为有向图 G 的完全关联矩阵,简记为 $\boldsymbol{M}$。

图 8.4.5 的完全关联矩阵为

$$\boldsymbol{M}(G)=\begin{pmatrix}-1 & 1 & 0 & 0 & 0 \\ 1 & -1 & 1 & 0 & 0 \\ 0 & 0 & 0 & 1 & 1 \\ 0 & 0 & -1 & -1 & -1\end{pmatrix}$$

设 $G=<V,E>$ 是有向图,G 的完全关联矩阵 $\boldsymbol{M}(G)$ 有以下性质。

(1) 每列有一个 1 和一个 -1,这说明每条有向边有一个始点和一个终点。

(2) 每行中 1 的个数是对应结点的出度,-1 的个数是对应结点的入度。

(3) 所有元素之和是 0,这说明所有结点出度的和等于所有结点入度的和。

(4) 两列相同,则对应的两条边是平行边。

习 题

1. 设无向图 G 有 16 条边,有 3 个 4 度结点、4 个 3 度结点,其余结点的度数均小于 3,问:G 中至少有几个结点?

2. 试证明习题 2 图中(a)和(b)两个图是同构的。

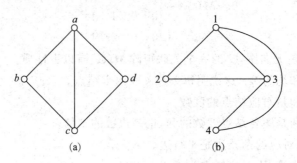

(a)   (b)

习题 2 图

3. 设 $G_1 = \langle V_1, E_1 \rangle$ 与 $G_2 = \langle V_2, E_2 \rangle$ 是两个无向图,其中:

$V_1 = \{a, b, c, d, e\}, E_1 = \{(a,b), (a,c), (a,c), (b,c), (b,d), (d,e), (c,e), (e,e)\}$

$V_2 = \{1, 2, 3, 4, 5\}, E_2 = \{(1,2), (1,3), (1,3), (2,3), (2,4), (4,5), (3,5), (4,4)\}$

E_1 和 E_2 中重复出现的无序对是图中的平行边,如 E_1 中的 (a,c) 和 E_2 中的 $(1,3)$ 都是平行边。

(1) 试画出 G_1 和 G_2 的图形。

(2) 证明 G_1 和 G_2 不同构。

4. 设 G 是 n 阶自补图,试证明 $n = 4k$ 或 $n = 4k+1$,其中 k 为正整数。画出 5 个结点的自补图。是否有 3 个结点或 6 个结点的自补图?

5. 若 G 是一个至少有三个结点的连通图,则下列命题是等价的。

(1) G 没有桥。

(2) G 的每两个结点在一条公共的简单回路上。

(3) G 的每一个结点和一条边在一条公共的简单回路上。

(4) G 的每两条边在一条公共的简单回路上。

(5) 对 G 的每一对结点和每一条边,有一条联结这两个结点而且含有这条边的简单路。

(6) 对 G 的每一对结点和每一条边,有一条联结这两个结点而不含有这条边的基本路。

(7) 对每三个结点,有一条联结任何两个结点而且含第三个结点的简单路。

6. 设 $G = \langle V, E \rangle$ 是一个简单有向图,$V = \{v_1, v_2, v_3, v_4\}$,其邻接矩阵如下:

$$A(G) = \begin{pmatrix} 0 & 1 & 0 & 0 \\ 0 & 0 & 1 & 1 \\ 1 & 1 & 0 & 1 \\ 1 & 1 & 0 & 0 \end{pmatrix}$$

(1) 求 v_1 的出度 $\deg^+(v_1)$。

(2) 求 v_4 的入度 $\deg^-(v_4)$。

(3) 由 v_1 到 v_4 长度为 2 的路有几条?

7. 有向图 G 如习题7图所示。

（1）写出 G 的邻接矩阵。

（2）根据邻接矩阵求各结点的出度和入度。

（3）求 G 中长度为3的路的总数，其中有多少条回路？

（4）求 G 的可达性矩阵。

（5）求 G 的完全关联矩阵。

（6）由完全关联矩阵求各结点的出度和入度。

8. 无向图 G 如习题8图所示。

（1）写出 G 的邻接矩阵。

（2）根据邻接矩阵求各结点的度数。

（3）求 G 中长度为3的路的总数，其中有多少条回路？

（4）求 G 的连通矩阵。

（5）求 G 的完全关联矩阵。

（6）由完全关联矩阵求各结点的度数。

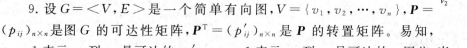

习题7图

习题8图

9. 设 $G=\langle V,E\rangle$ 是一个简单有向图，$V=\{v_1,v_2,\cdots,v_n\}$，$\boldsymbol{P}=(p_{ij})_{n\times n}$ 是图 G 的可达性矩阵，$\boldsymbol{P}^{\mathrm{T}}=(p'_{ij})_{n\times n}$ 是 $\boldsymbol{P}$ 的转置矩阵。易知，$p_{ij}=1$ 表示 v_i 到 v_j 是可达的；$p'_{ij}=p_{ji}=1$ 表示 v_j 到 v_i 是可达的。因此，当 $p_{ij}\wedge p'_{ij}=1$ 时，v_i 和 v_j 是互相可达的。由此可求得图 G 的强分图。例如，图 G 的可达性矩阵 $\boldsymbol{P}$ 为

$$\boldsymbol{P}=\begin{pmatrix}1&0&1&1&1\\0&1&1&1&1\\0&0&1&1&1\\0&0&1&1&1\\0&0&1&1&1\end{pmatrix}\quad \boldsymbol{P}^{\mathrm{T}}=\begin{pmatrix}1&0&0&0&0\\0&1&0&0&0\\1&1&1&1&1\\1&1&1&1&1\\1&1&1&1&1\end{pmatrix}\quad \boldsymbol{P}\wedge\boldsymbol{P}^{\mathrm{T}}=\begin{pmatrix}1&0&0&0&0\\0&1&0&0&0\\0&0&1&1&1\\0&0&1&1&1\\0&0&1&1&1\end{pmatrix}$$

其中：$\boldsymbol{P}\wedge\boldsymbol{P}^{\mathrm{T}}$ 定义为矩阵 $\boldsymbol{P}$ 和矩阵 $\boldsymbol{P}^{\mathrm{T}}$ 的对应元素的合取。

由此可知由 $\{v_1\}$，$\{v_2\}$，$\{v_3,v_4,v_5\}$ 导出的子图是 G 的强分图。

试用这种办法求题2的所有强分图。

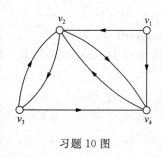

习题10图

10. 设 $G=\langle V,E\rangle$ 是一个简单有向图，$V=\{v_1,v_2,\cdots,v_n\}$，$\boldsymbol{A}$ 是 G 的邻接矩阵，G 的距离矩阵 $\boldsymbol{D}=(d_{ij})_{n\times n}$ 定义如下：

$d_{ij}=\infty$　如果 $d\langle v_i,v_j\rangle=\infty$。

$d_{ii}=0$　$i=1,\cdots,n$。

$d_{ij}=k$　k 是使 $a_{ij}^k\neq 0$ 的最小正整数。

试写出习题10图的距离矩阵 $\boldsymbol{D}$，并说明 $d_{ij}=1$ 是什么意义。

11. 试证明一个有向图 G 是单侧连通的当且仅当它有一条经过每一结点的路。

12. 设 $G=\langle V,E\rangle$ 是一个简单图，$|V|=n$，$|E|=m$，$m>1/2(n-1)(n-2)$。证明 G 是连通的。

Chapter 9　Euler Graph and Hamiltonian Graph

Can you start from a vertex and move forward along the edges of the graph, passing through each edge of the graph exactly once and returning to that vertex? Similarly, can we start from a vertex and move along the edges of the graph, passing through each vertex exactly once and returning to that vertex? This chapter conducts research and discussion based on this issue.

9.1　Euler Graph

Definition 9.1.1　In graph G (directed graph or undirected graph) without isolated nodes, if there is a path that passes through each edge once and only once, the path is called Euler path. If a loop passes through each edge once and only once, the loop is called an Euler cycle. A graph with Euler loop is called an Euler Graph.

Theorem 9.1.1　Undirected graph G has an Euler road if and only if G is connected and has zero or two odd degree nodes.

Proof　Suppose G has an Euler path, in the following we prove that G is connected and has zero or two odd degree nodes.

Suppose G has an Euler path $L: v_0 e_1 v_1 e_2 v_2 ... e_k v_k$, the road goes through each side of G. Because there are no isolated nodes in G, so this path goes through all the nodes of G, that is, all the nodes of G are on this path. Therefore, any two nodes in G are connected, so G is a connected graph.

Suppose v_i is any node of graph G. If v_i is not the endpoint of L, each time it passes through v_i along L, it passes through the two edges associated with this node, and the degree of this node is increased by 2. Since each edge of G is on the path and does not repeat, v_i must have an even degree. If $v_i = v_0$, the endpoint of L, when $v_0 = v_k$, then the degrees of v_0 and v_k are even, that is, the nodes with no black degree in G. When $v_0 \neq v_k$, the degrees of v_0 and v_k are odd, that is, there are two odd degree nodes in G.

Suppose G is connected and has zero or two odd degree nodes, in the following we prove that G has an Euler path.

Suppose G is a connected graph with zero or two odd degree nodes, an Euler path is constructed by the following method.

(1) If there are two oddity nodes in G, a simple path is constructed from one of them v_0.

Starting from v_0, passing through the associated edge e_1 and entering v_1. If v_1 is an even degree node, it must be possible to enter v_2 through the associated edge e_2 from v_1. And so on, you take each side only once. Since G is a connected graph, another odd node v_k must be reached, thus obtaining a simple path L_1: $v_0 e_1 v_1 e_2 v_2 \ldots e_k v_k$.

If there is no odd node in G, then starting from any node v_0, using the above method, it must return to node v_0 and get a simple loop L_1: $v_0 e_1 v_1 e_2 v_2 \ldots v_0$.

(2) If L_1 goes through all the edges of G, then L_1 is Euler path.

(3) Otherwise, L_1 is deleted from G to obtain the subgraph G', and the degree of each node in G' is even. Since G is a connected graph, L_1 and G' overlap at least one node v_i, and repeat from v_i in G' to obtain a simple loop L_2.

(4) If the combination of L_1 and L_2 is exactly G, then Euler path is obtained; otherwise, repeat to get simple loop L_3.

And so on until you get an Euler path that goes through all the edges of G.

Inference Undirected graph G has an Euler loop if and only if G is connected and all nodes are even.

This corollary states that an undirected graph G is an Euler graph if and only if G is connected and all nodes are even.

The degree of each node in Figure 9.1.1(a) is even 2, so there is an Euler loop in (a), which is an Euler graph; in Figure 9.1.1(b), there are two nodes whose degree is odd 3, so there is an Euler path in (b), but there is no Euler loop, which is not an Euler graph. In Figure 9.1.1(c), the degrees of all four nodes are odd 3, and there is no Euler path or Euler loop in (c), which is not an Euler graph.

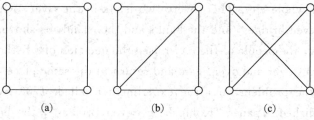

(a)   (b)   (c)

Figure 9.1.1

Theorem 9.1.2 A directed graph G has an Euler loop if and only if G is strongly connected (For each pair of v_i, v_j, $v_i \neq v_j$, there are paths from v_i to v_j and from v_j to v_i) and the indegree of each node is equal to the outdegree.

This theorem states that it is necessary and sufficient for a directed graph G to be an Euler graph if G is strongly connected and the indegree of each node is equal to the outdegree.

Theorem 9.1.3 A directed graph G has an Euler path if and only if G is unidirectional connected(for any node v_1 and v_2, there is at least one path from v_1 to v_2 and from v_2 to v_1) and the indegree of each node is equal to the outdegree except for two nodes. In these two nodes, the indegree of one node is 1 greater than the outdegree, and the indegree of the other node is 1 less than the outdegree.

These two theorems can be regarded as generalizations of Theorem 9.1.1 and corollaries. Because for any node of a directed graph, if the indegree is equal to the outdegree, the degree of the node is even. If the difference between indegree and outdegree is 1, then the degree of the node is odd. Therefore, the proofs of Theorem 9.1.2 and Theorem 9.1.3 are similar to the proofs of Theorem 9.1.1.

Figure 9.1.2(a) is strongly connected and the indegree of each node is equal to the outdegree, which is equal to 1. Therefore, there is an Euler loop in (a), which is the Euler graph. Figure 9.1.2 (b) is strongly connected, and there are two nodes with the same indegree and outdegree. There are two nodes with different indegree and outdegree, among which the indegree of one node is 1 greater than the outdegree, and the indegree of the other node is 1 less than the outdegree. So there is an Euler path, but there is no Euler loop, not an Euler graph. In Figure 9.1.2(c), the indegree and outdegree of the four nodes are not equal, and there is no Euler path, let alone Euler loop.

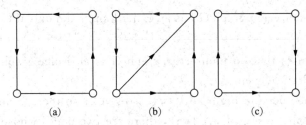

(a)   (b)   (c)

Figure   9.1.2

Konigsberg Seven Bridges

Once upon a time, there was a city called Konigsberg where there was a Pregel River that ran through the whole city. There were two islets in the river, which were connected with the island by seven bridges, and the banks and the island, as shown in Figure 9.1.3. In the mid-18th century, the people of the city raised the question of whether it was possible to walk the seven Bridges over and over again and return to the same place. Many people in the city tried to solve the problem, but the experimenter failed. 1736 Swiss mathematician Leonhard Euler published a paper "Konigsberg Seven Bridges", the basic content of this paper is Theorem 9.1.1. In this paper, it is proved for the first time that this problem is unsolvable. He abstracts the river bank and the island into the node of the picture, and draws the bridge as the corresponding connecting side, as shown in Figure 9.1.4. Therefore, it is equivalent to the existence of a loop in the figure that goes through each edge once and only

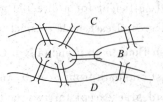

Figure   9.1.3

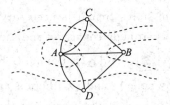

Figure   9.1.4

once, that is, there is an Euler loop in the figure. Because all four nodes in the graph have odd degrees. So there is no Euler loop.

9.2　Hamiltonian Graph

A problem very similar to Euler loops and Euler graph is the Hamiltonian loop and Hamiltonian graph problem. It was first proposed by the Irish mathematician Sir Willian Hamiltonian in 1859 first put forward.

Definition 9.2.1　In figure G (directed or undirected), if there is a path that passes through each node once and only once, the path is called Hamiltonian path. If there is a loop that passes through each node once and only once, the loop is called a Hamiltonian loop. A graph with a Hamiltonian loop is called a Hamiltonian graph.

Different from the Euler graph, there is not a simple sufficient and necessary condition to judge whether a graph is a Hamiltonian graph, only some necessary and sufficient conditions. The necessary and sufficient condition for determining a graph to be a Hamiltonian graph are one of the fundamental unsolved problems in graph theory. Here are some necessary and sufficient conditions for Hamiltonian graphs.

Gives a pact: $G=<V, E>$ is a graph, $S \subseteq V$, $G-S$ is used to represent the subgraph resulting from deleting all nodes in S from graph G.

Theorem 9.2.1　Undirected graph $G=<V, E>$ is a Hamiltonian graph, and S is any non empty subset of V, then $W(G-S) \leqslant |S|$.

Proof　Let C be a Hamiltonian loop of G, $\forall v_1 \in S$, then $C-\{v_1\}$ (delete node v_1 from C) is a path, in which all nodes are connected. If another node V_2 in S is deleted, then $W(C-\{v_1, v_2\}) \leqslant 2$, so $W(C-\{v_1, v_2\}) \leqslant |\{v_1, v_2\}|$. By induction, we get $W(C-S) \leqslant |S|$.

The condition of this theorem is a necessary condition of the Hamiltonian graph, not a sufficient condition. By this theorem, it can be proved that a graph is not a Hamiltonian graph, that is, a graph that satisfies the condition of this theorem is not necessarily a Hamiltonian graph, and a graph that does not satisfy the condition of this theorem is definitely not a Hamiltonian graph.

For example, Figure 9.2.1 is called the Peterson graph. The Peterson graph satisfies the condition of Theorem 9.2.1, but it can be verified that it is not a Hamiltonian graph. Another example is shown in Figure 9.2.2, $S=\{a, b\}$, then $W(G-S)=3$, $|S|=2$. Therefore, $W(G-S) \leqslant |S|$ does not hold. So it's not a Hamiltonian graph.

Example　9.2.1　If G is an undirected connected graph, please prove that if there are cut points or bridges in G (If a vertex is deleted and becomes a disconnected graph, the vertex is called a cut point, and the edges where all vertices are cut points are called bridges), then G must not be a Hamiltonian graph.

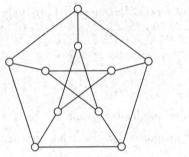 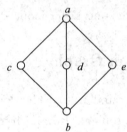

Figure 9.2.1   Figure 9.2.2

Proof Let v be the cutting point. Delete node v, and the graph becomes disconnected and has at least two connected branches. To $S=\{v\}$, $W(G-S)\geqslant 2>1=|S|$, namely $W(G-S)>|S|$. According to Theorem 9.2.1, G is not a Hamiltonian graph.

If there is a bridge in G, then one end of the bridge is a cut point. It comes down to the case where there is a cut point in G. So G is not a Hamiltonian graph either.

Theorem 9.2.2 If $G=<V, E>$ order n is a simple undirected graph, each pair of nodes in the G degree and greater than or equal to $n-1$, then there is a Hamiltonian path in G.

Inference $G=<V, E>$ is an n-order simple undirected graph, if the sum of the degrees of each pair of nodes in G is greater than or equal to n, then a Hamiltonian loop exists in G.

Theorem 9.2.2 and its corollary give sufficient but not necessary conditions for the existence of a Hamiltonian path and a Hamiltonian loop in the figure. For example, In Figure 9.2.3(a), there are 6 nodes, and the sum of the degrees of any two nodes is less than 5, but there is a Hamiltonian path in the figure. As shown in Figure 9.2.3 (b), there are also 6 nodes, and the sum of the degrees of any two nodes is less than 6. However, there is a Hamiltonian loop in the graph.

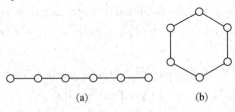

(a)   (b)

Figure 9.2.3

Example 9.2.2 $G=<V, E>$ is a simple undirected graph, $|V|=n$, $|E|=m$, suppose $m>(n-1)(n-2)$. Prove that there is a Hamiltonian path in G.

Proof First, it is proved that the sum of the degrees of any two different nodes in figure G is greater than or equal to $n-1$, and then the result to be proved in this example is obtained by using Theorem 9.2.2.

By contradiction, let there be two nodes v_1 and v_2 in graph G, $\deg(v_1)+\deg(v_2)<n-1$,

so $\deg(v_1)+\deg(v_2)\leqslant n-2$. After deleting nodes v_1 and v_2, the figure G' is obtained. G' is an undirected simple graph with $n-2$ nodes. Let the number of edges of G' be m', $m' \geqslant m-(n-2)>(n-1)(n-2)-(n-2)=(n-2)^2$, then $m'>(n-2)^2$. On the other hand, G' is an undirected simple graph with $n-2$ nodes, and it has $\dfrac{1}{2}(n-2)(n-3)$ edges at most. So $m'\leqslant(n-2)(n-3)$.

This is a contradiction. So $\deg(v_1)+\deg(v_2)\geqslant n-1$, according to Theorem 9.2.2, there is a Hamiltonian path in G.

Example 9.2.3　There are 5 scenic spots in a certain place. If there are two paths connecting each scenic spot with other points, can I go through each scenic spot to complete the 5 scenic spots exactly once?

Solution　Since there are 5 scenic spots, it can be regarded as an undirected graph with 5 nodes, in which each place has two paths connecting with other nodes, so $\deg(v_i)=2$, $i=1, \ldots, 5$. So for any two points v_i, v_j

$$\deg(v_i)+\deg(v_j)=2+2=4=5-1$$

Theorem 9.2.2 shows that there must be a Hamiltonian path in this figure, and the problem has a solution.

Definition 9.2.2　A given graph $G=<V, E>$ has n nodes, if we connect non-adjacent nodes in graph G whose degrees sum to at least n, we obtain graph G'. Repeat the above steps for graph G' until no such pairs of nodes exist anymore. The resulting graph is called the closure of the original graph G, denoted as $C(G)$.

For example, Figure 9.2.4 shows the process of constructing the closure of graph G with six nodes. In this case, $C(G)$ is the complete graph. In general, $C(G)$ may not be a complete graph.

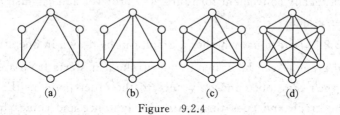

(a)　　　　(b)　　　　(c)　　　　(d)

Figure　9.2.4

Theorem 9.2.3　Let G be a simple graph and G be a Hamiltonian graph if and only if the closure of G is a Hamiltonian graph.

Example 9.2.4　Graph G is shown in Figure 9.2.5, proving that there is no Hamiltonian path in G.

Proof　Take any node a and label it with A, and all the nodes that are adjacent to it are labeled with B. And you keep marking all the nodes that are adjacent to B with A, and then you mark all the nodes that are adjacent to A with B, until you've marked all the nodes.

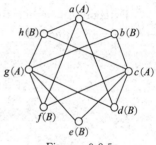

Figure　9.2.5

If there is a Hamiltonian path in the graph, then it must alternate between the node marked A and the node marked B, so either the node marked A is the same as the node marked B, or the difference between the two is 1. However, there are 3 nodes marked A and 5 nodes marked B in this example, and the difference between them is 2, so there is no Hamiltonian path in this figure.

9.3 Application of Euler Graph and Hamiltonian Graph

Example 9.3.1 Postman problem.

Professor Guan Meigu, a Mathematician in China, first posed a question: when a postman delivers mail, he has to go through all the streets within the scope of delivery each time, and then he returns to the post office. What route should he taken to make the distance the shortest? This is the postman problem. If the delivery location is represented by points, the street available to the postman is represented by edges, and the weight of each edge represents the distance between two adjacent locations, then a weight graph G is formed.

The postman problem is actually to find a loop passing through each edge on the weight graph G, and the weight sum of each edge is the smallest, so such a loop is called the optimal loop. If the graph of the street taken by the postman is an Euler graph, then the postman problem is easy to solve, because any Euler circuit in graph G is the optimal circuit. But if graph G is not an Euler graph, then the problem is complicated, and the optimal circuit of graph G will pass through some edges in the graph more than once. For example, in the design of the park scenic route, to consider the main factor is to view all the scenic spots as far as possible, at the same time as much as possible to take repeated paths, in order to save energy. How to design the path connection between various scenic spots can be scientifically and rationally convenient for tourists to browse, and can also be solved with the knowledge of Euler graph.

Example 9.3.2 Determine whether a graph can be drawn in one stroke.

An undirected graph G can be drawn in one stroke if it starts from a point in the graph and goes through each edge once and only once to the other point in the graph and starts from a point in the graph and goes through each edge once and comes back to that point again. To determine whether a graph can be drawn with one stroke, in fact, is to determine whether it contains Euler paths containing all the edges, that is, a graph is all even-degree nodes or if two nodes are odd and the rest are even.

Determine whether the following Figure 9.3.1 can be drawn in one stroke without repeating on either side.

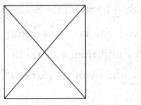

Figure 9.3.1

Solution The left figure cannot be drawn with one stroke, and there are 4 odd degree nodes. On the right, you can draw it in one stroke, and each node has an even degree.

Example 9.3.3 Travel salesman problem.

The traveling salesman problem is to go from a certain city to each city once and only once, and then to return to the starting city and ask for a loop that minimizes the sum of the loop.

The problem of traveling salesman is essentially to find a Hamiltonian loop on a complete graph(a graph with every pair of nodes connected by edges is called a complete graph) with edge weighting, so that the sum of weights on each edge of the loop is minimized, and the edge weights are the length of the traffic lines connecting the two cities.

An efficient algorithm to solve the traveling salesman problem has not been successful, there is a kind of approximate algorithm is the most adjacent algorithm, the basic idea is very simple: when the salesperson is in a city, the next step is to choose the nearest city that has not yet been to as the next stop until all the cities are finished.

Nearest neighbor algorithm:

(1) Select any point in the complete graph as the starting point, find a point closest to the starting point, form an initial path of the edge, and then use step (1) to expand the path point by point.

(2) Let x represent the latest vertex added to the path. From all vertices on the path, select a point closest to x and add the edge connecting x to this point of the path. Repeat this step until all vertices in the complete graph are included in the path.

(3) Put the edge between the starting point and the last added vertex into the loop.

Example 9.3.4 Seating arrangement.

Suppose you know the following facts: a speaks English, b speaks English and Chinese, c speaks English, Italian and Russian, d speaks Japanese and Chinese, e speaks German and Italian, f speaks French, Japanese and Russian, and g speaks French and German. How should the seven people be seated so that each of them could talk to the person next to him?

Solution Suppose undirected graph $G = <V, E>$. $V = \{a, b, c, d, e, f, g\}$, $E = \{<u, v>|u, v \in V$ and u and v have a common language}, as shown in Figure 9.3.2, G is a connected graph. The seven people are seated around a round table, so that each person can talk with both sides. That is, find the Hamiltonian loop in graph G, which is $abdfgec$ after observation.

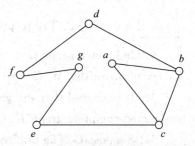

Figure　9.3.2

Example 9.3.5 A meeting is attended by 20 people, each of whom has at least 10 friends. These 20 people gather around a round table and sit down. Is it possible for each person to have two adjacent friends? On what basis?

Solution Using node to represent people. According to the question, when two people are friends, an edge is connected between corresponding nodes, then an undirected graph $G=<V, E>$, which can be converted to the problem of the Hamiltonian loop. For any node u, $v \in V$, $\deg(u) \geqslant 10$, $\deg(v) \geqslant 10$, so $\deg(u)+\deg(v) \geqslant 20$. According to the sufficient condition theorem of the Hamiltonian loop, it can be known that G is a Hamiltonian graph. There is a Hamiltonian loop in G.

Exercises

1. Is the following statement true or false?

(1) The complete graph $K_n (n \geqslant 3)$ is an Euler graph.

(2) A directed complete graph of order $n(n \geqslant 2)$ is an Euler graph.

(3) When r, s are positive even Numbers, the complete bipartite graph $K_{r,s}$ is an Euler graph.

2. Draw an undirected Euler graph satisfying the following conditions.

(1) It has odd number of nodes and odd number of edges.

(2) It has even number of nodes and even number of edges.

(3) It has odd number of nodes and even number of edges.

(4) It has even number of nodes and odd number of edges.

3. Draw a directed Euler diagram that meets the four requirements in problem 1.

4. If an undirected graph G is an Euler graph, is there a cut edge in G? Why is that?

5. What is the value of n, undirected perfect K_n has an Euler loop? Does this conclusion hold true for K_n, a directed complete graph? Why is that?

6. Complete the following questions.

(1) Draw a graph with an Euler loop and a Hamiltonian loop.

(2) Draw a graph with an Euler loop but no Hamiltonian loop.

(3) Draw a graph that has no Euler loop, but one Hamiltonian loop.

7. Proof: If G is directed Euler graph, then G is strongly connected.

8. Suppose $G=<V, E>$, It's an undirected simple graph, $|V|=n$, $|E|=m$, $m = \frac{1}{2}(n-1)(n-2)+2$. Try to prove that there is a Hamiltonian loop in G.

9. Undirected graph $G=<V, E>$ has a Hamiltonian path, S is any nonempty subset of V, try to prove that $W(G-S) \leqslant |S|+1$.

10. It is proved that the graph obtained after the arbitrary deletion of $n-3$ edges in the undirected complete graph $K_n (n \geqslant 6)$ is a Hamiltonian graph.

11. Suppose G is an undirected connected graph, it is proved that G is not a Hamiltonian graph if there is a bridge or cut point in G.

12. The Petersson graph is neither an Euler graph nor a Hamiltonian graph. At least how many new edges do I have to add to make it an Euler graph? At least how many new edges can be added to make it a Hamiltonian graph?

13. There are n of them, and we know that any two of them together know the other

$n-2$. Proof: When n is $\geqslant 3$, these n people can line up, so that any two adjacent people know each other. When n is $\geqslant 4$, these n people can form a circle, so that everyone knows the people on both sides.

14. A factory produces two-color cloth made of yarns of 6 colors. Each color is known to match at least 3 other colors in a batch of bichromatic cloth. It is proved that 3 kinds of double color cloth can be selected from this batch. They are made of 6 different colored yarns.

第 9 章　欧拉图与哈密顿图

能否从一个顶点出发沿着图的边前进,恰好经过图的每条边一次并且回到这个顶点? 同样,能否从一个顶点出发沿着图的边前进,恰好经过图中每个顶点一次并且回到这个顶点? 本章基于此问题进行研究和讨论。

9.1　欧拉图

定义 9.1.1 ▶ 在无孤立结点的图 G(有向图或无向图)中,如果存在一条路经过每一条边一次且仅一次,则称该路为欧拉路。如果存在一条回路经过每一条边一次且仅一次,则称该回路为欧拉回路。具有欧拉回路的图称为欧拉图。

定理 9.1.1 ▶ 无向图 G 具有一条欧拉路当且仅当 G 是连通的且有零个或两个奇度结点。

证明　设 G 具有一条欧拉路,下证 G 是连通的且有零个或两个奇度结点。

设 G 中有一条欧拉路 $L: v_0 e_1 v_1 e_2 v_2 \cdots e_k v_k$,该路经过 G 的每一条边。因为 G 中无孤立结点,所以该路经过 G 的所有结点,即 G 的所有结点都在该路上。故 G 中任意两个结点连通,从而 G 是连通图。

设 v_i 是图 G 的任意结点,若 v_i 不是 L 的端点,每当沿 L 经过 v_i 一次都经过该结点关联的两条边,为该结点的度数增加 2。由于 G 的每一条边都在该路上,且不重复,所以 v_i 的度数必为偶数。若 v_i 是 L 的端点 v_0,当 $v_0 = v_k$ 时,v_0 和 v_k 的度数也为偶数,即 G 中无奇度结点;当 $v_0 \neq v_k$ 时,v_0 和 v_k 的度数均为奇数,即 G 中有两个奇度结点。

设 G 是连通的且有零个或两个奇度结点,下证 G 具有一条欧拉路。

设 G 是连通图,有零个或两个奇度结点,用下述方法构造一条欧拉路。

(1) 若 G 中有两个奇度结点,则从其中的一个 v_0 开始构造一条简单路。

从 v_0 出发经关联边 e_1 进入 v_1,若 v_1 是偶数度结点,则必可以由 v_1 经关联边 e_2 进入 v_2。如此下去,每边只取一次。由于 G 是连通图,必可到达另一个奇度结点 v_k,从而得到一条简单路 $L_1: v_0 e_1 v_1 e_2 v_2 \cdots e_k v_k$。

若 G 中无奇度结点,则从任意结点 v_0 出发,用上述方法必可回到结点 v_0,得到一条简单回路 $L_1: v_0 e_1 v_1 e_2 v_2 \cdots v_0$。

(2) 若 L_1 经过 G 的所有边,L_1 就是欧拉路。

(3) 否则,在 G 中删除 L_1,得子图 G',则 G' 中的每一个结点的度数为偶数。因为 G 是连通图,故 L_1 和 G' 至少有一个结点 v_i 重合,在 G' 中由 v_i 出发重复(1),得到简单回路 L_2。

(4) 若 L_1 与 L_2 组合在一起恰是 G,则得到了欧拉路,否则,重复(3)可得到简单回路 L_3。以此类推,直到得到一条经过 G 中所有边的欧拉路。

推论 无向图 G 具有一条欧拉回路,当且仅当 G 是连通的且所有结点都是偶度的。

这个推论说明,无向图 G 是欧拉图,当且仅当 G 是连通的且所有结点都是偶度的。

图 9.1.1(a)的每一个结点的度数都是偶数 2,所以(a)中有一个欧拉回路,是欧拉图;图 9.1.1(b)中有两个结点的度数是奇数 3,故(b)中有一个欧拉路,但没有欧拉回路,不是欧拉图;图 9.1.1(c)中四个结点的度数都是奇数 3,(c)中没有欧拉路,更没有欧拉回路,不是欧拉图。

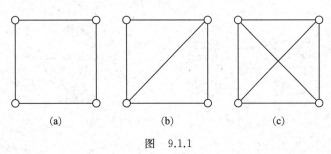

图   9.1.1

定理 9.1.2 ▶有向图 G 具有一条欧拉回路,当且仅当 G 是强连通(对于每一对 v_i 和 v_j,$v_i \neq v_j$,从 v_i 到 v_j 和从 v_j 到 v_i 都存在路径)的且每个结点的入度等于出度。

这个定理说明,有向图 G 是欧拉图的充分必要条件是 G 是强连通的且每个结点的入度等于出度。

定理 9.1.3 ▶有向图 G 具有一条欧拉路,当且仅当 G 是单向连通(对于任意节点 v_1 和 v_2,至少存在从 v_1 到 v_2 和从 v_2 到 v_1 的路径中的一条)的且除两个结点外,每个结点的入度等于出度,而在这两个结点中,一个结点的入度比出度大 1,另一个结点的入度比出度小 1。

这两个定理可以看作定理 9.1.1 和推论的推广。因为对于有向图的任意一个结点,如果入度与出度相等,则该结点的度数为偶数。若入度与出度之差为 1,则该结点的度数为奇数。因此定理 9.1.2 和定理 9.1.3 的证明与定理 9.1.1 的证明类似。

图 9.1.2(a)是强连通的且每一个结点的入度等于出度,都等于 1,所以(a)中有一个欧拉回路,是欧拉图;图 9.1.2 (b)是强连通的且有两个结点的入度与出度相等,有两个结点的入度与出度不相等,其中一个结点的入度比出度大 1,另一个结点的入度比出度小 1,故有一个欧拉路,但没有欧拉回路,不是欧拉图;图 9.1.2(c)中四个结点的入度与出度都不相等,没有欧拉路,更没有欧拉回路。

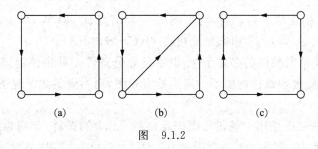

图   9.1.2

哥尼斯堡七桥问题

从前,一个称为哥尼斯堡的城市有一条横贯全市的普雷格尔河,河中有两个小岛,用七座桥将岛和岛、河岸和岛连接起来,如图 9.1.3 所示。18 世纪中叶,该城市的人们提出了一个问题:能否不重复地走完七座桥,最后回到原地。城中的很多人试图解决这个问题,然而试验者都以失败告终。1736 年,瑞士的数学家列昂哈德·欧拉发表了一篇论文《哥尼斯堡七桥问题》,这篇论文的基本内容就是定理 9.1.1。在这篇论文中第一次论证了这个问题是无解的。他将河岸和岛抽象成图的结点,把桥画成相应的连接边,如图 9.1.4 所示。于是,不重复地走完七座桥,最后回到原地,等价于图中存在一条回路,经过每一条边一次且仅一次,即图中存在欧拉回路。因为图中的四个结点的度数都是奇数,所以图中不存在欧拉回路。

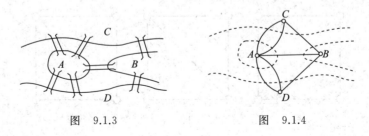

图　9.1.3　　　　　　　　　图　9.1.4

9.2　哈密顿图

一个与欧拉回路和欧拉图非常相似的问题是哈密顿回路和哈密顿图问题。它是由爱尔兰数学家威廉·哈密顿爵士于 1859 年首先提出的。

定义 9.2.1 ▶ 在图 G(有向图或无向图)中,如果存在一条路,经过每个结点一次且仅一次,则该路称为哈密顿路。如果存在一条回路,经过每个结点一次且仅一次,则称该回路为哈密顿回路。具有哈密顿回路的图称为哈密顿图。

与欧拉图不同,判断一个图是否为哈密顿图至今还没有一个简单的充分必要条件,只有一些必要条件和充分条件。判断一个图是哈密顿图的充分必要条件是图论中尚未解决的基本难题之一。下面介绍一些关于哈密顿图的必要条件和充分条件。

先给出一个约定:设 $G=<V,E>$ 是图,$S\subseteq V$,用 $G-S$ 表示从图 G 中删除 S 中的所有结点所得的子图。

定理 9.2.1 ▶ 设无向图 $G=<V,E>$ 是哈密顿图,S 是 V 的任意非空子集,则 $W(G-S)\leqslant|S|$。

证明 设 C 是 G 中的一条哈密顿回路,$\forall v_1\in S$,则 $C-\{v_1\}$(从 C 中删除结点 v_1)是一条路,路上各结点相互连通。再删除 S 中的另一个结点 v_2,则 $W(C-\{v_1,v_2\})\leqslant 2$,所以 $W(C-\{v_1,v_2\})\leqslant|\{v_1,v_2\}|$。由归纳法可得 $W(C-S)\leqslant|S|$。

本定理的条件是哈密顿图的必要条件,但不是充分条件。利用该定理可以证明一个图不是哈密顿图,即满足此定理条件的图不一定是哈密顿图,而不满足此定理条件的图一定不是哈密顿图。

例如,图 9.2.1 叫彼得松图。彼得松图满足定理 9.2.1 的条件,但可以验证它不是哈密顿图。又如,在图 9.2.2 中,令 $S=\{a,b\}$,则 $W(G-S)=3$,$|S|=2$,故 $W(G-S)\leqslant|S|$ 不成立。所以该图不是哈密顿图。

例 9.2.1 设 G 是无向连通图,证明:如果 G 中有割点或桥(如果删除某个顶点后变为非连通图,该顶点称为割点,顶点都为割点的边称为桥),则 G 一定不是哈密顿图。

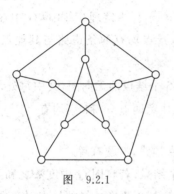

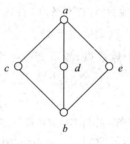

图 9.2.1   图 9.2.2

证明 设 v 是割点。删除结点 v，图变为不连通的，且至少有两个连通分支。令 $S = \{v\}$，则 $W(G-S) \geq 2 > 1 = |S|$，即 $W(G-S) > |S|$。依据定理 9.2.1，G 不是哈密顿图。

如果 G 中有桥，则该桥的一个端点是割点。可以归结为 G 中有割点的情况。所以 G 也不是哈密顿图。

定理 9.2.2 ▶ 设 $G = <V, E>$ 是 n 阶无向简单图，如果 G 中每一对结点度数的和大于或等于 $n-1$，则在 G 中存在一条哈密顿路。

推论 设 $G = <V, E>$ 是 n 阶无向简单图，如果 G 中每一对结点度数的和大于或等于 n，则在 G 中存在一条哈密顿回路。

定理 9.2.2 及其推论给出了图中存在一条哈密顿路和哈密顿回路的充分条件，而不是必要条件。例如，图 9.2.3(a) 中有 6 个结点，任意两个结点度数的和小于 5，但图中存在一条哈密顿路。又如，图 9.2.3 (b) 中也有 6 个结点，任意两个结点度数的和小于 6，但图中存在一条哈密顿回路。

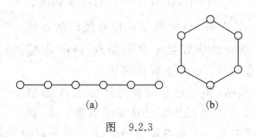

(a)   (b)

图 9.2.3

例 9.2.2 设 $G = <V, E>$ 是一个无向简单图，$|V| = n$，$|E| = m$，设 $m > (n-1)(n-2)$。试证明 G 中存在一条哈密顿路。

证明 先证明图 G 中任意两个不同的结点度数之和大于或等于 $n-1$，然后利用定理 9.2.2 即得本例要证的结果。

用反证法，设图 G 中存在两个结点 v_1 和 v_2，使 $\deg(v_1) + \deg(v_2) < n-1$，即 $\deg(v_1) + \deg(v_2) \leq n-2$。删去结点 v_1 和 v_2 后得到图 G'，G' 是具有 $n-2$ 个结点的无向简单图。设 G' 的边数为 m'，则 $m' \geq m - (n-2) > (n-1)(n-2) - (n-2) = (n-2)^2$，即 $m' > (n-2)^2$。另外，G' 是具有 $n-2$ 个结点的无向简单图，它最多有 $\dfrac{1}{2}(n-2)(n-3)$ 条边。所以 $m' \leq (n-2)(n-3)$。

由此得出矛盾。所以 $\deg(v_1)+\deg(v_2)\geqslant n-1$，由定理 9.2.2 得 G 中存在一条哈密顿路。

例 9.2.3 某地有 5 个风景点，若每个风景点均有两条道路与其他点相通，问是否可经过每个风景点恰好一次游完这 5 处？

解 因为有 5 个风景点，故可看成一个有 5 个结点的无向图，其中每处均有两条路与其他结点相通，所以 $\deg(v_i)=2,i=1,\cdots,5$。故对任意两点 v_i 和 v_j 均有

$$\deg(v_i)+\deg(v_j)=2+2=4=5-1$$

由定理 9.2.2 可知，此图中一定有一条哈密顿路，本题有解。

定义 9.2.2 ▶ 给定图 $G=<V,E>$ 有 n 个结点，若将图 G 中度数之和至少是 n 的非邻接结点连接起来得图 G'，对图 G' 重复上述步骤，直到不再有这样的结点对存在为止，所得到的图称为原图 G 的闭包，记作 $C(G)$。

例如，图 9.2.4 给出了有 6 个结点的图 G 构造其闭包的过程。在这个例子中，$C(G)$ 是完全图。一般情况下，$C(G)$ 也可能不是完全图。

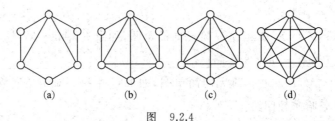

图 9.2.4

定理 9.2.3 ▶ 设 G 是简单图，G 是哈密顿图当且仅当 G 的闭包是哈密顿图。

例 9.2.4 图 G 如图 9.2.5 所示，证明 G 中没有哈密顿路。

证明 任取一结点 a 用 A 标记，所有与它相邻接的结点标记为 B。继续不断地用 A 标记所有邻接于 B 的结点，再用 B 标记所有邻接于 A 的结点，直到所有的结点标记完毕。

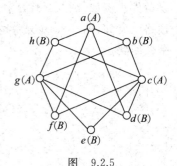

图 9.2.5

如果图中有一条哈密顿路，那么它必交替通过标记 A 的结点和标记 B 的结点，故标记 A 的结点与标记 B 的结点数一样，或者两者相差为 1。然而本例中有 3 个结点标记为 A，5 个结点标记为 B，它们相差为 2，所以该图不存在哈密顿路。

9.3 欧拉图与哈密顿图的应用

例 9.3.1 邮路问题。

我国数学家管梅谷教授首先提出一个问题：一个邮递员在投送邮件时，每次要走遍负责投递范围内的各条街道，然后再回到邮局，他应该按什么样的路线走才能使所走的路程最短？这就是邮路问题。如果将投递地点用点来表示，邮递员可供选择的街道用边来表示，每条边上的权表示两相邻地点的距离，这样就构成了一个赋权图 G。

邮路问题实际就是在赋权图 G 上找到一条通过各边的回路，且各边的权和最小，称这样的回路为最优回路。如果邮递员所走街道的图形是一个欧拉图，则邮路问题容易解决，因为图

G 中任何一条欧拉回路都是最优回路。但如果图 G 不是欧拉图,问题就比较复杂,图 G 的最优回路将通过图中某些边超过一次。如在设计公园景区路径时,要考虑的一个主要影响因素是尽可能观看全部景点,同时尽量少走重复路径,以节省体力。如何设计各景点之间的路径连接才能科学合理地便于游客的浏览? 也可用欧拉图的知识来解决。

例 9.3.2 判断一个图是否可以一笔画出。

一个无向图 G 可一笔画出的两种情况:从图的一点出发经过每条边一次且只一次到达图的另一点;从图的一点出发通过每条边一次且一次又回到该点。判断图形是否可以一笔画出,实际上即判断是否存在包含所有边的欧拉路,即一个图全为偶数度结点或恰有两个结点是奇数其余都为偶数。

判定图 9.3.1 是否可一笔连续画出而不在任何一边上重复画过。

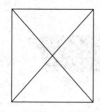 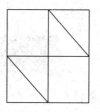

图   9.3.1

解 左图不可以一笔画出,因为图中有 4 个奇度结点。右图可以一笔画出,其中每个结点的度数均为偶数。

例 9.3.3 旅行售货员问题。

售货员从某城市出发到各个城市去一次并且只去一次,然后回到出发城市,要求出一条巡回路线,使该巡回路线的总和最小,这就是旅行售货员问题。

旅行售货员问题实质上是在一个边赋权的完全图(每一对结点之间都有边相连的图称为完全图)上找到一条哈密顿回路,使回路上各边的权之和最小,边上的权即为连接两城市交通线路的长度。

求解旅行售货员问题的有效算法至今尚未成功,有一种近似算法叫最邻近算法,它的基本思想非常简单:当售货员在某一城市时,下一步就选择与这个城市最邻近的、还没有去过的城市作为下一站,如此进行直到走完所有城市为止。

最邻近算法如下。

(1) 在完全图中任选一点作为起始点,找出一个与起始点最近的点,形成一条边的初始路,然后用步骤(1)逐点扩充这条路。

(2) 设 x 表示最新加入这条路上的顶点,从不在路上的所有顶点中选一个与 x 最邻近的点,把连接 x 与此点的边加到这条路上。重复这一步,直到完全图中所有顶点都包含在路中。

(3) 把起始点和最后加入的顶点间的边放入,得到回路。

例 9.3.4 座位安排问题。

设已知下列事实:a 会讲英语,b 会讲英语和汉语,c 会讲英语、意大利语和俄语,d 会讲日语和汉语,e 会讲德语和意大利语,f 会讲法语、日语和俄语,g 会讲法语、德语。问这 7 个人应如何排座位,才能使每个人和他身边的人交谈?

解 设无向图 $G=<V,E>$。其中,$V=\{a,b,c,d,e,f,g\}$,$E=\{<u,v>|u,v\in V$ 且 u

和 v 有共同语言}，如图 9.3.2 所示，图 G 是连通图，将这 7 个人安排围圆桌而坐，使每个人能与两边交谈，即在图 G 中找哈密顿回路，经观察，该回路是 $abdfgec$。

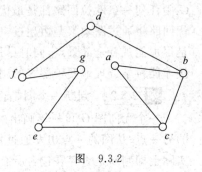

例 9.3.5 某次会议有 20 人参加，其中每人都至少有 10 个朋友，这 20 人围成一圆桌入席，是否可能使每人相邻的两位都是朋友？根据是什么？

解 用结点表示人，根据题意，两人是朋友时相应结点间连一条边，得到一个无向图 $G = \langle V, E \rangle$，可转化为求哈密顿回路问题。由于对任意结点 $u, v \in V$，有 $\deg(u) \geqslant 10, \deg(v) \geqslant 10$，因而 $\deg(u) + \deg(v) \geqslant 20$。根据求哈密顿回路的充分条件定理可知，图 G 是哈密顿图，图 G 中存在哈密顿回路，按此回路中的各点位置入席即为所求。

图 9.3.2

习 题

1. 判断下列命题是真是假。

(1) 完全图 $K_n (n \geqslant 3)$ 是欧拉图。

(2) $n(n \geqslant 2)$ 阶有向完全图是欧拉图。

(3) 当 r, s 为正偶数时，完全二部图 $K_{r,s}$ 是欧拉图。

2. 画出一个满足下述条件的无向欧拉图。

(1) 奇数个结点，奇数条边。

(2) 偶数个结点，偶数条边。

(3) 奇数个结点，偶数条边。

(4) 偶数个结点，奇数条边。

3. 画出满足题 1 中的四个要求的有向欧拉图。

4. 若无向图 G 是欧拉图，图 G 中是否存在割边？为什么？

5. n 取怎样的值，无向完全图 K_n 有一条欧拉回路？这个结论对有向完全图 K_n 成立吗？为什么？

6. 完成下列各题。

(1) 画一个有一条欧拉回路和一条哈密顿回路的图。

(2) 画一个有一条欧拉回路但没有哈密顿回路的图。

(3) 画一个没有欧拉回路但有一条哈密顿回路的图。

7. 证明：若 G 是有向欧拉图，则 G 是强连通的。

8. 设 $G = \langle V, E \rangle$ 是一个无向简单图，$|V| = n$，$|E| = m$，设 $m = \frac{1}{2}(n-1)(n-2) + 2$。

试证明 G 中存在一条哈密顿回路。

9. 设无向图 $G = \langle V, E \rangle$ 具有哈密顿路，S 是 V 的任意非空子集，试证明 $W(G-S) \leqslant |S| + 1$。

10. 试证明在无向完全图 $K_n (n \geqslant 6)$ 中任意删除 $n-3$ 条边后所得的图是哈密顿图。

11. 设 G 是无向连通图，证明：若 G 中有桥或割点，则 G 不是哈密顿图。

12. 彼得松图既不是欧拉图，也不是哈密顿图。至少加几条新边才能使它成为欧拉图？至少加几条新边才能使它成为哈密顿图？

13. 今有 n 个人,已知他们中的任何两人合起来认识其余的 $n-2$ 个人。证明:当 $n \geqslant 3$ 时,这 n 个人能排成一列,使任何两个相邻的人都相互认识。而当 $n \geqslant 4$ 时,这 n 个人能排成一个圆圈,使每个人都认识左右两侧的人。

14. 某工厂生产由 6 种颜色的纱织成的双色布。已知在一批双色布中,每种颜色至少与其他 3 种颜色相搭配。证明可以从这批双色布中挑出 3 种,它们由 6 种不同颜色的纱织成。

Chapter 10   Planar Graph

The planar graph is an extremely important part of graph theory. Its properties and applications play an important role in the field of computers. In this chapter, we will introduce the related knowledge of planar graphs.

10.1   Basic Concepts of the Planar Graph

Definition 10.1.1   Let $G=<V$, $E>$ be an undirected graph, if all the nodes and edges of G can be drawn on a plane so that there is no intersection point between any two edges except the endpoints, G is called a planar graph; otherwise, G is called a nonplanar graph.

If the vertex set V of an undirected graph $G=<V$, $E>$ can be divided into two subsets V_1 and V_2 ($V_1 \cap V_2 = \varnothing$), so that one of the two endpoints of any edge in G belongs to V_1 and the other belongs to V_2, then G is called a bipartite graph. If $|V_1|=n$, $|V_2|=m$, then let G be a complete bipartite graph $K_{n,m}$.

In Figure 10.1.1, (a) is an undirected complete graph K_3, which is a planar graph, (b) is an undirected complete graph K_4. Although it has intersecting edges, it can obtain (c) by properly adjusting the position of the edges. In (c), there is no intersection point between any two sides except the endpoint, so K_4 is a planar graph (d) is a bipartite graph, and $K_{1,3}$ is a planar graph. (e) is a bipartite graph $K_{2,3}$, and (f) is obtained by changing the drawing. According to the Definition 10.1.1, $K_{2,3}$ is a planar graph (g) is K_5 and (h) is $K_{3,3}$. No matter how to adjust the position of the edge, it can not make any two sides have no intersection except the endpoint, so it is not a planar graph.

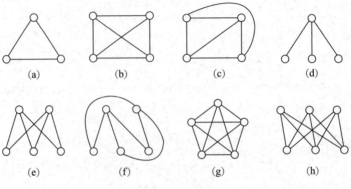

(a)   (b)   (c)   (d)

(e)   (f)   (g)   (h)

Figure   10.1.1

Undirected complete K_5 and complete bipartite graph $K_{3,3}$ are two important nonplanar graphs, which play an important role in the study of plane graph theory.

The following three conclusions are obvious.

Theorem 10.1.1　If G is a planar graph, then any subgraph of G is a planar graph.

According to this theorem, any subgraph of an undirected complete graph K_4 is a planar graph, and any subgraph of a bipartite graph $K_{2,3}$ is also a planar graph.

Theorem 10.1.2　If G is a nonplanar graph, then any supergraph of G is a nonplanar graph.

Inference　It is concluded that undirected complete graph $K_n (n \geqslant 5)$ and bipartite graph $K_{3,n} (n \geqslant 3)$ are non-planar graphs.

Theorem 10.1.3　If G is a planar graph, the graph obtained by adding parallel edges or selfcycles in G is still a planar graph.

This theorem shows that parallel edges and selfcycles do not affect the planarity of graphs, therefore, parallel edges and selfcycles are not considered when studying the planarity of graphs.

Definition 10.1.2　If G is a planar graph, the region bounded by the edges of the graph contains neither the nodes nor the edges of the graph, such an area is called a face of G. The surface with the infinite area is called infinite surface or outer surface, and the face with the finite area is called finite face or inner face. The cycle formed by the edges surrounding the surface is called the boundary of the face, the number of edges on the boundary is called the degree of the face, and the degree of the face r is denoted as $\deg(r)$.

The infinite face is often denoted as r_0 and the finite face as r_1, r_2, ...

Figure 10.1.2 is a planar graph with four faces: r_0, r_1, r_2, r_3, where r_0 is an infinite face and the other three are finite. The boundary of r_0 is composed of cycle $abcda$ and cycle $ffghijgf$, and its degree $\deg(r_0)=11$; the boundary of r_1 is composed of cycle $abcdeda$ with degree $\deg(r_1)=6$; the boundary of r_2 is composed of cycle $ghijg$ with degree $\deg(r_2)=4$; the boundary of r_3 is selfcycle on f, followed by $\deg(r_3)=1$.

Theorem 10.1.4　Let $G=<V, E>$ be a finite planar graph with n faces, then the sum of the times of all faces is equal to twice the number of edges. It is $\sum_{i=1}^{n} \deg(r_i)=2|E|$.

Proof　For any edge of G, or the common boundary of two faces, the number of times of each face is increased by 1; or if it is calculated twice as a boundary in a face, the number of times of the face is increased by 2. In either case, any edge of G increases the sum of the times of the faces in the graph by 2, so the sum of the times of all the faces is equal to twice the number of edges.

In Figure 10.1.2, there are 11 edges, and the sum of the times of all faces is:

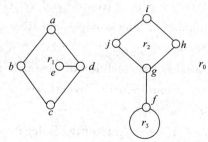

Figure　10.1.2

$$\deg(r_0)+\deg(r_1)+\deg(r_2)+\deg(r_3)=11+6+4+1=22=2\times11$$

Definition 10.1.3 If G is a simple planar graph, if an edge (v, u) is added between any pair of nonadjacent nodes v and u of G, the resulting graph is a nonplanar graph, then G is called a maximal planar graph.

There are several conclusions about maximal planar graphs.

(1) Maximal planar graphs are connected graphs.

(2) It is impossible for a maximal planar graph with a node number greater than or equal to 3 to have cut points and bridges.

(3) Let G be a simple connected planar graph with a node number greater than or equal to 3, and G be a maximal planar graph if and only if the degree of each face of G is 3.

10.2 Euler Formula and Judgment of Planar Graph

Theorem 10.2.1 (**Euler formula**) Any connected planar graph G, if G has n nodes, m edges and r faces, then the Euler formula $n-m+r=2$ holds.

Proof Inductive proof of the number of edges m.

(1) Suppose $m=0$, because G is a connected graph, so G can only be a trivial graph. In this case, $n=1$, $m=0$, $r=1$, $n-m+r=2$ holds.

(2) Suppose $m=k$ $(k\geqslant1)$, Euler's formula holds. If $m=k+1$, Euler's formula also holds.

An undirected graph that is connected without a loop is called a tree. Each connected branch is an undirected graph of a tree called a forest. In an undirected tree, nodes with degree 1 are called leaves.

If G is a tree and there are at least two leaves in G, then G is a nontrivial tree. Let v be one of the leaves. Let $G'=G-\{v\}$ (delete node v from G and get G'), then G' is still a connected graph. Let n', m' and r' be the number of nodes, edges and faces of G', respectively. Then $n'=n-1$, $m'=m-1=k$, $r'=r$. So $n=n'+1$, $m=m'+1$, $r=r'$. Because G' is a connected graph and $m'=k$, G' satisfies the condition of the inductive hypothesis. According to the inductive hypothesis:

$$n'-m'+r'=2$$

So, $n-m+r=(n'+1)-(m'+1)+r'=n'-m'+r'=2$.

If G is not a tree, then G contains a cycle. Let e be on a cycle of G. Let $G'=G-\{e\}$ (remove edge e from G and get G'), then G' is still connected. Let n', m' and r' be the number of nodes, edges and faces of G', respectively. Then $n'=n$, $m'=m-1=k$, $r'=r-1$. So that $n=n'$, $m=m'+1$, $r=r'+1$. Because G' is a connected graph and $m'=k$, so G' satisfies the condition of the inductive hypothesis.

From inductive hypothesis: $n'-m'+r'=2$, so $n-m+r=n'-(m'+1)+(r'+1)=n'-m'+r'=2$.

The existence of the Euler formula is an important property of the planar graph, and the condition is the connectivity of the planar graph. Euler's formula can also be extended to unconnected planar graphs. Write this result as the following theorem.

Theorem 10.2.2   Let G be a planar graph with k connected branches, n nodes, m edges and r faces, then the formula $n-m+r=k+1$ holds.

Proof Let the connected branch of G be G_i, where G_i, n_i, m_i and r_i are the numbers of nodes, edges and faces of G_i respectively. According to Theorem 10.1.2, $n_i-m_i+r_i=2$, $i=1, \ldots, k$. For $\sum_{i=1}^{k} m_i=m$, $\sum_{i=1}^{k} n_i=n$, since each G_i has an outer face and G has only one outer face, the number of faces of G is $r=\sum_{i=1}^{k} r_i-k+1$, that is $\sum_{i=1}^{k} r_i=r+k-1$. So

$$2k=\sum_{i=1}^{k}(n_i-m_i+r_i)=\sum_{i=1}^{k}n_i-\sum_{i=1}^{k}m_i+\sum_{i=1}^{k}r_i=n-m+r+k-1$$

After finishing, there is $n-m+r=k+1$.

Some properties of planar graphs can be deduced by the Euler formula.

Theorem 10.2.3   Let G be a connected planar graph with n nodes and m edges, and the degree of each face is at least k $(k\geqslant 3)$, then the relation $m\leqslant\frac{k}{k-2}(n-2)$ holds.

Proof Let G have r faces, which is known by Theorem 10.1.4, $2m=\sum_{i=1}^{r}\deg(r_i)$.

Because the degree of each face is greater than or equal to k, the sum of the degree of all faces is greater than or equal to kr, therefore $2m=\sum_{i=1}^{r}\deg(r_i)\geqslant kr$. That is $2m\geqslant kr$.

Substituting Euler formula $r=2+m-n$, $2m\geqslant kr=k(2+m-n)$ is obtained, and $m\leqslant\frac{k}{k-2}(n-2)$.

Example 10.2.1   Using the above theorem, it is proved that neither K_5 nor $K_{3,3}$ is a planar graph.

Solution K_5 is shown in Figure 10.1.1 (g), the number of nodes is $n=5$ and the number of edges is $m=10$. Let's assume that K_5 is a planar graph. Obviously, K_5 is connected because there is no selfloop and parallel edge in K_5, so the degree of each face is at least 3, that is, $k=3$ in Theorem 10.2.3, and K_5 satisfies the condition of Theorem 10.2.3. So there should be

$$m\leqslant\frac{k}{k-2}(n-2)$$

However, substituting $n=5$, $m=10$ and $k=3$ gives $10\nleqslant 9=\frac{3}{3-2}\times(5-2)$, which is contradictory. So K_5 is not a planar graph.

$K_{3,3}$ is shown in Figure 10.1.1 (h), the number of nodes is $n=6$, and the number of edges is $m=9$. Suppose $K_{3,3}$ is a planar graph. Obviously, $K_{3,3}$ is connected, because two of any three nodes in $K_{3,3}$ are not adjacent, so the degree of each face is greater than or equal to 4, that is, $k=4$ in Theorem 10.2.3, $K_{3,3}$ satisfies the condition of Theorem 10.2.3. So,

$$m \leqslant \frac{k}{k-2}(n-2)$$

However, substituting $n=6$, $m=9$ and $k=4$ give $9 \nleqslant 8 = \frac{4}{4-2} \times (6-2)$ which is contradictory. So $K_{3,3}$ is not a planar graph.

Theorem 10.2.4   Let G be a planar graph with s connected branches, n nodes and m edges, and the degree of each face is at least $k(k \geqslant 3)$, then $m \leqslant \frac{k}{k-2}(n-s-1)$ holds.

Theorem 10.2.5   Let G be a simple planar graph with $n(n \geqslant 3)$ nodes and m edges, then $m \leqslant 3n-6$.

Proof Let G have s $(s \geqslant 1)$ connected branches.

If G is a tree or forest,

$$m = n-s \leqslant n = n+6-6 = n+3+3-6 \leqslant n+n+n-6 = 3n-6$$

Then $m \leqslant 3n-6$.

If G is not a tree or a forest, then G must contain a cycle. Because G is a simple planar graph, the length of the loops in it is greater than or equal to 3. Therefore, the number of faces $k \geqslant 3$ and $\frac{k}{k-2} = 1 + \frac{2}{k-2}$ reaches the maximum 3 when $k=3$, According to Theorem 10.2.4:

$$m \leqslant \frac{k}{k-2}(n-s-1) \leqslant 3(n-s-1) \leqslant 3(n-1-1) \leqslant 3n-6$$

Theorem 10.2.6   Let G be a maximal planar graph with $n(n \geqslant 3)$ nodes and m edges, then $m = 3n-6$.

Proof Let G have r faces. Since the maximal plane graph is connected, G is connected. According to Euler formula $r = 2+m-n$, and because G is a maximal planar graph, the degree of every face of G is 3, so $2m = \sum\limits_{i=1}^{r} \deg(r_i) = 3r = 3(2+m-n)$.

After finishing, $m = 3n-6$.

In a planar graph, the smallest degree of each node is called the minimum degree, denoted as $\delta(G)$. The largest degree of each node is called the maximum degree, denoted as $\Delta(G)$.

Theorem 10.2.7   Let G be a simple phanar graph, then the minimum degree of G is $\delta(G) \leqslant 5$.

Proof Let G have n nodes and m edges.

When $n \leqslant 6$, because G is a simple graph,

$$\delta(G) \leqslant \Delta(G) \leqslant 5$$

In the case of $n \geqslant 7$, if $\delta(G) \geqslant 6$, the degree of each node is greater than or equal to 6, and the sum of degrees of all nodes in G is greater than or equal to $6n$. Therefore

$$2m = \sum\limits_{i=1}^{n} \deg(v_i) \geqslant 6n, \ m \geqslant 3n > 3n-6, \text{ then } m > 3n-6$$

It is in contradiction with Theorem 10.2.5.

Insert a new degree 2 nodes on the edge of a given graph G, so that an edge is divided into two edges, as shown in Figure 10.2.1 (a), is called inserting a 2-degree node; for a node of degree 2, removing this node makes its associated two edges become one edge, as shown in Figure 10.2.1 (b), it is called removing a 2-degree node.

Definition 10.2.1   Given two graphs G_1 and G_2, if they are isomorphic, or G_1 and G_2 are isomorphic by repeatedly inserting a 2-degree node or removing a 2-degree node, G_1 and G_2 are said to be isomorphic in a 2-degree node or G_1 and G_2 are homeomorphism.

Theorem 10.2.8   A graph is planar if and only if it does not contain subgraphs isomorphic to $K_{3,3}$ or K_5 in 2-degree nodes.

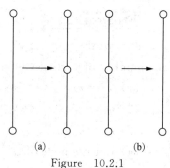

(a)   (b)

Figure   10.2.1

Definition 10.2.2   Let $G = <V, E>$ be an undirected graph, $e \in E$, delete e in G, merge the two endpoints u and v of e into a new endpoint, denoted by w. w relates all the edges of u and v except e, which is called the contraction of edge e. If some edges of a graph G are contracted to obtain a graph G', then G is said to be contracted to G'.

Theorem 10.2.9   A graph is a planar graph if and only if it does not shrink to a subgraph of K_5 or $K_{3,3}$.

10.3   Dual Graph and Properties

The dual graph is a kind of graph associated with a planar graph, which can find the maximum flow (minimum cut) in the original graph. This section will introduce the dual graph.

Definition 10.3.1   Let $G = <V, E>$ be a planar graph with faces r_1, r_2, ..., r_n. If there is a graph $G^* = <V^*, E^*>$ satisfies the following conditions.

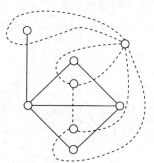

(1) For any face r_i of graph G, there is only one node $v_i^* \in V^*$.

(2) For the face r_i of graph G, the common boundary $e_k \in E$ of r_j exists and only exists one edge $e_k^* \in E^*$, such that $e_k^* = (v_i^*, v_j^*)$ and e_k^* intersect with e_k.

(3) If and only if e_k is the boundary of a face r_i, there is a selfloop e_k^* intersecting with e_k.

Then a graph G^* is said to be a dual graph of a planar graph G.

Figure   10.3.1

In Figure 10.3.1, if the graph with a solid line edge is G, then the graph with a dotted line edge is dual graph G^* of G.

From the definition, we can see that the dual graph G^* of planar graph G has the following properties.

(1) G^* is a connected planar graph.

(2) G^* is the dual graph of graph G, and G is also the dual graph of graph G^*.

(3) If edge e is a selfloop in G^*, then in G^*, the edge e^* corresponding to e is a bridge; if edge e is a bridge, then in G^* the edge e^* corresponding to e is a selfloop.

Theorem 10.3.1 Let G be a connected planar graph and G^* be a dual graph of G. m, n and r are the number of vertices, edges and faces of G, respectively. m^*, n^* and r^* are the number of vertices, edges and faces of G^*, respectively. Then

(1) $n^* = r$

(2) $m^* = m$

(3) $r^* = n$

(4) Let the node of G^* lie in the face r_i of G, then $\deg(v_i^*) = \deg(r_i)$. Where $\deg(v_i^*)$ is the degree of the node v_i^* and $\deg(r_i)$ is the degree of the face r_i.

Proof According to Definition 10.3.1, (1), (2), obviously. (3) since G and G^* are connected planar graphs, they satisfy Euler's formula, $n - m + r = 2$, $n^* - m^* + r^* = 2$. Therefore, $r^* = 2 - n^* + m^* = 2 - r + m = 2 + m - r = n$. (4) Let the boundary of the face r_i of G be C_i, and let there be k_1 bridges and k_2 non-bridge edges in C_i, so the length of C_i is $2k_1 + k_2$, that is, $\deg(r_i) = 2k_1 + k_2$, while there are k_1 selfloops at the nodes of r_i, v_i^* corresponding to k_1 bridges, and k_2 non-bridge edges correspond to k_2 edges emanating from v_i^*, namely $\deg(v_i^*) = 2k_1 + k_2$; and $\deg(v_i^*) = \deg(r_i)$.

Theorem 10.3.2 Let G be a planar graph with k $(k \geq 2)$ connected branches and G^* be the dual graph of G. m, n and r are the number of nodes, edges and faces of G, respectively. m^*, n^* and r^* are the number of nodes, edges and faces of G^*, respectively. Then

(1) $n^* = r$

(2) $m^* = m$

(3) $r^* = n - k + 1$

(4) Let the node of G^*, v_i^* lie in the face r_i of G, then $\deg(v_i^*) = \deg(r_i)$. Where $\deg(v_i^*)$ is the degree of the node v_i^* and $\deg(r_i)$ is the degree of the face r_i.

Definition 10.3.2 If the dual graph G^* of G is isomorphic to G, then G is called a self dual graph.

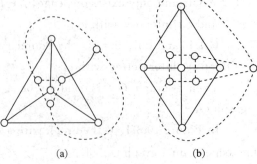

(a)　　　　　　　　(b)

Figure 10.3.2

In (a), (b) of Figure 10.3.2, if the solid line edge graph is G, then the dotted line edge graph is the dual graph G^* of G. We can see that graph G and its dual graph G^* are isomorphic. Therefore, the solid line edge graphs in Figure 10.3.2 (a) and (b) are self dual graphs.

10.4　Application of the Planar Graph

The application of the planar graph is used in many fields, but it is often combined with other knowledge to solve problems, and rarely appears in a single way.

Definition 10.4.1　Let G be a graph without selfloop, and assign a color to each node of G so that adjacent nodes have different colors. This is called normal coloring of graph G, or coloring for short. Graph G is said to be n-colored if n colors are used. When coloring a graph G, the minimum number of colors required is called the coloring number of G, which is recorded as $X(G)$.

There is a simple and effective method for the normal coloring of graph G, which is called the Welch Powell method, its basic ideas are as follows.

(1) The nodes in graph G are arranged in descending order of degree (This arrangement may not be unique because some nodes have the same degree).

(2) The first node is colored with the first color, and each node that is not adjacent to the previous coloring node is colored with the same color according to the arrangement order.

(3) Repeat the second color for the uncolored nodes, and continue with the third color until all nodes are colored.

Example 10.4.1　Using the Welch Powell method to color Figure 10.4.1.

Solution　(1) The nodes are arranged in descending order: b, f, a, c, e, g, d, h. Their degrees are 4, 4, 3, 3, 3, 3, 3, 2, 2.

(2) b is colored with the first color, and the node g that is not adjacent to b is also colored with this color.

(3) The second color is applied to the nodes f and c.

(4) The third color is applied to the nodes a, d and h.

(5) The fourth color is applied to the node e.

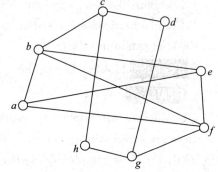

Figure　10.4.1

So G is four colors. Note that G can't be a tricolor, because a, b, e, f are adjacent to each other, so they must have four colors. So $X(G)=4$.

From Definition 10.4.1, we can see that the following theorem holds.

Theorem 10.4.1　$X(G)=1$ if and only if G is a zero graph.

Theorem 10.4.2　Let G contain at least one edge, then $X(G)=2$ if and only if G is a

Theorem 10.4.3 $X(K_n)=n$.

Proof Because every node of a complete graph with n nodes is adjacent to other $n-1$ nodes, the coloring number of n nodes should not be less than n, and the coloring number of n nodes should be at most n, $X(K_n)=n$.

Definition 10.4.2 It is called a kind of normal coloring of map G by coloring each country of map G with a different color for neighboring countries. If n colors are used in coloring G, the map G is called n-colored. When coloring a graph G, the minimum number of colors required is called the coloring number of G, which is recorded as $X^*(G)$.

The coloring of a map can be transformed into the node coloring of its dual graph. See the following theorem.

Theorem 10.4.4 A map G is n-colored if and only if its dual graph G^* is n colored.

Proof Let G be n-colored and we prove that its dual graph G^* is n-colored.

Because the map G is connected, according to Theorem 10.4.1, $n^*=r$, that is, each face of G contains a node of G^*. Let lie in the plane r_i of G. It is easy to know that if the nodes of G^*, v_i^*, v_j^* are adjacent to each other, the color of r_i and r_j is different. Therefore, the color of v_i^* and v_j^* is also different, so G^* is n-color.

If the dual graph G^* of G is n-colored, it can be proved that G is also n-colored.

It can be seen from Theorem 10.4.4 that studying the coloring of maps (face coloring) is equivalent to studying the node coloring of planar graphs. For the point coloring of planar graphs, the following theorem has been proved so far.

Theorem 10.4.5 Any planar graph G is at most 5-colored.

Theorem 10.4.5 is called the Five Color Theorem, also known as the Hewood theorem.

The geometric characteristics of the planar graph are widely used in various fields, such as the application in informatics.

Example 10.4.2 Count the number of triangles in Figure 10.4.2.

Solution 1

Enumerate all triples to determine whether the three vertices are adjacent to each other. Because there are C_n^3 triplets in total, the time complexity is $O(n^3)$.

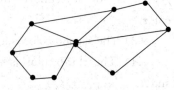

Solution 2

Figure 10.4.2

Enumerate an edge, and then enumerate the third vertex to determine whether it is adjacent to the two endpoints on the edge. According to Definition 10.1.1, Figure 10.4.2 is a planar graph, and the number of edges in Figure 10.4.2 is $O(n)$, so the complexity of solution 2 is $O(n^2)$.

Comparison between solution 1 and solution 2.

Solution 1 is just a simple enumeration, did not pay attention to the actual situation of the problem, but in fact the number of triangles is very small, solution 1 does a lot of useless enumeration, so the efficiency is very low. Solution 2 enumerates the third vertex from the edge, which is in line with the actual situation of the problem and avoids many unnecessary enumerations, so solution 2 is more efficient than solution 1.

The most typical application of planar graph coloring is map coloring.

Map coloring is a kind of combination configuration. It is a partition of the map face set. It assigns a color to each face of the map so that adjacent faces (which have common boundary edges) have different colors. This kind of color assignment is called a coloring of the map. In other words, the map face set is divided into several subsets, so that no two sides of each subset are adjacent. In this way, the faces in each subset can be colored with one color, so that different subsets use different colors. Among all the coloring of map M, the number of colors that use the least color is called the color number of map M. The vertex coloring of a map, or, to do normal coloring for the vertices of its isomorphic graph, is the map coloring of its dual map.

Exercises

1. Let G be a connected planar graph with n nodes and m edges, and has k faces, and k is equal to(　　).

　　A. $m-n+2$　　　B. $n-m-2$　　　C. $n+m-2$　　　D. $m+n+2$

2. In a connected simple planar graph with 6 vertices and 12 edges, each face is composed of (　　) edges.

　　A. 2　　　　　B. 4　　　　　C. 3　　　　　D. 5

3. Let G be a connected planar graph with v nodes, e edges and r faces, then $r =$(　　).

　　A. $e-v+2$　　　B. $v+e-2$　　　C. $e-v-2$　　　D. $e+v+2$

4. Let G be a connected planar graph with 5 vertices and 6 faces, then the number of edges of G is (　　).

　　A. 9　　　　　B. 5　　　　　C. 6　　　　　D. 11

5. Let $G=<V, E>$, then the following conclusion (　　) is true.

　　A. $\deg(V)=2|E|$　　　　　B. $\deg(V)=|E|$

　　C. $\sum_{v\in V}\deg(v)=2|E|$　　　　D. $\sum_{v\in V}\deg(v)=|E|$

6. Any planar graph G is at most (　　) chromatic.

　　A. 3　　　　　B. 4　　　　　C. 5　　　　　D. 6

7. For a complete graph K_{10} with 10 nodes, the minimum number of colors $X(K_{10})$ is (　　).

　　A. 10　　　　　B. 9　　　　　C. 11　　　　　D. 12

8. Judge whether the following statements are correct.

(1) In the plan view, any two edges can have intersections other than the endpoints. (　　)

(2) K_5 is a planar graph. (　　)

(3) Let G_1, G_2 be planar graphs. If $G_1\cong G_2$, then their dual graphs $G_1^*\cong G_2^*$.(　　)

(4) K_4 is a planar graph. (　　)

9. Determine whether graph G is a planar graph and explain the reasons.

10. Verify that K_5 and $K_{3,3}$ are minimal non-planar graphs.

11. Let G be a simple planar graph with m edges of order n, and let $m < 30$ be known, proof that $\delta(G) \leqslant 4$.

12. Draw all non-isomorphic connected simple nonplanar graphs of order 6.

13. Verify that the graph in Exercise 13 Figure is a planar graph.

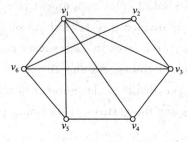

Exercise 9 Figure

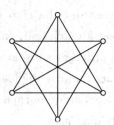

Exercise 13 Figure

14. Verify that the graph in Exercise 14 Figure is a maximum planar graph.

15. Verify that the planar graph in Exercise 15 Figure satisfies the Euler formula.

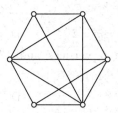

Exercise 14 Figure

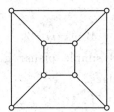

Exercise 15 Figure

第 10 章 平 面 图

平面图是图论中极其重要的组成部分,它的性质及应用都在计算机领域内发挥了重要作用,本章我们将介绍平面图的相关知识。

10.1 平面图的基本概念

定义 10.1.1 ▶ 设 $G=<V,E>$ 是无向图,如果能把 G 的所有结点和边画在一个平面上,使任何两边除端点外没有其他交点,则称 G 为平面图,否则称 G 为非平面图。

若能将无向图 $G=<V,E>$ 的顶点集 V 划分为两个子集 V_1 和 $V_2(V_1 \bigcap V_2 = \varnothing)$,使 G 中任何一条边的两个端点一个属于 V_1,另一个属于 V_2,则称 G 为二部图。若 $|V_1|=n$,$|V_2|=m$,则记完全二部图 G 为 $K_{n,m}$。

在图 10.1.1 中,(a)是无向完全图 K_3,它是平面图;(b)是无向完全图 K_4,它虽有相交边,但适当调整边的位置得到(c);(c)中任何两边除端点外没有其他交点,所以 K_4 为平面图;(d)是二部图 $K_{1,3}$,$K_{1,3}$ 是平面图;(e)是二部图 $K_{2,3}$,经改画以后得到(f),由定义 10.1.1 知,$K_{2,3}$ 是平面图;(g),(h)分别是 K_5 和 $K_{3,3}$,无论怎样调整边的位置都不能使任何两边除端点外没有其他交点,所以不是平面图。

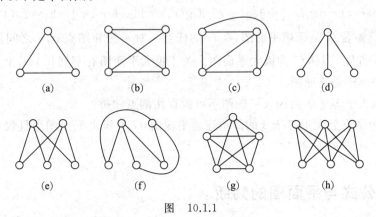

图 10.1.1

无向完全图 K_5 和完全二部图 $K_{3,3}$ 是两个重要的非平面图,它们在平面图理论的研究中有非常重要的作用。

下列三个结论是显然的。

定理 10.1.1 ▶ 若 G 是平面图,则 G 的任何子图都是平面图。

由此定理可知,无向完全图 K_4 的任何子图都是平面图,二部图 $K_{2,3}$ 的任何子图也是平面图。

定理 10.1.2 ▶ 若 G 是非平面图,则 G 的任何母图都是非平面图。

推论 无向完全图 $K_n(n \geq 5)$ 和二部图 $K_{3,n}(n \geq 3)$ 都是非平面图。

定理 10.1.3 ▶ 若 G 是平面图,则 G 中增加平行边或自回路后所得的图还是平面图。

本定理说明平行边与自回路不影响图的平面性,因此研究图的平面性时,不考虑平行边和自回路。

定义 10.1.2 ▶ 若 G 是平面图,由图中的边所包围的区域内既不包含图的结点,也不包含图的边,这样的区域称为图 G 的一个面。其中面积无限的面称为无限面或外部面,面积有限的面称为有限面或内部面。包围面的诸边组成的回路称为该面的边界,边界的边数称为该面的次数,面 r 的次数记为 $\deg(r)$。

常将无限面记为 r_0,有限面记为 $r_1, r_2, \cdots$。

图 10.1.2 是一个平面图,它有四个面:r_0, r_1, r_2, r_3。其中 r_0 是无限面,其他三个面是有限面。r_0 的边界是由回路 $abcda$ 和回路 $ffghijgf$ 组成的,其次数 $\deg(r_0) = 11$;r_1 的边界是由回路 $abcdeda$ 组成的,其次数 $\deg(r_1) = 6$;r_2 的边界是由回路 $ghijg$ 组成的,其次数 $\deg(r_2) = 4$;r_3 的边界是 f 上的自回路,其次数 $\deg(r_3) = 1$。

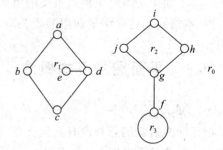

图 10.1.2

定理 10.1.4 ▶ 设 $G = <V, E>$ 是有限平面图,有 n 个面,则所有面的次数之和等于边数的 2 倍,即 $\sum_{i=1}^{n} \deg(r_i) = 2 \mid E \mid$。

证明 G 的任何一边,或者是两个面的公共边界,为每一个面的次数增加 1;或者在一个面中作为边界重复计算两次,为该面的次数增加 2。无论哪种情况,G 的任何一边为图中面的次数和增加 2,故所有面的次数之和等于边数的 2 倍。

在图 10.1.2 中有 11 条边,所有面的次数之和为

$$\deg(r_0) + \deg(r_1) + \deg(r_2) + \deg(r_3) = 11 + 6 + 4 + 1 = 22 = 2 \times 11$$

定义 10.1.3 ▶ 若 G 是简单平面图,在 G 的任意一对不相邻结点 v, u 之间加入边 (v, u),所得的图为非平面图,则称 G 为极大平面图。关于极大平面图有下列几个结论。

(1) 极大平面图是连通图。

(2) 结点数大于等于 3 的极大平面图不可能存在割点和桥。

(3) 设 G 为结点数大于等于 3 的简单连通平面图,G 为极大平面图当且仅当 G 的每个面的次数均为 3。

10.2 欧拉公式与平面图的判断

定理 10.2.1 ▶ (欧拉公式)任意连通平面图 G,若有 n 个结点、m 条边和 r 个面,则欧拉公式 $n - m + r = 2$ 成立。

证明 对边数 m 进行归纳证明。

(1) 设 $m = 0$,由于 G 是连通图,所以 G 只能是平凡图。这时,$n = 1, m = 0, r = 1, n - m + r = 2$ 成立。

(2) 设 $m = k(k \geq 1)$ 时欧拉公式成立,当 $m = k + 1$ 时欧拉公式也成立。

连通无回路的无向图称为树,每个连通分支都是树的无向图称为森林,在无向树中,度数为1的结点称为树叶。

若 G 是树,G 中至少有两片树叶,则 G 是非平凡树。设 v 为其中一片树叶。令 $G'=G-\{v\}$(从 G 中删除结点 v,得到 G'),则 G' 仍然是连通图。设 n',m' 和 r' 分别是 G' 的结点数、边数和面数,则 $n'=n-1,m'=m-1=k,r'=r$。于是 $n=n'+1,m=m'+1,r=r'$。因为 G' 是连通图且 $m'=k$,所以 G' 满足归纳假设的条件。由归纳假设知:

$$n'-m'+r'=2$$

故 $n-m+r=(n'+1)-(m'+1)+r'=n'-m'+r'=2$。

若 G 不是树,则 G 中含有回路。设边 e 在 G 的某个回路上。令 $G'=G-\{e\}$(从 G 中删除边 e,得到 G'),则 G' 仍然是连通图。设 n',m' 和 r' 分别是 G' 的结点数、边数和面数,则 $n'=n,m'=m-1=k,r'=r-1$。于是 $n=n',m=m'+1,r=r'+1$。因为 G' 是连通图且 $m'=k$,所以 G' 满足归纳假设的条件。

由归纳假设知:$n'-m'+r'=2$,所以 $n-m+r=n'-(m'+1)+(r'+1)=n'-m'+r'=2$。

欧拉公式成立是平面图的重要性质,条件是平面图的连通性。欧拉公式还可以推广到非连通的平面图中。将这个结果写成下面的定理。

定理 10.2.2 ▶ 设 G 是平面图,有 k 个连通分支、n 个结点、m 条边、r 个面,则公式 $n-m+r=k+1$ 成立。

证明 设 G 的连通分支为 G_i,n_i,m_i 和 r_i 分别是 G_i 的结点数、边数和面数,根据定理 10.1.2,$n_i-m_i+r_i=2,i=1,\cdots,k$。而 $\sum_{i=1}^{k}m_i=m$,$\sum_{i=1}^{k}n_i=n$,由于每一个 G_i 有一个外部面,而 G 只有一个外部面,所以 G 的面数 $r=\sum_{i=1}^{k}r_i-k+1$,即 $\sum_{i=1}^{k}r_i=r+k-1$。所以

$$2k=\sum_{i=1}^{k}(n_i-m_i+r_i)=\sum_{i=1}^{k}n_i-\sum_{i=1}^{k}m_i+\sum_{i=1}^{k}r_i=n-m+r+k-1$$

整理后有 $n-m+r=k+1$。

由欧拉公式可以推出平面图的一些性质。

定理 10.2.3 ▶ 设 G 是连通的平面图,有 n 个结点、m 条边,每个面的次数至少为 $k(k\geqslant3)$,则关系式 $m\leqslant\dfrac{k}{k-2}(n-2)$ 成立。

证明 设 G 中有 r 个面,由定理 10.1.4 知,$2m=\sum_{i=1}^{r}\deg(r_i)$。

因为每个面的次数大于等于 k,所以所有面的次数之和大于等于 kr,于是,$2m=\sum_{i=1}^{r}\deg(r_i)\geqslant kr$。即 $2m\geqslant kr$。代入欧拉公式 $r=2+m-n$,得 $2m\geqslant kr=k(2+m-n)$,整理得:$m\leqslant\dfrac{k}{k-2}(n-2)$。

例 10.2.1 利用上述定理证明 K_5 和 $K_{3,3}$ 都不是平面图。

解 K_5 如图 10.1.1(g)所示,结点数 $n=5$,边数 $m=10$。假设 K_5 是平面图。显然 K_5 是连通的,因为 K_5 中无自回路和平行边,所以每个面的次数至少为3,即定理 10.2.3 中的 $k=3$,

K_5 满足定理 10.2.3 的条件。所以应有

$$m \leqslant \frac{k}{k-2}(n-2)$$

但是,代入 $n=5,m=10$ 和 $k=3$ 得 $10 \leqslant 9 = \frac{3}{3-2} \times (5-2)$,矛盾。所以 K_5 不是平面图。

$K_{3,3}$ 如图 10.1.1(h)所示,结点数 $n=6$,边数 $m=9$。假设 $K_{3,3}$ 是平面图。显然 $K_{3,3}$ 是连通的,因为在 $K_{3,3}$ 中任何三个结点中必有两个是不相邻接的,所以每个面的次数大于等于 4,即定理 10.2.3 中的 $k=4$,$K_{3,3}$ 满足定理 10.2.3 的条件。所以应有

$$m \leqslant \frac{k}{k-2}(n-2)$$

但是,代入 $n=6,m=9$ 和 $k=4$ 得 $9 \leqslant 8 = \frac{4}{4-2} \times (6-2)$,矛盾。所以 $K_{3,3}$ 不是平面图。

定理 10.2.4 ▶ 设 G 是平面图,有 s 个连通分支、n 个结点、m 条边,每个面的次数至少为 $k(k \geqslant 3)$,则 $m \leqslant \frac{k}{k-2}(n-s-1)$ 成立。

定理 10.2.5 ▶ 设 G 是简单平面图,有 $n(n \geqslant 3)$ 个结点、m 条边,则 $m \leqslant 3n-6$。

证明 设 G 有 s $(s \geqslant 1)$ 个连通分支。

若 G 为树或森林,

$$m = n-s \leqslant n = n+6-6 = n+3+3-6 \leqslant n+n+n-6 = 3n-6$$

即 $m \leqslant 3n-6$。

若 G 不是树,也不是森林,则 G 中必含有回路。又因为 G 是简单平面图,所以其中回路的长度均大于等于 3,因此各面次数 $k \geqslant 3$。又 $\frac{k}{k-2} = 1 + \frac{2}{k-2}$,当 $k=3$ 时,达到最大值 3,由定理 10.2.4 有:

$$m \leqslant \frac{k}{k-2}(n-s-1) \leqslant 3(n-s-1) \leqslant 3(n-1-1) \leqslant 3n-6$$

定理 10.2.6 ▶ 设 G 是极大平面图,有 $n(n \geqslant 3)$ 个结点、m 条边,则 $m=3n-6$。

证明 设 G 有 r 个面。由于极大平面图是连通图,所以 G 是连通图。由欧拉公式得 $r = 2+m-n$,又因为 G 是极大平面图,G 的每个面的次数均为 3,所以 $2m = \sum_{i=1}^{r} \deg(r_i) = 3r = 3(2+m-n)$。

整理后,$m=3n-6$。

在平面图中,各结点度数中的最小者称为最小度,记为 $\delta(G)$;各结点度数中的最大者称为最大度,记为 $\Delta(G)$。

定理 10.2.7 ▶ 设 G 是简单平面图,G 的最小度 $\delta(G) \leqslant 5$。

证明 设 G 有 n 个结点、m 条边。

当 $n \leqslant 6$ 时,因为 G 是简单图,有

$$\delta(G) \leqslant \Delta(G) \leqslant 5$$

以下证 $n \geqslant 7$ 的情况,若 $\delta(G) \geqslant 6$,即每个结点的度数大于等于 6,G 中所有结点度数之和大于等于 $6n$。于是

$$2m = \sum_{i=1}^{n} \deg(v_i) \geqslant 6n, m \geqslant 3n > 3n - 6, \text{即 } m > 3n - 6$$

与定理 10.2.5 矛盾。

在给定图 G 的边上插入一个新的度数为 2 的结点,使一条边分成两条边,如图 10.2.1(a) 所示,叫作插入一个 2 度结点;对于度数为 2 的结点,去掉这个结点,使它关联的两条边变成一条边,如图 10.2.1(b)所示,叫作去掉一个 2 度结点。

定义 10.2.1 ▶给定两个图 G_1 和 G_2,如果它们是同构的,或者通过反复插入一个 2 度结点或去掉一个 2 度结点后使 G_1 和 G_2 同构,则称 G_1 和 G_2 在 2 度结点内同构或 G_1 和 G_2 同胚。

定理 10.2.8 ▶一个图是平面图,当且仅当它不包含与 $K_{3,3}$ 或 K_5 在 2 度结点内同构的子图。

定义 10.2.2 ▶设 $G = <V, E>$ 是无向图,$e \in E$,在 G 中删除 e,将 e 的两个端点 u 和 v 合并成一个新端点,记为 w。w 关联除 e 以外 u 和 v 关联的所有边,称为边 e 的收缩。如果对图 G 的某些边进行收缩,得到图 G',则称将 G 收缩到 G'。

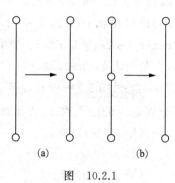

图　10.2.1

定理 10.2.9 ▶一个图是平面图,当且仅当它既没有收缩到 K_5 的子图,也没有收缩到 $K_{3,3}$ 的子图。

10.3　对偶图及其性质

对偶图是平面图相伴的一种图,可以求出原图中的最大流(最小割),本节将对对偶图进行相关介绍。

定义 10.3.1 ▶设 $G = <V, E>$ 是平面图,它具有面 $r_1, r_2, \cdots, r_n$。若有图 $G^* = <V^*, E^*>$ 满足下述条件。

(1) 对于图 G 的任何一个面 r_i,内部有且仅有一个结点 $v_i^* \in V^*$。

(2) 对于图 G 的面 r_i, r_j 的公共边界 $e_k \in E$,存在且仅存在一个边 $e_k^* \in E^*$,使 $e_k^* = (v_i^*, v_j^*)$,且 e_k^* 与 e_k 相交。

(3) 当且仅当 e_k 只是一个面 r_i 的边界时,存在一个自回路 e_k^* 与 e_k 相交。

则称图 G^* 是平面图 G 的对偶图。

在图 10.3.1 中,设实线边的图是 G,则虚线边的图是 G 的对偶图 G^*。

从定义可以看出平面图 G 的对偶图 G^* 有下列性质。

(1) G^* 是连通平面图。

(2) G^* 是图 G 的对偶图,G 也是图 G^* 的对偶图。

(3) 若边 e 为 G 中的自回路,则在 G^* 中,与 e 对应的边 e^* 为桥;若边 e 为桥,则在 G^* 中,与 e 对应的边 e^* 为自回路。

定理 10.3.1 ▶设 G 是连通平面图,G^* 是图 G 的对偶图,m, n 和 r 分别是 G 的结点数、边数和面数,m^*, n^* 和 r^* 分别是 G^* 的结点数、边数和面数,则

(1) $n^* = r$

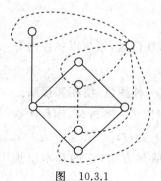

图　10.3.1

(2) $m^* = m$

(3) $r^* = n$

(4) 设 G^* 的结点 v_i^* 位于 G 的面 r_i 中,则 $\deg(v_i^*) = \deg(r_i)$。其中,$\deg(v_i^*)$ 是结点 v_i^* 的度数,$\deg(r_i)$ 是面 r_i 的次数。

证明 根据定义 10.3.1,(1)、(2)显然成立。(3)由于 G 与 G^* 都是连通平面图,因此满足欧拉公式,$n - m + r = 2$,$n^* - m^* + r^* = 2$。故 $r^* = 2 - n^* + m^* = 2 - r + m = 2 + m - r = n$。(4)设 G 的面 r_i 的边界为 C_i,设 C_i 中有 k_1 个桥,k_2 个非桥边,于是 C_i 的长度为 $2k_1 + k_2$,即 $\deg(r_i) = 2k_1 + k_2$;而 k_1 个桥对应面 r_i 中的结点 v_i^* 处有 k_1 个自回路,k_2 个非桥边对应从 v_i^* 处引出 k_2 条边,即 $\deg(v_i^*) = 2k_1 + k_2$;从而有 $\deg(v_i^*) = \deg(r_i)$。

定理 10.3.2 ▶ 设 G 是平面图,具有 $k(k \geqslant 2)$ 个连通分支,G^* 是图 G 的对偶图,m,n 和 r 分别是 G 的结点数、边数和面数,m^*,n^* 和 r^* 分别是 G^* 的结点数、边数和面数,则

(1) $n^* = r$

(2) $m^* = m$

(3) $r^* = n - k + 1$

(4) 设 G^* 的结点 v_i^* 位于 G 的面 r_i 中,则 $\deg(v_i^*) = \deg(r_i)$。其中,$\deg(v_i^*)$ 是结点 v_i^* 的度数,$\deg(r_i)$ 是面 r_i 的次数。

定义 10.3.2 ▶ 如果 G 的对偶图 G^* 与 G 同构,则称 G 为自对偶图。

在图 10.3.2 的(a)、(b)中,设实线边图是 G,则虚线边图是 G 的对偶图 $G*$。可以看出,图 G 和它的对偶图 $G*$ 是同构的。所以,图 10.3.2(a)、(b)中的实线边图都是自对偶图。

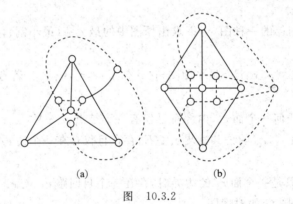

图　10.3.2

10.4　平面图的应用

平面图在很多领域都有用到,但是它往往和其他知识结合起来解决问题,很少以单独的方法出现在问题中。

定义 10.4.1 ▶ 设 G 是一个无自回路图,对 G 的每一个结点指定一种颜色,使相邻接结点具有不同颜色,称为对图 G 的正常着色,简称着色。如果图 G 着色时用了 n 种颜色,则称图 G 为 n 色的。对图 G 着色时,需要的最少颜色数称为 G 的着色数,记为 $X(G)$。

对图 G 的正常着色有一个简单而有效的方法,叫韦尔奇·鲍威尔方法,它的基本思想如下。

　　(1) 将图 G 中的结点按照度数的递减次序进行排列(这种排列可能不唯一,因为有些结点有相同的度数)。

　　(2) 用第一种颜色对第一个结点着色,并且按排列次序,对与前面着色结点不邻接的每一个结点着上同样的颜色。

　　(3) 用第二种颜色对尚未着色的结点重复(2),用第三种颜色继续这种做法,直到所有结点全部着上色为止。

例 10.4.1　用韦尔奇·鲍威尔法对图 10.4.1 着色。

解　(1) 按照度数递减次序排列各结点:$b,f,a,c,$
e,g,d,h。它们的度数分别是 4,4,3,3,3,3,2,2。

　　(2) 用第一种颜色对 b 着色,并对与 b 不相邻的结点 g 也着这种颜色。

　　(3) 对结点 f 和 c 着第二种颜色。

　　(4) 对结点 a、d 和 h 着第三种颜色。

　　(5) 对结点 e 着第四种颜色。

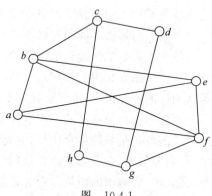

因此 G 是四色的。注意 G 不可能是三色的,因为
a,b,e,f 是相互邻接的,故必须着四种颜色。所以
$X(G)=4$。

图　　10.4.1

　　从定义 10.4.1 可以看出下列定理成立。

定理 10.4.1　$X(G)=1$,当且仅当 G 是零图。

定理 10.4.2　设 G 中至少含有一条边,则 $X(G)=2$,当且仅当 G 是二部图。

定理 10.4.3　$X(K_n)=n$。

证明　因为 n 个结点的完全图的每一个结点与其他 $n-1$ 个结点都邻接,故 n 个结点的着色数不能少于 n,又因为 n 个结点的着色数至多为 n,故 $X(K_n)=n$。

定义 10.4.2　对地图 G 的每个国家涂上一种颜色,使相邻的国家涂不同的颜色,称为对地图 G 的一种面正常着色,简称着色。如果对图 G 着色时用了 n 种颜色,则称地图 G 为 n 色的。对图 G 着色时,需要的最少颜色数称为 G 的着色数,记为 $X^*(G)$。

　　研究地图的着色可以转化成对它的对偶图的结点着色,见下面的定理。

定理 10.4.4　地图 G 是 n 色的,当且仅当它的对偶图 G^* 是 n 色的。

证明　设地图 G 是 n 色的,下证它的对偶图 G^* 是 n 色的。

　　因为地图 G 连通,由定理 10.4.1 可知,$n^*=r$,即 G 的每个面中含 G^* 的一个结点,设位于 G 的面 r_i 内。将 G^* 的结点 v_i^* 涂 r_i 的颜色,易知,若 v_i^* 与 v_j^* 相邻,则由于 r_i 与 r_j 的颜色不同,所以 v_i^* 与 v_j^* 的颜色也不同,因而 G^* 是 n 色的。

　　当 G 的对偶图 G^* 是 n 色的,类似地可以证明 G 也是 n 色的。

　　由定理 10.4.4 可知,研究地图的着色(面着色)等价于研究平面图的结点着色。对于平面图的结点着色问题,到目前为止,人们证明了下面的定理。

定理 10.4.5　任何平面图 G 最多是 5 色的。

定理 10.4.5 称为五色定理,也称为希伍德定理。

平面图的几何特点被广泛应用于各种领域,如在信息学中的应用。

例 10.4.2 统计图 10.4.2 中三角形的数目。

解法 1

枚举所有的三元组,判断三个顶点是否两两相邻。由于总共有 C_n^3 个三元组,因此时间复杂度为 $O(n^3)$。

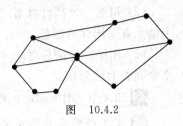

图 10.4.2

解法 2

枚举一条边,再枚举第三个顶点,判断其是否与边上的两个端点相邻。根据定义 10.1.1 可知,图 10.4.2 为平面图,图 10.4.2 的边数为 $O(n)$,故解法 2 的复杂度为 $O(n^2)$。

解法 1 与解法 2 的比较如下。

解法 1 只是单纯地枚举,没有注意到问题的实际情况,而实际上三角形的数目是很少的,解法 1 做了很多无用的枚举,因此效率很低。解法 2 则是从边出发,枚举第三个顶点,这正好符合问题的实际情况,避开了许多不必要的枚举,所以解法 2 比解法 1 更加高效。

平面图着色最典型的应用就是地图着色。

地图着色是一种组合构形,它是对于地图面集的一种分划,分配地图的每一个面一种颜色,使相邻的面(指有公共边界边)具有不同的颜色,称这样一种色的分配为这个地图的一个着色,或者说,将地图的面集分划为若干个子集,使每个子集中的任何两面均不相邻,这样就可以将每个子集中的面用一种颜色着染,使不同子集用的颜色不同。在地图 M 的所有着色中,使用颜色最少的着色的颜色数目称为地图 M 的色数。地图的顶点着色,或者说,对于与它同构的图的顶点做正常着色,就是其对偶地图的地图着色。

习 题

1. 设 G 是有 n 个结点、m 条边的连通平面图,且有 k 个面,则 k 等于(　　)。

    A. $m-n+2$　　　　B. $n-m-2$　　　　C. $n+m-2$　　　　D. $m+n+2$

2. 在具有 6 个顶点、12 条边的连通简单平面图中,每个面是由(　　)条边组成。

    A. 2　　　　　　　B. 4　　　　　　　C. 3　　　　　　　D. 5

3. 设 G 是连通平面图,有 v 个结点、e 条边、r 个面,则 $r=$(　　)。

    A. $e-v+2$　　　　B. $v+e-2$　　　　C. $e-v-2$　　　　D. $e+v+2$

4. 设 G 是连通平面图,有 5 个顶点、6 个面,则 G 的边数是(　　)。

    A. 9　　　　　　　B. 5　　　　　　　C. 6　　　　　　　D. 11

5. 设图 $G=<V,E>$,则下列结论中成立的是(　　)。

    A. $\deg(V)=2|E|$　　　　　　　　　　B. $\deg(V)=|E|$

    C. $\sum\limits_{v\in V}\deg(v)=2|E|$　　　　　　D. $\sum\limits_{v\in V}\deg(v)=|E|$

6. 任意平面图 G 最多是(　　)色的。

    A. 3　　　　　　　B. 4　　　　　　　C. 5　　　　　　　D. 6

7. 对于 10 个结点的完全图 K_{10},对其着色时,需要的最少颜色数 $X(K_{10})$ 为(　　)。

    A. 10　　　　　　B. 9　　　　　　　C. 11　　　　　　D. 12

8. 判断下列说法是否正确。

(1) 平面图中,任何两条边除端点外可以有其他交点。(　　　)

（2）K_5 是平面图。（　　）

（3）设 G_1，G_2 是平面图，若 $G_1 \cong G_2$，则它们的对偶图 $G_1^* \cong G_2^*$。（　　）

（4）K_4 是平面图。（　　）

9. 判断习题 9 图 G 是不是平面图，并说明理由。

10. 验证 K_5 和 $K_{3,3}$ 都是极小非平面图。

11. 设 G 是 n 阶 m 条边的简单平面图，已知 $m < 30$，证明：$\delta(G) \leqslant 4$。

12. 画出 6 阶的所有非同构的连通的简单的非平面图。

13. 验证习题 13 图是平面图。

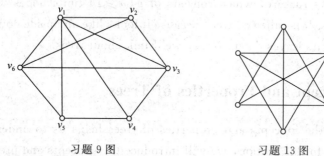
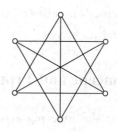

习题 9 图　　　　　　　　习题 13 图

14. 验证习题 14 图是极大平面图。

15. 验证习题 15 图平面图满足欧拉公式。

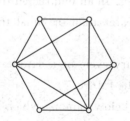
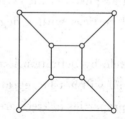

习题 14 图　　　　　　　　习题 15 图

Chapter 11　Tree

A tree is a data structure, which consists of $n\,(n \geqslant 1)$ finite nodes to form a set with hierarchical relations. It is called a "tree" because it looks like an upside down tree, that is, it has its roots up and its leaves down. Next, we'll talk about trees.

11.1　The Concept and Properties of Trees

Understanding the concepts and properties of trees helps us to understand the related knowledge of trees. In this section, we will introduce the concepts and properties of trees.

Definition 11.1.1　An undirected graph that is connected without a cycle is called an undirected tree, or simply a tree. Each connected branch is an undirected graph of a tree called a forest. Trivial graphs are called trivial trees. In an undirected tree, nodes of degree 1 are called leaves, and vertices with a degree greater than or equal to 2 are called branch points.

Description: A cycle by definition is a primary cycle or a simple cycle. It is so agreed upon in this chapter and will not be repeated in the future.

Some important properties of trees are given below, among which Theorem 11.1.1 gives several necessary and sufficient conditions for trees.

Theorem 11.1.1　Let $G = <V, E>$ is an undirected graph with n order m edges, then the following propositions are equivalent.

(1) G is a tree.

(2) There is a unique path between any two vertices in G.

(3) There is no cycle in G and $m = n - 1$.

(4) G is connected and $m = n - 1$.

(5) G is connected and any edge is a bridge.

(6) There is no cycle in G, but the graph resulting from adding a new edge between any two different vertices has a unique cycle with a new edge.

Proof　(1)$\Rightarrow$(2). According to the connectivity of G, $\forall u, v \in V$, there is a path between u and v. If the path is not unique, and let Γ_1 and Γ_2 are the path from u to v, then there must be a cycle formed by the edge of Γ_1 and Γ_2, which is contradictory to the non-loop in G.

(2)$\Rightarrow$(3). First, we prove that there is no cycle in G. If there is a loop associated with a vertex v in G, then there are two paths of length 0 and 1 from v to v (note that the primary loop is a special case of the path), which contradicts the known conditions. If there is a cycle of length greater than or equal to 2 in G, then there are two different paths between any two vertices on the circle, which also leads to a contradiction. Now, the induction method is used to prove that $m=n-1$.

When $n=1$, C is a trivial graph, and the conclusion is obviously true. Let when $n \leqslant k(k \geqslant 1)$ the conclusion is valid. When $n=k+1$, let $e=(u, v)$ be an edge in G. Since there is no cycle in G, $G-e$ is two connected branches G_1, G_2. Let n_i, m_i is the number of vertices and edges in G_i respectively, then $n_i \leqslant k$. By the induction hypothesis, $m_i=n_i-1$, $i=1, 2$. So, $m=m_1+m_2+1=n_1+n_2-2+1=n-1$. It turns out that this is also true when $n=k+1$.

(3)$\Rightarrow$(4). Just to prove that G is connected. Suppose not, let's say that G has $s(s \geqslant 2)$ connected branches G_1, G_2, ..., G_i. There are no cycles in each G, so G_i are all trees. From (1)$\Rightarrow$(2)$\Rightarrow$(3) known, $m_i=n_i-1$. So $m=\sum_{i=1}^{s} m_i=\sum_{i=1}^{s} n_i-s=n-s$. Since $s \geqslant 2$, this contradicts with $m=n-1$.

(4)$\Rightarrow$(5). Just to prove that every edge in G is a bridge. $\forall e \in E$, all have $|E(G-e)|=n-1-1=n-2$. $G-e$ is not a connected graph, so e is a bridge.

(5)$\Rightarrow$(6). Since each edge in G is a bridge, there is no cycle in G. And because G is connected, G is a tree. We can know from (1)$\Rightarrow$(2), let e be the new edge added between u and v, then $\Gamma \cup e$ is a cycle, and it is obviously unique.

(6)$\Rightarrow$(1). Just to prove that G is connected. For any two different vertices u and v, the addition of a new edge e between u and v produce a unique cycle C containing e. Obviously, $C-e$ is the path from u to v in G, so $u \sim v$ can be known from the arbitrariness of u and v that G is connected.

Theorem 11.1.2 Let T be a nontrivial undirected tree of order n, then there are at least two leaves in T.

Proof Let T have x leaves, it can be known from Definition 11.1.1 that,
$$2(n-1)=\sum d(v_i) \geqslant x+2(n-x)$$
Solve for $x \geqslant 2$.

Example 11.1.1 Draw all the undirected trees of order 6 that are not isomorphic.

Solution Let T be an undirected tree of order 6. According to Theorem 11.1.1, the number of edges of T is $m=5$. The handshake theorem tells us that the sum of the degrees of the six vertices of T is equal to 10. So, the degree column of T must be one of the following:

(1) 1,1,1,1,1,5
(2) 1,1,1,1,2,4
(3) 1,1,1,1,3,3
(4) 1,1,1,2,2,3
(5) 1,1,2,2,2,2

Their corresponding trees are shown in Figure 11.1.1, where T_1 corresponds to (1), T_2 corresponds to (2), T_3 corresponds to (3), T_4, T_5 correspond to (4), T_6 corresponds to (5). (4) corresponds to two non-isomorphic trees, two 2-degree vertices are adjacent in one tree, not adjacent in the other tree, other conditions corresponding to a tree are not adjacent; all other cases correspond to a nonisomorphic tree.

An $n(n \geqslant 3)$ undirected tree with only one branch vertex and the degree of the branch vertex is $n-1$ is often called a star graph, and its only branch vertex is called the star center. T_1 of Figure 11.1.1 shows a star of order 6.

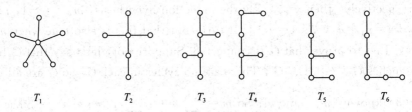

$$T_1 \qquad T_2 \qquad T_3 \qquad T_4 \qquad T_5 \qquad T_6$$

Figure 11.1.1

Example 11.1.2 Draw all undirected trees of order 7 that have 3 leaves, 1 vertex of degree 3, and other vertices whose degrees are not equal to 1 and 3.

Solution Let T be an undirected tree satisfying the requirements, and the number of edges $m = 6$. So, $1 \times 3 + 3 + d(v_4) + d(v_5) + d(v_6) = 12$. Because $d(v_j)(j = 4, 5, 6)$ is not equal to 1 and 3, and it's greater than or equal to 1, less than or equal to 6, so $d(v_j) = 2(j = 4, 5, 6)$. So the degree of T is listed as

$$1, 1, 1, 2, 2, 2, 3$$

Thus, the degree of the three vertices adjacent to a 3 degree vertex is only possible in the following three ways.

$$1, 1, 2 \quad 1, 2, 2 \quad 2, 2, 2$$

The corresponding trees are shown in Figure 11.1.2, and there is a non-isomorphic tree of order 7 for each possible tree.

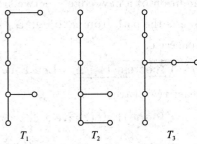

$$T_1 \qquad T_2 \qquad T_3$$

Figure 11.1.2

11.2 Spanning Tree

Spanning tree is widely used in the computer networks, which can avoid the endless cycle of frames, and the storage structure of s tree, which is convenient for data searching and sorting.

Definition 11.2.1 If the spanning subgraph T of undirected graph G is a tree, T is said to be a spanning tree of G. Let T be the spanning tree of G, and the edges of C in T are called branches of T, strings that are not in T are called T. Call the derived graph of all strings of T is the remainder of T, let me call that $\overline{T}$.

Note that $\overline{T}$ doesn't have to be connected, and it doesn't have to have a cycle. In Figure 11.2.1, the real edge is a spanning tree T of the figure, while the remainder tree is shown as the virtual edge, which is not connected and contains cycles.

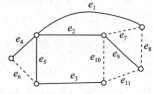

Theorem 11.2.1　Undirected graph G has a living adult tree if and only if G is a connected graph.

Figure　11.2.1

Proof　The necessity is obvious. Now let's prove sufficiency. If there is no loop in G, then G is its own spanning tree. If G contains a cycle, take any cycle and delete an edge on the cycle at will, you have cycle connectivity. If there are still cycles, take another cycle and delete one edge of the cycle. Repeat until there are no cycles left. The resulting graph is acyclic (of course acyclic), connected and a spanning subgraph of G, so it is a spanning tree of G.

The proof of Theorem 11.2.1 is a constructive proof, and this method of generating spanning trees is called the break method.

From Theorem 11.2.1 and the number of edges in the tree is equal to the number of vertices minus 1, the following inference can be immediately obtained.

Inference　Let G be an undirected connected graph with m edges of order n, then $m \geqslant n-1$.

Theorem 11.2.2　Let T be a spanning tree in the undirected connected graph G and e be any chord of T, then the other sides which contain only one chord e in G are cycles of branches, and the cycles corresponding to different chords are also different.

Proof　Let $e=(u, v)$, it can be seen from Theorem 11.1.1 that in T, there is a unique path $\Gamma_{(u, v)}$ between u and v, then the requirement $\Gamma_{(u, v)} \cup e$ is satisfied. It is obvious that different chords have different cycles.

The following definitions can be given by Theorem 11.2.2.

Definition 11.2.2　Let T be a spanning tree of an undirected connected graph G of order n with m edges, let $e'_1, e'_2, \ldots, e'_{m-n+1}$ be the chord of T, $C_r(r=1, 2, \ldots, m-n+1)$ is the cycle formed by chord e'_r and branches in G produced by adding chord e'_r to T, and C_r is called the basic cycle or basic cycle of the corresponding chord e'_r of G. $\{C_1, C_2, \ldots, C_{m-n+1}\}$ is the basic cycle system of G corresponding to T, and $m-n+1$ is the coil rank of G, denoted as $\xi(G)$.

In Figure 11.2.1, the basic cycle of the chord corresponding to the spanning tree e_6, e_7, e_8, e_{10}, e_{11} is

$$C_1 = e_6 e_4 e_5$$
$$C_2 = e_7 e_2 e_5$$
$$C_3 = e_8 e_9 e_2 e_1$$
$$C_4 = e_{10} e_3 e_5 e_2$$
$$C_5 = e_{11} e_3 e_5 e_2 e_9$$

The cycle rank of this figure is 5, and the basic cycle system is $\{C_1, C_2, C_3, C_4, C_5\}$.

It is not difficult to see that the ring rank of undirected connected graph G has nothing to do with the selection of the spanning tree, but different spanning trees may correspond to different basic cycle systems.

Theorem 11.2.3 Let T be a spanning tree of connected graph G and e is a branch of T, then there exists only branch e in G, and the rest edges are the cut sets of chords, and the corresponding cut sets of different branches are also different.

Proof It can be seen from Theorem 11.1.1 that e is a bridge of T, so T-e has two connected branches T_1, T_2, let $S_e = \{e' \mid e' \in E(G)$ and the vertex e' belonging to $V(T_1)$ and $V(T_2)\}$. Obviously, S_e is the cut set of G, where $e \in S_e$ and all but e are chords in S_e. Since each cut set S_e contains only one branch, different branches have different cut sets.

Definition 11.2.3 Let T be a spanning tree of n order connected graph G, and e_1, e_2, ..., e_{n-1} is the branch of T, $S_i (i=1, 2, ..., n-1)$ is the cut set composed of twig e_i and chord, and S_i is the basic cut set of the corresponding twig e_i of G. $\{S_1, S_2, ..., S_{n-1}\}$ is called the basic cut-set system of G corresponding to T, and $n-1$ is called the cut-set rank of G, denoted as $\eta(G)$.

In Figure 11.2.1, the basic cut sets of e_1, e_2, e_3, e_4, e_5, e_9 corresponding branches are respectively

$$S_1 = \{e_1, e_7, e_8\}$$
$$S_2 = \{e_2, e_7, e_8, e_{10}, e_{11}\}$$
$$S_3 = \{e_3, e_{10}, e_{11}\}$$
$$S_4 = \{e_4, e_6\}$$
$$S_5 = \{e_5, e_6, e_{10}, e_{11}\}$$
$$S_6 = \{e_9, e_8, e_{11}\}$$

The basic cut set system is $\{S_1, S_2, S_3, S_4, S_5, S_6\}$, the rank of the cut set is 6.

The cut rank $\eta(G)$ of connected graph G does not change with different spanning trees, but the basic cut set system corresponding to different spanning trees may be different.

The minimum spanning tree in a connected weighted graph is discussed below.

Definition 11.2.4 Let the undirected connected belt weight graph $G = <V, E, W> (W$ is the weight of each edge), T be a spanning tree of G, and the sum of all edge weights of T be called the weight of T, denoted as $W(T)$. The least weighted spanning tree of all spanning trees of G is called the minimum spanning tree of G.

There are many algorithms to find the generated spanning tree. Here is the method of avoiding circles(Kruskal algorithm).

Let n order undirected connected weighted graph $G = <V, E, W>$ have m edges. Suppose there are no loops in G (otherwise, all loops can be deleted first), arrange m edges in order of weight from smallest to largest, and set as e_1, e_2, ..., e_m.

Take e_1 in T, and then check e_2, e_3, ..., e_m in turn. If $e_j (j \geqslant 2)$ and the edge already in T cannot form a loop, then take e_j in T, otherwise discard e_j.

When the algorithm stops, T is the minimum spanning tree of G (proof omitted).

Example 11.2.1 Find the minimum spanning tree in the two figures shown in Figure 11.2.2.

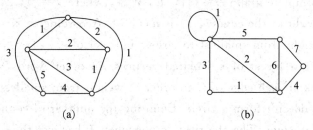

(a)   (b)

Figure 11.2.2

Solution The minimum spanning tree obtained by avoiding the circle method is shown in Figure 11.2.3, and their weights are 6 and 12 respectively.

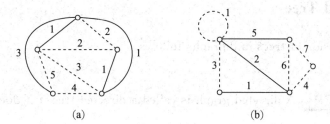

(a)   (b)

Figure 11.2.3

Example 11.2.2 Single chain clustering in data analysis. Various clustering operations are often used in data analysis. The so-called clustering operation is to aggregate the data in dataset D into several subclasses according to their similarity. This operation in Data mining, image processing, circuit design, system division is often used. Let's consider the simplest single chain clustering.

Suppose there is a set of discrete data $D = \{a_1, a_2, \ldots, a_n\}$. D defines a similarity function d. For any two data a_i, $a_j \in D$, a_i and a_j, the value of the similarity function is $d(i, j)$, usually $0 \leqslant d(i, j) \leqslant 1$, and d has the property of symmetry, namely $d(i, j) = d(j, i)$.

Given a positive integer $k (1 < k < n)$, a k cluster of D is a k partition of D $\pi = \{C_1, C_2, \ldots, C_k\}$. We want data in the same subclass to be as "close" as possible, and data in different subclasses to be as "far away" as possible. Divide the minimum interval π by $D(\pi)$ for this purpose as defined below.

For any two different subclasses C_i, C_j, define the distance between them $D(C_i, C_j)$ to be the minimum value of the similarity between the data in C_i and the data in C_j, there is

$$D(C_i, C_j) = \min\{d(i, j) \mid a_i \in C_i, a_j \in C_j\}$$

The minimum interval of k cluster $\pi = \{C_1, C_2, \ldots, C_k\}$ is

$$D(\pi) = \min\{D(C_i, C_j) \mid C_i, C_j \in \pi, 1 \leqslant i < j \leqslant k\}$$

Our question is: given the similarity function D and the positive integer k on datasets D and D, how to find the clustering π that makes the $D(\pi)$ maximum?

The Kruskal algorithm of the minimum spanning tree can be used to solve this problem. Define a weighted complete graph $G=<V, E, W>$, where $V=\{1, 2, ..., n\}$, for any i, $j \in V$, $i \neq j$, the weight of the edge(i, j) is $d(i, j)$. According to Kruskal algorithm, the edge is sorted by weight from smallest to largest in order of $e_1, e_2, ..., e_{n(n-1)/2}$. The initial T has no edges, it's a forest of n isolated vertices. In other words, T has n connected branches. Next, look at each edge of G in order of weight from smallest to largest, and add it to T as long as it doesn't form a circle. Counting the number of connected branches of T during the addition of edges. The algorithm stops when T has exactly k connected branches. At this time, the k connected branches obtained are exactly k subclasses of the desired cluster $C_1, C_2, ..., C_k$, whose minimum interval reaches the maximum.

11.3　Directed Tree

A brief discussion of trees in digraphs follows.

1. Root tree

Definition 11.3.1　A directed graph is called a directed tree if it does not consider the direction of the arc.

A directed tree is shown in Figure 11.3.1.

Definition 11.3.2　A directed tree is called a rooted tree if just one vertex has an entry degree of 0 and all the other vertices have an entry degree of 1. Vertices with a degree of entry of 0 are called roots, vertices with a degree of exit of 0 are called leaves or hanging points, and vertices with a degree of exit of 0 are called branch points or inner points.

Figure 11.3.2 shows a root tree, where v_1 is the root, v_1, v_2, v_4, v_8, v_9 is the branch vertex, the rest of the vertices are leaves.

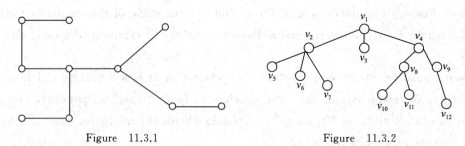

Figure　11.3.1　　　　　　Figure　11.3.2

Definition 11.3.3　There can only be a unique path from the root to any vertex v_i, and the length of this path is called the series of v_i, also known as the length of v_i the vertex path. In Figure 11.3.2, the series for the vertices v_1 is 0, the series for the vertices v_2, v_3, v_4 is 1, the series for the vertices v_5, v_6, v_7, v_8, v_9 is 2, and the series for the vertices v_{10},

v_{11}, v_{12} is 3. The maximum value of the series for all vertices in a rooted tree is called the height of the tree.

In computer science, there are some technical terms for root trees, which are described as follows.

Let u be a branch point of the root tree. If there is an arc (u, w) from u to w, then w is called the son of u or the father of w. If a vertex has two sons, call them brothers; If there is a directed way from u to z, call z a descendant of u or call u an ancestor of z.

Let u be a branch point of root tree T, with u as the root, all descendants of u constitute vertex set V', and all arcs on the directed path from u to these descendants constitute arc set E', then subfigure $T' = (V', E')$ of T is called a subtree with u as the root.

2. Ordered Tree

For a rooted tree, there can be two different ways of drawing roots below or above, as shown in Figure 11.3.3.

(a)   (b)   (c)

Figure   11.3.3

Figure 11.3.3(a) is the natural representation of the root tree, that is, the tree grows upward from its root.

Figure 11.3.3(b) and Figure 11.3.3(c) both grow from the root downwards. They are of the same composition and differ only in the order in which the vertices at each level appear from left to right.

Definition 11.3.4   The order of vertices or arcs in a rooted tree is specified in a definite way. This tree is called an ordered tree.

If the vertices are in order, then:

(1) Root order. ①Access to the root; ②Traverse the left subtree; ③Traverse the right subtree.

(2) Middle root order. ①Traverse the left subtree; ②Access to the root; ③Traverse the right subtree.

(3) Post root order. ① Traverse the left subtree; ② Traverse the right subtree; ③Access to the root.

3. Binary Tree

Definition 11.3.5   In the root tree, if the degree of each vertex is less than or equal to m, the tree is called an m-cross tree. If the degree of the output of each vertex is exactly equal to m or zero, the tree is called a complete m-cross tree. A complete m-tree with the same series of leaves is also called a regular m-tree. In particular, when m is equal to 2, it's called a binary tree.

There are a lot of practical problems that can be represented by a binary tree or an m-cross tree.

For example, in a tennis match between M and E, if one player wins two or three sets in a row, he or she wins. The game is over. Figure 11.3.4 shows the various possible scenarios in which the race is played. It has ten leaves, and each path from the root to a leaf corresponds to the race, such that MM, MEMM, MEMEM, MEMEE, MEE, EMM, EMEMM, EMEME, EMEE, EE.

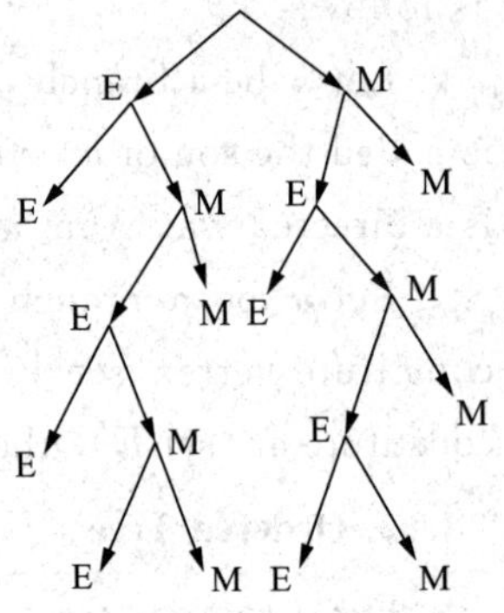

Figure 11.3.4

Theorem 11.3.1 If the complete binary tree T has n branch points, and the total length of all branches is I, and the total length of all leaves is E, then

$$E = I + 2n$$

Proof The number of branch points n is summarized.

Let's say it's true for $n-1$. The number of branch points is n investigated as follows.

If you delete the edge incident with a branch vertex v and its sons, whose path length is l, and whose two sons are leaves, then you get a tree T', E reduces by $2(l+1)$; And by deleting the son of v, we increase E by l, so the change value of E is $-l-2$, the change value of I is $-l$. From the induction hypothesis we know, in T', $E' = I' + 2n'$, here $n' = n-1$, and then we add v and its two sons to T', and we get T. Because $E' = I' + 2(n-1)$ and $E' = E - l - 2$, $I' = I - l$, $n' = n-1$, so $E = I + 2n$.

11.4 Root Trees and Their Applications

The concept of root trees is simple and has a wide range of applications. Let's briefly introduce root trees and their applications.

Definition 11.4.1 Let T be a rooted tree, $\forall v \in V(T)$, and the derived subgraph T' of v and its descendants is called a rooted tree with v as the root of T.

The root trees derived by the two sons of each branch point of a binary regular ordered tree are called the left and right subtrees of the branch point respectively.

The following introduces the application of binary tree.

Definition 11.4.2 Let a Binary tree T have t leaves v_1, v_2, ..., v_t, with weights of w_1, w_2, ..., w_t respectively, and call $W(T) = \sum w_i l(v_i)$ the weights of T, wherein $l(v_i)$ is the number of layers of v_i. Among all binary trees with t leaves and weight w_1, w_2, ..., w_t, the least weight binary tree is called the optimal binary tree.

The three trees shown in Figure 11.4.1 are all binary trees with weights of 2, 2, 3, 3, 5. They have the following rights.

$$W(T_1)=2\times2+2\times2+3\times3+5\times3+3\times2=38$$
$$W(T_2)=3\times4+5\times4+3\times3+2\times2+2\times1=47$$
$$W(T_3)=3\times3+3\times3+5\times2+2\times2+2\times2=36$$

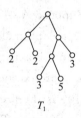 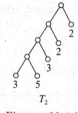 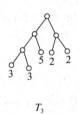

Figure 11.4.1

The following is the introduction of the optimal binary tree algorithm—Huffman algorithm.

Given real number w_1, w_2, ..., w_t,

(1) Make t leaves, with w_1, w_2, ..., w_t as the weight.

(2) Select the two vertices (not necessarily leaves) with the lowest weight among all the vertices of degree 0, and add a new branch vertex, whose weight is equal to the sum of the weights of the two sons.

(3) Repeat (2) until there is only one vertex with an entry degree of 0.

$W(T)$ is equal to the sum of the weights of all the branching vertices.

Example 11.4.1 Find the optimal binary tree with weights 2, 2, 3, 3, 5.

Solution Figure 11.4.2 shows the process of calculating the optimal tree with Huffman algorithm. (e) is the optimal tree, $W(T)=34$. This indicates that none of the three trees shown in Figure 11.4.1 are optimal.

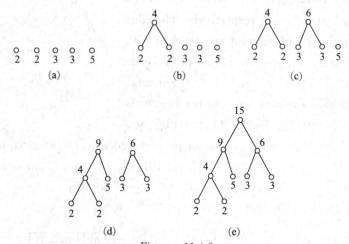

Figure 11.4.2

In communication, binary codes are commonly used to represent symbols. For example, the binary codes 00, 01, 10, 11 of length 2 can represent A, B, C, D respectively. This notation is called isometric code notation. If A, B, C and D appear at roughly the same frequency during transmission, it is a good idea to use the unequal length encoding code. However, when the frequency of their appearance is very different, in order to save binary digits and achieve the purpose of improving efficiency, it is necessary to use non-equal-length codes.

Definition 11.4.3 Let $\alpha_1\alpha_2...\alpha_{n-1}\alpha_n$ be a symbol string of length n and its substring α_1, $\alpha_1\alpha_2$, ..., $\alpha_1\alpha_2...\alpha_{n-1}$ be the prefix of the symbol string. Suppose $A=\{\beta_1, \beta_2, ..., \beta_m\}$ is a set of symbol strings. If any two symbol strings of A are not prefixed, A is called prefix code. A prefix code consisting of a 0-1 symbol string is called a binary prefix code.

For example, {1, 00, 011, 0101, 01001, 01000} is a prefix code and {1, 00, 011, 0101, 0100, 01001, 01000} is not a prefix code because 0100 is the prefix for 01001.

Binary tree can be used to generate binary prefix codes. Suppose T is a binary tree with n leaves, then T has one or two sons per branch point. Let v be the branching point of T, if v has two sons, on the two sides of the derivation of v, the left side is 0, and the right side is 1. If v has only one son, the edge from it can be marked 0 or 1. Let v_i be any leaf of T, the symbol strings of the labels (0 or 1) on each edge of the path from the root to v_i are placed v_i in the order above the path, and the t symbol strings of the leaves of t constitute a binary code. It can be seen from the practice that the prefix of symbol string at the leaf v_i is reached at the vertex (excluding v_i) on the path from the root to v_i, so the symbol string set obtained must be prefix code. If T is a binary regular tree, the prefix code generated by T is unique. In other words, a binary regular tree can generate a unique binary prefix code.

Example 11.4.2 Find the prefix code generated by two binary trees as shown in Figure 11.4.3.

Solution Figure 11.4.3 (a) is a binary tree, but is not regular. The two edges derived from each branch vertex were marked with 0 and 1 respectively. The edge derived from the right son of the root was marked with 1, as shown in Figure 11.4.4 (a). The prefix code generated was {11, 01, 000, 0010, 0011}. If the edge mark of the right son of the root is 0, the prefix code is {10, 01, 000, 0010, 0011}. Figure 11.4.3 (b) is a

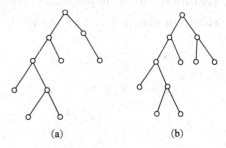

Figure 11.4.3

binary regular tree, which can only generate unique prefix codes. The calibrated binary regular tree is shown in Figure 11.4.4 (b), with prefix codes of {01,10, 11, 000, 0010, 0011}.

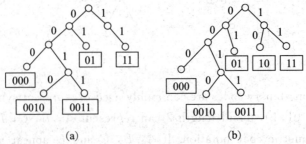

Figure 11.4.4

Any prefix code generated above can be used to transmit five symbols, such as A, B,

C, D and E. But when these letters appear at different frequencies in the text, the binary bits used to transmit the text are different. Suppose there be t symbols in total, the frequency of symbols represented by the binary string at the leaf v_i is v_i, and the length of the binary string at the leaf c_i, v_i is equal to the number of layers, so the binary digit used for transmitting m symbols is $m \sum_{i=1}^{i} c_i l(v_i)$. Obviously, the prefix code generated by the optimal binary tree weighted by the frequency of each symbol USES the least number of binary digits, and is called the optimal prefix code by the optimal binary tree. The best prefix code transmits the least number of binary digits.

The following describes the binary tree travel and application.

Accessing each vertex of a rooted tree once and only once is called traversing or touring a tree.

There are three ways for binary ordered regular trees to travel.

(1) Inorder traversal method. The order of access is left subtree, root, and right subtree.

(2) Preorder traversal method. The order of access is root, left subtree, and right subtree.

(3) Postorder traversal method. The order of access is left subtree, right subtree, and root.

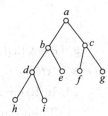

Figure 11.4.5

For binary ordered regular tree T as shown in Figure 11.4.5, the traversal results in the inorder, preorder, and postorder are as follows:

$$((h\ \underline{d}i)\underline{b}e)\ \underline{a}(f\underline{c}g)$$

$$\underline{a}(\underline{b}(\underline{d}hi)e)(\underline{c}fg)$$

$$((hi\ \underline{d})\ e\ \underline{b})\ (fg\underline{c})\underline{a}$$

In the above equation, $\underline{v}$ means v is the root of the root tree. The binary ordered regular tree can be used to represent the equations of four operations, and then different algorithms can be obtained according to different access methods. Orderly regular use binary tree representation formula method is as follows:

In computing the number on the leaves, and then the sequence according to the operation will be operators $(+, -, *, \div)$ on the branch point, each branch point said an operation, also said the results of this operation, it is the son with its two operands, and dividend, the minuend on the left.

Exercises

1. Draw all undirected trees of order 5 and order 7 that are not isomorphic.

2. An undirected tree T has 5 leaves, 3 branch vertices of 2 degrees, and the other branch vertices are all vertices of 3 degrees. How many vertices does T have?

3. Proof: Any undirected tree T is a bipartite graph.

4. Under what conditions, the undirected tree T is a semi-Eulerian graph.

5. Under what conditions, the undirected tree T is a semi-Hamiltonian diagram.

6. Let T be the spanning tree of undirected tree G and the cotree of T, and the edge cut set without G is proved.

7. In the undirected graph shown in Exercise 7 Figure, how many nonisomorphic spanning trees with edges as branches are there? Let me draw them.

8. How many non-isomorphic spanning trees are in the undirected graph shown in Exercise 8 Figure? Draw the trees (hint: pick from all 6 nonisomorphic trees).

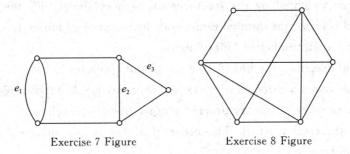

Exercise 7 Figure   Exercise 8 Figure

9. Given that the undirected graph G of n order with m edges is a forest composed of $k(h \geqslant 2)$ trees, it is proved that $m = n - k$.

10. In Exercise 10 Figure, the real edges form a spanning tree, denoted as T.

(1) Point out the chord of T, the basic loop corresponding to each chord and the basic loop system corresponding to T.

(2) Point out all branches of T and the basic cut set corresponding to each branch and the basic cut set system corresponding to T.

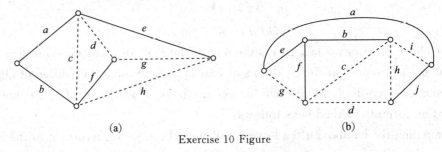

(a)   (b)

Exercise 10 Figure

11. Find the minimum spanning tree of the two weighted graphs in Exercise 11 Figure.

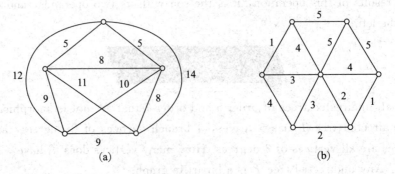

(a)   (b)

Exercise 11 Figure

12. Draw a fully regular binary tree of height 3.

13. In the digraph shown in Exercise 13 Figure, is there a spanning subgraph of the root tree? If there are, how many are non-isomorphic?

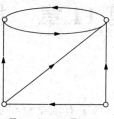

Exercise 13 Figure

14. Draw an optimal binary tree with weights of 3, 4, 5, 6, 7, 8, 9 and calculate its weights.

15. Use the binary tree shown in Exercise 15 Figure to generate a binary prefix code.

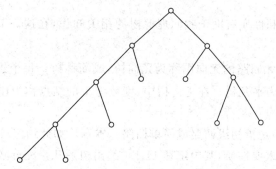

Exercise 15 Figure

第 11 章　树

　　树是一种数据结构,它是由 $n(n \geq 1)$ 个有限结点组成一个具有层次关系的集合。把它叫作"树"是因为它看起来像一棵倒挂的树,也就是说它是根在上而叶在下的。接下来我们将介绍树的相关知识。

11.1　树的概念与性质

　　了解树的相关概念和性质有助于我们理解树的相关知识,在这一节我们将介绍树的概念和性质。

　　定义 11.1.1 ▶ 连通无回路的无向图称为无向树,或简称树。每个连通分支都是树的无向图称为森林。平凡图称为平凡树。在无向树中,度数为 1 的结点称为树叶,度数大于或等于 2 的顶点称为分支点。

　　说明: 定义中的回路是指初级回路或简单回路。本章均如此约定,以后不再重复说明。

　　下面给出树的一些重要性质,其中定理 11.1.1 给出树的几个充分必要条件。

　　定理 11.1.1 ▶ 设 $G = <V, E>$ 是 n 阶 m 条边的无向图,则下面各命题是等价的。

　　(1) G 是树。

　　(2) G 中任意两个顶点之间存在唯一的路径。

　　(3) G 中无回路且 $m = n - 1$。

　　(4) G 是连通的且 $m = n - 1$。

　　(5) G 是连通的且任何边均为桥。

　　(6) G 中没有回路,但在任何两个不同的顶点之间加一条新边后所得图中有唯一的一个含新边的圈。

　　证明 (1)⇒(2)。由 G 的连通性可知,$\forall u, v \in V$,u 与 v 之间存在一条路径。若路径不是唯一的,设 Γ_1 和 Γ_2 都是 u 到 v 的路径,则必存在由 Γ_1 和 Γ_2 上的边构成的回路,这与 G 中无回路矛盾。

　　(2)⇒(3)。首先证明 G 中无回路。若 G 中存在关联某顶点 v 的环,则 v 到 v 存在长为 0 和 1 的两条路径(注意初级回路是路径的特殊情况),这与已知条件矛盾。若 G 中存在长度大于或等于 2 的圈,则圈上任何两个顶点之间都存在两条不同的路径,这也引出矛盾,下面用归纳法证明 $m = n - 1$。

　　$n = 1$ 时,C 为平凡图,结论显然成立。设 $n \leq k(k \geq 1)$ 时结论成立。当 $n = k + 1$ 时,设 $e = (u, v)$ 为 G 中的一条边。由于 G 中无回路,所以 $G-e$ 为两个连通分支 G_1, G_2。设 n_i, m_i 分别为 G_i 中的顶点数和边数,则 $n_i \leq k$。由归纳假设,$m_i = n_i - 1, i = 1, 2$。于是,$m = m_1 + m_2 + 1 =$

$n_1+n_2-2+1=n-1$。得证,当 $n=k+1$ 时结论也成立。

(3)⇒(4)。只需证明 G 是连通的。假设 G 是不连通的,设 G 有 $s(s \geqslant 2)$ 个连通分支 G_1, $G_2,\cdots,G_i$。每个 G 中均无回路,因而 G_i 全为树。由(1)⇒(2)⇒(3)可知,$m_i=n_i-1$。于是 $m=\sum\limits_{i=1}^{s} m_i=\sum\limits_{i=1}^{s} n_i-s=n-s$。由于 $s \geqslant 2$,这与 $m=n-1$ 矛盾。

(4)⇒(5)。只需证明 G 中每条边均为桥。$\forall e \in E$,均有 $|E(G-e)|=n-1-1=n-2$,则 $G-e$ 不是连通图,故 e 为桥。

(5)⇒(6)。由于 G 中每条边均为桥,因此 G 中无圈。又由于 G 连通,所以 G 为树。由 (1)⇒(2)可知,设 e 是在 u,v 之间添加的新边,则 $\Gamma \cup e$ 是一个圈,且显然是唯一的。

(6)⇒(1)。只需证明 G 是连通的。对任意两个不同的顶点 u 和 v,在 u,v 之间添加一条新边 e 后产生唯一的一个含 e 的圈 C。显然,$C-e$ 为 G 中 u 到 v 的通路,故 $u \sim v$ 由 u,v 的任意性可知,G 是连通的。

定理 11.1.2 ▶ 设 T 是 n 阶非平凡的无向树,则 T 中至少有两片树叶。

证明 设 T 中有 x 片树叶,由定义 11.1.1 可知,

$$2(n-1)=\sum d(v_i) \geqslant x+2(n-x)$$

解得 $x \geqslant 2$。

例 11.1.1 画出所有非同构的 6 阶无向树。

解 设 T 是 6 阶无向树。由定理 11.1.1 可知,T 的边数 $m=5$。由握手定理可知,T 的 6 个顶点的度数之和等于 10。于是,T 的度数列必为以下情况之一:

(1) 1,1,1,1,1,5

(2) 1,1,1,1,2,4

(3) 1,1,1,1,3,3

(4) 1,1,1,2,2,3

(5) 1,1,2,2,2,2

它们对应的树如图 11.1.1 所示,其中 T_1 对应(1),T_2 对应(2),T_3 对应(3),T_4 和 T_5 对应(4),T_6 对应(5)。(4)对应两棵非同构的树,在一棵树中两个 2 度顶点相邻,在另一棵树中则不相邻,其他均对应一棵树中不相邻;其他情况均对应一棵非同构的树。

常称只有一个分支点且分支点的度数为 $n-1$ 的 $n(n \geqslant 3)$ 阶无向树为星形图,称其唯一的分支点为星心。图 11.1.1 中的 T_1 是 6 阶星形图。

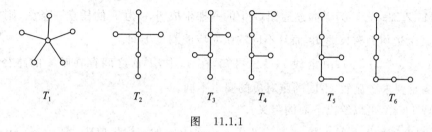

图　11.1.1

例 11.1.2 画出所有有 3 片树叶、1 个 3 度顶点、其余顶点度数不等于 1 和 3 的 7 阶非同构的无向树。

解 设 T 为满足要求的无向树,边数 $m=6$。于是,$1\times 3+3+d(v_4)+d(v_5)+d(v_6)=12$。由于 $d(v_j)(j=4,5,6)$ 不等于 1 和 3,且大于等于 1,小于等于 6,因此 $d(v_j)=2(j=4,5,6)$。于是 T 的度数列为

$$1,1,1,2,2,2,3$$

则与 3 度顶点相邻的 3 个顶点的度数只有下述 3 种可能:

$$1,1,2 \qquad 1,2,2 \qquad 2,2,2$$

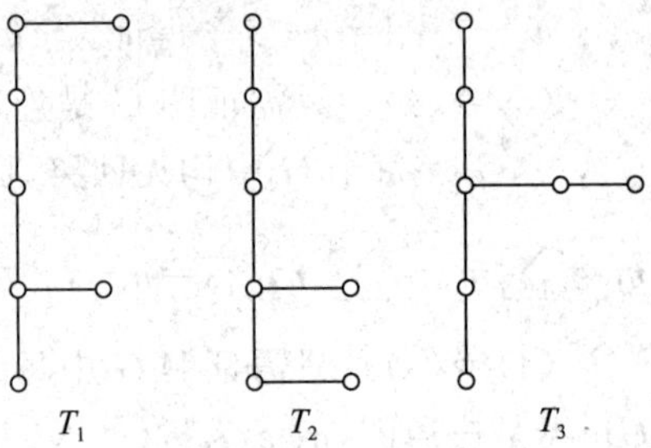

其对应的树如图 11.1.2 所示,每种可能对应一棵非同构的 7 阶树。

图 11.1.2

11.2 生成树

生成树的应用十分广泛,在计算机网络中可以避免帧发生死循环,而 s 树形的存储结构可方便数据查找和排序。

定义 11.2.1 ▶ 如果无向图 G 的生成子图 T 是树,则称 T 是 G 的生成树。设 T 是 G 的生成树,G 在 T 中的边称为 T 的树枝,不在 T 中的边称为 T 的弦。称 T 的所有弦的导出子图为 T 的余树,记作 $\overline{T}$。

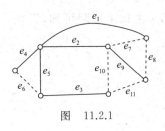

图 11.2.1

注意:$\overline{T}$ 不一定连通,也不一定不含回路。在图 11.2.1 中,实边图为该图的一棵生成树 T,余树 $\overline{T}$ 如虚边所示,它不连通且含有回路。

定理 11.2.1 ▶ 无向图 G 有生成树,当且仅当 G 是连通图。

证明 必要性显然成立。下面证明充分性。若 G 中无回路,则 G 为自己的生成树。若 G 中含圈,任取一圈,随意地删除圈上的一条边,便得到圈连通;若仍有圈,再任取一个圈并删去这个圈上的一条边,重复进行,直到最后无圈为止。最后得到的图无圈(当然无回路)、连通且是 G 的生成子图,因此是 G 的生成树。

定理 11.2.1 的证明是构造性证明,这个产生生成树的方法称为破圈法。

由定理 11.2.1 和树的边数等于顶点数减 1 可立即得到下述推论。

推论 设 G 为 n 阶 m 条边的无向连通图,则 $m\geqslant n-1$。

定理 11.2.2 ▶ 设 T 为无向连通图 G 中的一棵生成树,e 为 T 的任意一条弦,则 G 中只含一条弦 e,其余边均为树枝的圈,而且不同的弦对应的圈也不同。

证明 设 $e=(u,v)$,由定理 11.1.1 可知,在 T 中,u,v 之间存在唯一的路径 $\Gamma_{(u,v)}$,则 $\Gamma_{(u,v)}\cup e$ 满足要求。显然,不同的弦对应的圈也不同。

由定理 11.2.2 可以给出下面的定义。

定义 11.2.2 ▶ 设 T 是 n 阶 m 条边的无向连通图 G 的一棵生成树,设 $e_1',e_2',\cdots,e_{m-n+1}'$ 为 T 的弦,$C_r(r=1,2,\cdots,m-n+1)$ 为 T 添加弦 e_r' 产生的 G 中由弦 e_r' 和树枝构成的圈,称 C_r 为 G 的对应弦 e_r' 的基本回路或基本圈。称 $\{C_1,C_2,\cdots,C_{m-n+1}\}$ 为 G 对应 T 的基本回路系统,称 $m-n+1$ 为 G 的圈秩,记作 $\xi(G)$。

在图 11.2.1 中，对应生成树的弦 $e_6,e_7,e_8,e_{10},e_{11}$ 的基本回路为

$$C_1=e_6e_4e_5$$
$$C_2=e_7e_2e_5$$
$$C_3=e_8e_9e_2e_1$$
$$C_4=e_{10}e_3e_5e_2$$
$$C_5=e_{11}e_3e_5e_2e_9$$

此图的圈秩为 5，基本回路系统为 $\{C_1,C_2,C_3,C_4,C_5\}$。

不难看出，无向连通图 G 的圈秩与生成树的选取无关，但不同生成树对应的基本回路系统可能不同。

定理 11.2.3 ► 设 T 是连通图 G 的一棵生成树，e 为 T 的树枝，则 G 中存在只含树枝 e，其余边都是弦的割集，且不同的树枝对应的割集也不同。

证明 由定理 11.1.1 可知，e 是 T 的桥，因此 $T\text{-}e$ 有两个连通分支 T_1,T_2。令 $S_e=\{e'|e'\in E(G)$ 且 e' 的两个端点分别属于 $V(T_1)$ 和 $V(T_2)\}$。显然，S_e 为 G 的割集，$e\in S_e$ 且 S_e 中除 e 外都是弦。因为每个割集 S_e 只含一条树枝，故不同树枝对应的割集是不同的。

定义 11.2.3 ► 设 T 是 n 阶连通图 G 的一棵生成树，$e_1,e_2,\cdots,e_{n-1}$ 为 T 的树枝，$S_i(i=1,2,\cdots,n-1)$ 是由树枝 e_i 和弦构成的割集，则称 S_i 为 G 的对应树枝 e_i 的基本割集。称 $\{S_1,S_2,\cdots,S_{n-1}\}$ 为 G 对应 T 的基本割集系统，称 $n-1$ 为 G 的割集秩，记作 $\eta(G)$。

在图 11.2.1 中，对应树枝 e_1,e_2,e_3,e_4,e_5,e_9 的基本割集分别为

$$S_1=\{e_1,e_7,e_8\}$$
$$S_2=\{e_2,e_7,e_8,e_{10},e_{11}\}$$
$$S_3=\{e_3,e_{10},e_{11}\}$$
$$S_4=\{e_4,e_6\}$$
$$S_5=\{e_5,e_6,e_{10},e_{11}\}$$
$$S_6=\{e_9,e_8,e_{11}\}$$

基本割集系统为 $\{S_1,S_2,S_3,S_4,S_5,S_6\}$，割集秩为 6。

连通图 G 的割集秩 $\eta(G)$ 不因生成树的不同而改变，但不同生成树对应的基本割集系统可能不同。

下面讨论连通带权图中的最小生成树。

定义 11.2.4 ► 设无向连通带权图 $G=<V,E,W>$（W 为每条边上的权重），T 是 G 的一棵生成树，T 的各边权之和称为 T 的权，记作 $W(T)$。G 的所有生成树中权最小的生成树称为 G 的最小生成树。

求最小生成树有许多种算法，这里介绍避圈法（Kruskal 算法）。

设 n 阶无向连通带权图 $G=<V,E,W>$ 有 m 条边。不妨设 G 中没有环（否则，可以将所有的环先删去），将 m 条边按权从小到大顺序排列，设为 $e_1,e_2,\cdots,e_m$。

取 e_1 在 T 中，然后依次检查 $e_2,e_3,\cdots,e_m$。若 $e_j(j\geqslant2)$ 与已在 T 中的边不能构成回路，则取 e_j 在 T 中，否则弃去 e_j。

算法停止时得到的 T 为 G 的最小生成树（证明略）。

例 11.2.1 求如图 11.2.2 所示两个图中的最小生成树。

解 用避圈法，求出的最小生成树如图 11.2.3 所示，它们的权分别是 6 和 12。

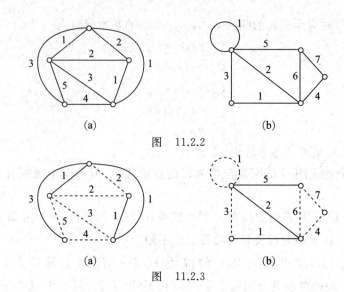

图　11.2.2

图　11.2.3

例 11.2.2　数据分析中的单链聚类。在数据分析中经常用到各种不同的聚类操作。所谓聚类操作,就是把数据集 D 中的数据按照它们之间的相似程度聚集成若干个子类。这种操作在数据挖掘、图像处理、电路设计、系统划分中经常用到。下面考虑一种最简单的单链聚类。

设有一组离散数据 $D=\{a_1,a_2,\cdots,a_n\}$。在 D 上定义一个相似度函数 d。对于任何两个数据 $a_i,a_j\in D,a_i$ 与 a_j 的相似度函数的值为 $d(i,j)$,通常 $0\leqslant d(i,j)\leqslant 1$,并且 d 具有对称性质,即 $d(i,j)=d(j,i)$。

给定正整数 $k(1<k<n)$,D 的一个 k 聚类是 D 的一个 k 划分 $\pi=\{C_1,C_2,\cdots,C_k\}$。我们希望同一子类中的数据尽可能地“接近”,而不同子类中的数据尽可能地“远离”。为此,可定义划分 π 的最小间隔 $D(\pi)$。

对任何两个不同的子类 C_i,C_j,定义它们之间的距离 $D(C_i,C_j)$ 是 C_i 中的数据与 C_j 中的数据的相似度的最小值,即

$$D(C_i,C_j)=\min\{d(i,j)\,|\,a_i\in C_i,a_j\in C_j\}$$

k 聚类 $\pi=\{C_1,C_2,\cdots,C_k\}$ 的最小间隔为

$$D(\pi)=\min\{D(C_i,C_j)\,|\,C_i,C_j\in\pi,1\leqslant i<j\leqslant k\}$$

我们的问题是:给定数据集 D 和 D 上的相似度函数 d 及正整数 k,如何求使 $D(\pi)$ 达到最大值的聚类 π?

可以利用最小生成树的 Kruskal 算法解决这个问题。定义带权完全图 $G=<V,E,W>$,其中 $V=\{1,2,\cdots,n\}$,对于任意 $i,j\in V,i\neq j$,边 (i,j) 的权为 $d(i,j)$。根据 Kruskal 算法,先将边按照权从小到大的顺序排序为 $e_1,e_2,\cdots,e_{n(n-1)/2}$。初始 T 中没有边,是由 n 个孤立顶点构成的森林。换句话说,T 有 n 个连通分支。接着,依次按照权从小到大的顺序考察 G 的每条边,只要不构成圈,就把它加到 T 中。在加入边的过程中计数 T 的连通分支个数。直到 T 恰好含有 k 个连通分支时算法停止。这时所得到的 k 个连通分支恰好就是所求聚类的 k 个子类 $C_1,C_2,\cdots,C_k$,它的最小间隔达到最大。

11.3　有向树

下面简单讨论有向图中的树。

1. 根树

定义 11.3.1 ▶ 有向图不考虑弧的方向时是一棵树,那么该有向图称为有向树。

如图 11.3.1 所示为一棵有向树。

定义 11.3.2 ▶ 一棵有向树,如果恰有一个顶点的入度为 0,其余所有顶点的入度都为 1,则称此有向树为根树。入度为 0 的顶点称为树根,出度为 0 的顶点称为树叶或悬挂点,出度不为 0 的顶点称为分支点或内点。

如图 11.3.2 所示为一棵根树,其中 v_1 为根,v_1,v_2,v_4,v_8,v_9 为分支点,其余顶点为树叶。

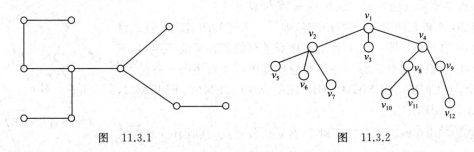

图　11.3.1　　　　　　　　　　　　　图　11.3.2

定义 11.3.3 ▶ 从根开始到任何顶点 v_i 只能有唯一的一条路,这条路的长度称为 v_i 的级数,也称为顶点 v_i 的路的长度。在图 11.3.2 中,顶点 v_1 的级数是 0,顶点 v_2,v_3,v_4 的级数是 1,顶点 v_5,v_6,v_7,v_8,v_9 的级数是 2,顶点 v_{10},v_{11},v_{12} 的级数是 3。一棵根树中所有顶点的级数的最大值称为树的高度。

在计算机科学中,关于根树还有一些专业术语,现介绍如下。

设 u 是根树的分支点,如果从 u 到 w 有一条弧 (u,w),则称 w 为 u 的儿子,或称 u 为 w 的父亲;如果一个顶点有两个儿子,称它们是兄弟;如果从 u 到 z 有一条有向路,称 z 是 u 的子孙或称 u 是 z 的祖辈。

设 u 是根树 T 的一个分支点,以 u 为根,u 的所有子孙组成顶点集 V',u 到这些子孙的有向路上的所有弧组成弧集 E',则 T 的子图 $T' = (V', E')$ 称为以 u 为根的子树。

2. 有序树

对于一棵根树,可以有根在下或根在上的两种不同画法,如图 11.3.3 所示。

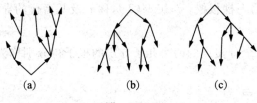

(a)　　　　　　(b)　　　　　(c)

图　11.3.3

图 11.3.3(a)是根树的自然表示法,即树从它的根向上生长。

图 11.3.3(b)和图 11.3.3(c)都是由根向下生长。它们是同构图,其差别仅在每一级上的顶点从左到右出现的次序不同。

定义 11.3.4 ▶ 根树中顶点或弧的次序用明确的方式指明,这种树称为有序树。

若以顶点为次序,则有以下三种情况。

（1）先根次序。①访问根；②遍历左子树；③遍历右子树。

（2）中根次序。①遍历左子树；②访问根；③遍历右子树。

（3）后根次序。①遍历左子树；②遍历右子树；③访问根。

3. 二叉树

定义 11.3.5 ▶ 在根树中,若每一个顶点的出度小于或等于 m,则称这棵树为 m 叉树。如果每一个顶点的出度恰好等于 m 或零,则称这棵树为完全 m 叉树。所有树叶级数相同的完全 m 叉树又称为正则 m 叉树。特别地,当 $m=2$ 时,称为二叉树。

有很多实际问题可用二叉树或 m 叉树表示。

例如,M 和 E 两人进行网球比赛,如果一人连胜两盘或共胜三盘就获胜,比赛结束。图 11.3.4 表示了比赛进行的各种可能情况,它有十片树叶,从根到树叶的每一条路对应比赛中可能发生的一种情况,即 MM, MEMM, MEMEM, MEMEE, MEE, EMM, EMEMM, EMEME, EMEE, FF。

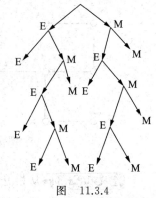

图 11.3.4

定理 11.3.1 ▶ 若完全二叉树 T 有 n 个分支点,且所有分支点的路长度总和为 I,所有树叶的路总长度为 E,则

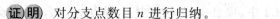

$$E = I + 2n$$

证明 对分支点数目 n 进行归纳。

假设对 $n-1$ 时结论成立。现考察分支点数为 n 的情况。

如果删去与一个分支点 v 相关联的边及其儿子,它的路长度为 l,并且它的两个儿子是树叶,那么得到树 T',E 减少 $2(l+1)$;又因为删去 v 的儿子而使 E 增大 l,所以 E 的变化值为 $-l-2$,I 的变化值为 $-l$。由归纳假设知,在 T' 中,$E' = I' + 2n'$,这里 $n' = n-1$,于是再将 v 及其两个儿子加到 T' 中,由此可以得到 T,$E' = I' + 2(n-1)$ 而且 $E' = E-l-2$,$I' = I-l$,$n' = n-1$,所以 $E = I + 2n$。

11.4 根树及其应用

根树的概念简单且应用广泛,接下来我们就简单介绍根树及其应用。

定义 11.4.1 ▶ 设 T 为一棵根树,$\forall v \in V(T)$,称 v 及其后代的导出子图 T' 为 T 的以 v 为根的根子树。

二叉正则有序树的每个分支点的两个儿子导出的根子树分别称为该分支点的左子树和右子树。

下面介绍二叉树的应用。

定义 11.4.2 ▶ 设二叉树 T 有 t 片树叶 $v_1, v_2, \cdots, v_t$,权分别为 $w_1, w_2, \cdots, w_t$,称 $W(T) = \sum w_i l(v_i)$ 为 T 的权,其中 $l(v_i)$ 是 v_i 的层数。在所有有 t 片树叶,带权 $w_1, w_2, \cdots, w_t$ 的二叉树中,权最小的二叉树称为最优二叉树。

如图 11.4.1 所示的三棵树 T_1, T_2, T_3 都是带权为 2,2,3,3,5 的二叉树。它们的权分别为

$$W(T_1)=2\times2+2\times2+3\times3+5\times3+3\times2=38$$
$$W(T_2)=3\times4+5\times4+3\times3+2\times2+2\times1=47$$
$$W(T_3)=3\times3+3\times3+5\times2+2\times2+2\times2=36$$

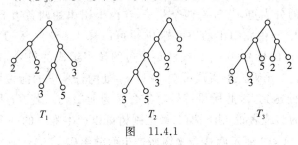

图 11.4.1

下面介绍求最优二叉树的算法——Huffman 算法。

给定实数 $w_1,w_2,\cdots,w_t$，

(1) 作 t 片树叶,分别以 $w_1,w_2,\cdots,w_t$ 为权。

(2) 在所有入度为 0 的顶点(不一定是树叶)中选出两个权最小的顶点,添加一个新分支点,它以这两个顶点为儿子,其权等于这两个儿子的权之和。

(3) 重复(2),直到只有一个入度为 0 的顶点为止。

$W(T)$ 等于所有分支点的权之和。

例 11.4.1 求带权 $2,2,3,3,5$ 的最优二叉树。

解 图 11.4.2 给出用 Huffman 算法计算最优树的过程。(e)为最优树,$W(T)=34$。这表明图 11.4.1 中的三棵树都不是最优树。

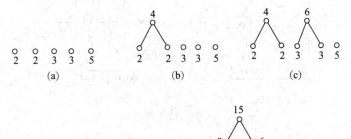

图 11.4.2

在通信中常用二进制编码表示符号。例如,可用长为 2 的二进制编码 $00,01,10,11$ 分别表示 A,B,C,D。称这种表示法为等长码表示法。若在传输中 A,B,C,D 出现的频率大体相同,用等长码表示是很好的方法。但当它们出现的频率相差悬殊时,为节省二进制数位,以达到提高效率的目的,就要采用非等长的编码。

定义 11.4.3 设 $\alpha_1\alpha_2\cdots\alpha_{n-1}\alpha_n$ 是长为 n 的符号串,称其子串 $\alpha_1,\alpha_1\alpha_2,\cdots,\alpha_1\alpha_2\cdots\alpha_{n-1}$ 为该符号串的前缀。设 $A=\{\beta_1,\beta_2,\cdots,\beta_m\}$ 是一个符号串集合,若 A 的任意两个符号串都互不为前缀,则称 A 为前缀码。由 0-1 符号串构成的前缀码称作二元前缀码。

例如，$\{1,00,011,0101,01001,01000\}$ 是前缀码，而 $\{1,00,011,0101,0100,01001,$ $01000\}$ 不是前缀码，因为 0100 是 01001 的前缀。

可用二叉树产生二元前缀码。设 T 是具有 n 片树叶的二叉树，则 T 的每个分支点有 1 或 2 个儿子。设 v 为 T 的分支点，若 v 有两个儿子，在由 v 引出的两条边上，左边的标 0，右边的标 1。若 v 只有一个儿子，由它引出的边可标 0，也可以标 1。设 v_i 是 T 的任意一片树叶，从树根到 v_i 的通路上各边的标号(0 或 1)按通路上边的顺序组成的符号串放在 v_i 处，t 片树叶的 t 个符号串组成一个二元码。由实践可知，树叶 v_i 处的符号串的前缀均在从树根到 v_i 的通路上的顶点(不含 v_i)处达到，因此所得符号串集合必为前缀码。若 T 是二叉正则树，则由 T 产生的前缀码是唯一的。或者说，由一棵二叉正则树可以产生唯一的一个二元前缀码。

例 11.4.2 求图 11.4.3 所示两棵二叉树所产生的前缀码。

解 图 11.4.3(a)是二叉树，但不是正则的。将每个分支点引出的两条边分别标 0 和 1，树根右儿子引出的边标 1，如图 11.4.4(a)所示，产生的前缀码为 $\{11,01,$ $000,0010,0011\}$。若将树根右儿子引出的边标 0，则产生的前缀码为 $\{10,01,000,0010,0011\}$。图 11.4.3(b)是二叉正则树，它只能产生唯一的前缀码，标定后的二叉正则树如图 11.4.4(b)所示，产生的前缀码为 $\{01,10,11,$ $000,0010,0011\}$。

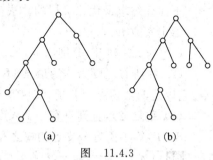

图 11.4.3

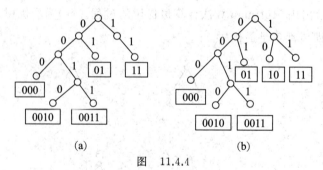

图 11.4.4

上面产生的任一个前缀码都可以用来传输 5 个符号，比如 A,B,C,D,E。但当在文本中这些字母出现的频率不同时，传输这个文本所用的二进制位数是不同的。设共有 t 个符号，用树叶 v_i 处的二进制串表示的符号出现的频率为 c_i，v_i 处的二进制串的长度等于 v_i 的层数 $l(v_i)$，因此传输 m 个符号使用的二进制位数为 $m\sum\limits_{i=1}^{i}c_il(v_i)$。显然，用以各个符号出现的频率为权的最优二叉树产生的前缀码所用的二进制位数最少，称这个由最优二叉树产生的前缀码为最佳前缀码。用最佳前缀码传输的二进制位数最省。

下面介绍二叉树的周游及应用。

对一棵根树的每个顶点都访问一次且仅一次称为行遍或周游一棵树。

对二叉有序正则树有以下三种周游方式。

(1) 中序行遍法。访问的次序为左子树、树根、右子树。

(2) 前序行遍法。访问的次序为树根、左子树、右子树。

(3) 后序行遍法。访问的次序为左子树、右子树、树根。

对图 11.4.5 所示的二叉有序正则树 T，按中序、前序、后序行遍的周游

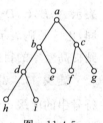

图 11.4.5

结果如下：

$$((h\ \underline{d}i)\underline{b}e)\ \underline{a}(f\underline{c}g)$$

$$\underline{a}(\underline{b}(\underline{d}hi)e)(\underline{c}fg)$$

$$((hi\ \underline{d})\ e\ \underline{b})\ (fg\underline{c})\underline{a}$$

上式中，$\underline{v}$ 表示 v 为根子树的树根。利用二叉有序正则树可以表示四则运算的算式，然后根据不同的访问方法得到不同的算法。用二叉有序正则树表示算式的方法如下。

参加运算的数都放在树叶上，然后按照运算的顺序依次将运算符（＋、－、＊、÷）放在分支点上，每个分支点表示一个运算，同时也表示这个运算的结果，它以它的两个运算对象为儿子，并规定被除数、被减数放在左边。

习　题

1. 画出所有 5 阶和 7 阶非同构的无向树。

2. 一棵无向树 T 有 5 片树叶、3 个 2 度分支点，其余的分支点都是 3 度顶点，问 T 有几个顶点？

3. 证明：任何无向树 T 都是二部图。

4. 在什么条件下，无向树 T 为半欧拉图？

5. 在什么条件下，无向树 T 为半哈密顿图？

6. 设 T 为无向树 G 的生成树，$\overline{T}$ 为 T 的余树，证明 $\overline{T}$ 中不含 G 的边割集。

7. 习题 7 图所示的无向图中，含边 e_1,e_2,e_3 作为树枝的非同构的生成树共有几棵？画出它们。

8. 习题 8 图所示的无向图中，有几棵非同构的生成树？画出这些树（提示：从所有 6 阶非同构的树中挑选）。

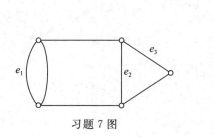

习题 7 图　　　　　　　　　　习题 8 图

9. 已知 n 阶 m 条边的无向图 G 是 $k(h\geqslant 2)$ 棵树组成的森林，证明：$m=n-k$。

10. 在习题图 10 中，实边构成一棵生成树，记为 T。

(1) 指出 T 的弦及每条弦对应的基本回路和对应 T 的基本回路系统。

(2) 指出 T 的所有树枝及每条树枝对应的基本割集和对应 T 的基本割集系统。

11. 求习题 11 图所示的两个带权图的最小生成树。

12. 画一棵树高为 3 的完全正则二叉树。

13. 在习题 13 图所示的有向图中存在是根树的生成子图吗？若存在，有几棵是非同构的？

14. 画一棵权为 3,4,5,6,7,8,9 的最优二叉树，并计算出它的权。

15. 用习题 15 图的二叉树产生一个二元前缀码。

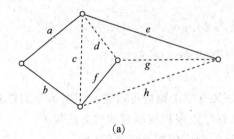

(a)　　　　　　　(b)

习题 10 图

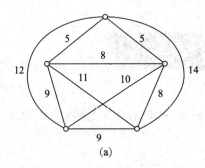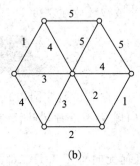

(a)　　　　　　　(b)

习题 11 图

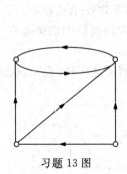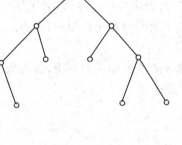

习题 13 图　　　　习题 15 图

Reference

参 考 文 献

[1] 邹晓红,刘天歌. 离散数学[M]. 秦皇岛:燕山大学出版社,2021.

[2] 朱保平. 离散数学[M]. 北京:清华大学出版社,2019.

[3] 黄亚群,蒋慕蓉,赵春娜. 离散数学[M]. 2 版. 北京:科学出版社,2020.

[4] 章炯民. 离散数学[M]. 4 版. 上海:华东师范大学出版社,2021.

[5] 杨文国,高华. 离散数学[M]. 2 版. 北京:清华大学出版社,2021.

[6] 罗熊. 离散数学[M]. 2 版. 北京:高等教育出版社,2021.

[7] DIJKSTRA E W, SCHOLTER C S. Predicate calculus and program semantics[M]. Berlin:Springer Science & Business Media,2012.

[8] BIGGS N L. Discrete mathematics[M]. Oxford:Oxford University Press,2002.

[9] EPP S S. Discrete mathematics with applications[M]. Stanford:Cengage learning,2010.

[10] CHANDRASEKARAN N, UMAPARVATHI M. Discrete mathematics[M]. Delhi:PHI Learning Pvt. Ltd.,2022.

[11] O'REGAN G. Guide to discrete mathematics[M]. Berlin:Springer International Publishing,2021.